ISNM
International Series of
Numerical Mathematics
Vol. 112

Edited by
K.-H. Hoffmann, München
H. D. Mittelmann, Tempe
J. Todd, Pasadena

ISNM
International Series of
Numerical Mathematics
Vol. 113

Edited by
K.-H. Hoffmann, München
H. D. Mittelmann, Tempe
J. Todd, Pasadena

Numerical Solution of the Incompressible Navier-Stokes Equations

L. Quartapelle

Springer Basel AG

Author

L. Quartapelle
Dipartimento di Fisica
Politecnico di Milano
Piazza Leonardo da Vinci, 32
I-20133 Milano
Italy

A CIP catalogue record for this book is available from the Library of Congress, Washington D.C., USA

Deutsche Bibliothek Cataloging-in-Publication Data
Quartapelle, Luigi:
Numerical solution of the incompressible Navier-Stokes equations / L. Quartapelle. – Basel ; Boston ; Berlin : Birkhäuser, 1993
(International series of numerical mathematics ; Vol. 113)
ISBN 978-3-0348-9689-4 ISBN 978-3-0348-8579-9 (eBook)
DOI 10.1007/978-3-0348-8579-9
NE: GT

Originally published by Birkhäuser Verlag in 1993
Softcover reprint of the hardcover 1st edition 1993
Camera-ready copy prepared by the author
Printed on acid-free paper produced from chlorine-free pulp
Cover design: Heinz Hiltbrunner, Basel

ISBN 978-3-0348-9689-4

9 8 7 6 5 4 3 2 1

a Giuliana

Contents

Le mani di Tou-Ema diventarono uccelli.
[And Tou-Ema's hands became birds.]

Motu-Iti: L'isola dei gabbiani, Roberto Piumini
Einaudi, Torino, 1989

Preface

This work is a revised and expanded version of the VKI lecture notes for the Lecture Series in Computational Fluid Dynamics held in the spring of 1992.

My aim was to provide graduate students and young researchers in computational fluid dynamics with a brief introduction to the numerical methods for the solution of the unsteady incompressible Navier–Stokes equations. Rather than presenting a great variety of numerical techniques, I preferred to give a unitary view of the methods which reduce the equations for viscous incompressible flows to a system of second-order equations of parabolic and elliptic type.

Several methods of this class have been developed for different formulations of the governing equations, exploiting various spatial discretization techniques, namely, finite differences, finite elements and spectral methods. They represent a significant subset of the schemes presently in use in CFD to solve the incompressible Navier–Stokes equations, and are closely related to my own reseach. However, restricting the analysis to them by no means intends to diminish the importance of several other numerical methods which rely on different formulations of the problem. For instance, just to recall a few very valuable numerical methods which are at the extremes of the range from general purpose techniques to highly specialized algorithms, I can mention the finite element penalty method for dealing with arbitrarily shaped domains (Bercovier and Engelman, *JCP* **30**, 181–201, 1979) and the spectral projection methods for the simulation of flows within geometrically simple boundaries (Moser, Moin and Leonard, *JCP* **52**, 524–544, 1983, and Dumas and Leonard, *JCP* 1993).

The various chapters are as self-contained as possible, so that the reader can study or consult each of them in an almost completely independent manner. To this purpose, the proof of very similar theorems establishing the equivalence of various formulations of the incompressible equations has been repeated, instead of indicating only the minor differences.

I am greatly indebted to Fausto Valz-Gris and Stanley Dennis, for the unique privilege of sharing their original ideas on the Navier–Stokes equations and incompressible viscous flows. I have also benefitted from the encouragements by Henri Cabannes whose intellectually generous attention to my research has been a nonminor motivation for persevering in it.

This work would have never been completed without the continuing contributions of my friends and colleagues Adriano Muzzio, Michele Napolitano and Pierre Pegon. Their suggestions, comments and criticisms have greatly influenced the writing and rewriting of the text up to the present form. I am also very indebted to Jean-Luc Guermond who gave the manuscript its few touches of mathematical elegance.

While I have tried with such excellent help to make the text as free as possible from errors and obscurities, of course some of them are still present. For those remaining I am the only responsible; corrections, comments and suggestions will be greatly appreciated: they can be e-mailed at `rfisic01@imipoli.it`.

I would like to thank Herman Deconinck, whose kind invitation to the von Karman Institute for Fluid Dynamics made me to start to write these pages and thus allowed my previous studies on the incompressible Navier–Stokes equations to arrive at a more definite settlement. Finally, I cannot forget my dear friend, the late Cornelis van den Muyzenberg, who introduced me to the realm of computers, giving me the chance of discovering the beauty of well designed computer algorithms.

L. Q.
Milano, Italy
June 1993

Chapter 1

The incompressible Navier–Stokes equations

1.1 Introduction

Methods for the approximate solution of the incompressible Navier–Stokes equations have been investigated rather extensively in the last two decades. Several numerical schemes have been developed for the calculation of both steady and unsteady viscous flows using the various possible formulations of the problem. An important aspect of the numerical methods for solving this class of problems is the proper specification of the boundary conditions, especially when solid walls are present. Of course, the appropriate treatment depends on which variables are chosen as unknowns and, from a numerical viewpoint, the specific form assumed by the boundary conditions may also depend on the method adopted for discretizing space and time.

The aim of the present study is to address the boundary condition issue in incompressible viscous problems and to attempt to provide a unitary framework for its analysis. The attention will be focussed mainly on the *time-dependent* equations and on formulations in which the incompressibility condition is imposed by means of a *Poisson equation* for a scalar or vector unknown. In other terms, the work addresses the theme of enforcing the continuity equation in numerical fluid dynamics of incompressible flows by solving an equivalent substitute equation which involves the Laplace operator. In particular, the full set of conditions required to supplement the parabolic and elliptic equations occurring in several different representations of the Navier–Stokes equations will be determined, with the aim of deriving *uncoupled* formulations of the incompressible equations.

The interest in uncoupled formulations of this kind is both theoretical and computational. From the theoretical viewpoint, writing the Navier–Stokes equations with each unknown provided with its own conditions is a necessary step for arriving at a correct and complete description of the behaviour of viscous incompressible flows in many circumstances. The development of several interesting fluid patterns is determined by the presence of solid boundaries, where the value of the velocity is prescribed. Unfortunately, this variable does not appear any more as a basic unknown in some useful formulations of the Navier–Stokes equations, as, for instance, the formulation based on the variables vorticity and stream function. It is therefore necessary to uncover the precise form which the boundary conditions originally attached to the velocity–pressure equations assume when passing to a different representation of the governing equations.

From the viewpoint of numerical methods, uncoupled formulations present an important advantage over coupled solution schemes: the discrete approximations of the unknown variables are not subject to conditions of compatibility which must be satisfied instead by the pair of variables, as, for instance, velocity and pressure, in coupled solution schemes.

The development of the uncoupled formulations derived in this study leads to the following important result: the vorticity field is found to be subject to conditions of an *integral* character, whenever no-slip velocity conditions are prescribed at the boundaries of the fluid domain. A similar result is shown to hold also for the pressure field in the uncoupled primitive variable representation of the Navier–Stokes equations. These integral conditions are used here to derive methods for the approximate solution of incompressible viscous problems as well as to recognize the common structure underlying several apparently unrelated numerical schemes presently in use.

1.2 Incompressible Navier–Stokes equations

The motion of a viscous incompressible fluid is governed by the Navier–Stokes equations

$$\frac{\partial \boldsymbol{u}}{\partial t} + (\boldsymbol{u} \cdot \boldsymbol{\nabla})\boldsymbol{u} = -\boldsymbol{\nabla} P + \nu \nabla^2 \boldsymbol{u}, \tag{1.1}$$

$$\boldsymbol{\nabla} \cdot \boldsymbol{u} = 0, \tag{1.2}$$

where $\boldsymbol{u}(\boldsymbol{x}, t)$ is the velocity, $P(\boldsymbol{x}, t)$ is the pressure divided by the density and ν is the (constant) coefficient of kinematic viscosity. The inclusion of a body force term $\boldsymbol{f}(\boldsymbol{x}, t)$ on the right-hand side of (1.1) would not change anything of all the subsequent analysis. The vector differential operators are indicated using standard notations, see, *e.g.*, Adams [2].

The statement of the problem is made complete by the specification of suitable boundary and initial conditions. A typical boundary condition consists in prescribing the value of the velocity $\boldsymbol{b}$ on the boundary,

$$\boldsymbol{u}|_S = \boldsymbol{b}(\boldsymbol{x}_S, t), \tag{1.3}$$

where S is the boundary of the domain V occupied by the fluid and $\boldsymbol{x}_S \in S$. When the boundary is a solid wall in contact with the fluid, the velocity boundary value $\boldsymbol{b}$ is equal to the velocity of the wall. The condition on the tangential components of velocity is known as the *no-slip condition*.

It is important to note that no boundary condition is prescribed for the pressure on no-slip boundaries and that it would be incorrect to impose one together with the velocity boundary condition (1.3). On the other hand, in some applications, velocity conditions different from (1.3) can be encountered, such as on inflow or outflow boundaries, and on plane of symmetry or antisymmetry. In these situations, the pressure can be supplemented by boundary conditions of Dirichlet or Neumann type. Since the no-slip condition is the most difficult kind of boundary condition for incompressible viscous problems, in the following the attention will be limited to the velocity boundary condition (1.3), with the only exception of section 5.7.

In the following, the fluid domain V is assumed to be open, bounded and simply connected, which implies that it is of finite extent and contains no holes. Furthermore, for simplicity, the boundary S is assumed to be a conveniently smooth closed surface.

The initial condition consists in the specification of the velocity field $\boldsymbol{u}_0$ at the initial time, $t = 0$, namely,

$$\boldsymbol{u}\big|_{t=0} = \boldsymbol{u}_0(\boldsymbol{x}). \tag{1.4}$$

The boundary velocity $\boldsymbol{b}$ must satisfy, for all $t \geq 0$, the global condition

$$\oint \boldsymbol{n} \cdot \boldsymbol{b}\, dS = 0, \tag{1.5}$$

which follows from integrating the continuity equation over V and using the divergence theorem. Here and throughout $\boldsymbol{n}$ denotes the outward unit normal to the boundary S. The symbol $\oint$ indicates here the boundary integral over a closed surface, but the same symbol is also used in the following to denote the line integral along a closed curve. Volume integrals and integrals over nonclosed surfaces will be always indicated by a single integration symbol $\int$ (with the only exception of Appendix D).

The initial velocity field $\boldsymbol{u}_0$ is assumed to be solenoidal, *i.e.*,

$$\boldsymbol{\nabla} \cdot \boldsymbol{u}_0 = 0. \tag{1.6}$$

Finally, the boundary and initial data $\boldsymbol{b}$ and $\boldsymbol{u}_0$ are assumed to satisfy the following compatibility condition:

$$\boldsymbol{n} \cdot \boldsymbol{b}\big|_{t=0} = \boldsymbol{n} \cdot \boldsymbol{u}_0\big|_S, \tag{1.7}$$

where, of course, $\boldsymbol{n} \cdot \boldsymbol{b}(\boldsymbol{x}_S, t)$ is taken to be a continuous function of time as $t \to 0^+$. Condition (1.7) is absent in the steady-state version of the problem.

Such a compatibility condition together with the solenoidality condition for the initial velocity field allows for an optimal choice of the linear space for the initial data. Therefore condition (1.7) affords solutions of the time-dependent equations with a *minimal* regularity, *viz.* H^1 for the velocity, as in most of current numerical schemes relying on spatial discretizations of local type.

This optimal setting has been provided by Ladyzhenskaya [24, p. 88] and Temam [35, p. 253], and includes the optimal condition of compatibility between the data specified for the (normal component of) velocity on the boundary and at the initial time.[1]

It is interesting to note that the same conditions of solenoidality and of compatibility between the boundary and initial data are necessary also to prove the existence and uniqueness of classical solutions of the time-dependent 2D Euler equations for an incompressible ideal fluid of zero viscosity (Kato 1967). Thus, the presence of the compatibility condition is due only to the incompressibility, irrespective of the viscous or nonviscous character of the fluid.

Additional compatibility conditions concerning the tangential components of the initial velocity field and the boundary condition have been considered for the viscous equations. However, the condition on the tangential components of the initial velocity is unnecessarily stringent. In fact, the solenoidality of the initial velocity and the compatibilty of the normal component of the boundary value of the velocity with that of the initial velocity are all that is needed for ensuring existence of a solution with some minimal regularity, whereas compatibilty of the tangential components of the initial and boundary data is required only if higher regularity is desired (Heywood and Rannacher 1982).

It can be also noted that there are computational fluid dynamicists who believe that no compatibility condition exists between the initial and boundary data for the incompressible Navier–Stokes problem, exactly like none exists for the parabolic equation governing the diffusion of the temperature in a heat-conducting medium. After all, the equations governing the motion of a viscous fluid define a parabolic problem so that no basic difference is expected in the mathematical structure with respect to the diffusion equation. However, this argument is not completely correct, because it neglects the rôle played by the incompressibility in the mathematical theory of the Navier–Stokes equations. In fact, the argument denying the existence of any compatibility condition between initial and boundary data is correct only as far as the tangential components of the velocity are concerned, while it is false when referred to the component normal to the boundary and to a vector field which must be solenoidal. This misunderstanding may explain why the importance of the aforementioned compatibility condition in viscous incompressible flows has not been fully recognized

[1]References to textbooks and monographs are indicated within brackets, giving usually the name of the author(s). References to articles are instead denoted always by the name of the author(s) and the year of publication.

so far in the CFD community. In this connection, it may be worthwhile to remind that the unsteady incompressible Navier–Stokes equations do define a parabolic problem, but only *after* it has been projected onto the space of solenoidal vector fields tangential to the boundary, and this means taking an initial velocity field which satisfies the compatibility condition.

In the next chapters the compatibility condition (1.7) will be used repeatedly to establish the equivalence of the equations governing incompressible flows with various different formulations involving the vorticity variable.

The compatibility condition (1.7) is *not* satisfied for problems characterized by an impulsive initial motion of bodies or walls in conctact with the fluid. However, in these situations, the proper initial condition for the flow around the body is given by a potential flow that establishes itself, by virtue of the incompressibility, in response to the sudden motion of the boundary (cf., *e.g.*, the discussion by Telionis [34, p. 81] and by Lighthill [27, p. 80]). (If V is not simply connected, the additional assumption of zero circulation around the body is necessary in order that the correction to initial velocity can be represented by the gradient of a scalar function.)

The initial potential velocity field is caused by the "jump" between the different values prescribed on the normal velocity by the boundary condition and by the initial condition. More precisely, if $\boldsymbol{n}\cdot\boldsymbol{b}|_{t=0} \neq \boldsymbol{n}\cdot\boldsymbol{u}_0|_S$, one introduces a velocity potential Φ_0 solution of the following Neumann problem

$$-\nabla^2\Phi_0 = 0, \qquad \boldsymbol{n}\cdot\boldsymbol{\nabla}\Phi_0|_S = \boldsymbol{n}\cdot\boldsymbol{b}|_{t=0} - \boldsymbol{n}\cdot\boldsymbol{u}_0|_S, \tag{1.8}$$

whose solvability condition $\oint \boldsymbol{n}\cdot(\boldsymbol{b}|_{t=0} - \boldsymbol{u}_0|_S)\,dS = 0$ is satisfied by conditions (1.5) and (1.6). Then, the initial velocity is replaced by a modified initial field $\boldsymbol{u}_0^\star$, defined as follows:

$$\boldsymbol{u}_0^\star = \boldsymbol{u}_0 + \boldsymbol{\nabla}\Phi_0. \tag{1.9}$$

With this modified initial field, the compatibility condition (1.7) is automatically satisfied, since

$$\begin{aligned} \boldsymbol{n}\cdot\boldsymbol{u}_0^\star|_S &= \boldsymbol{n}\cdot(\boldsymbol{u}_0 + \boldsymbol{\nabla}\Phi_0)|_S \\ &= \boldsymbol{n}\cdot\boldsymbol{u}_0|_S + \boldsymbol{n}\cdot\boldsymbol{b}|_{t=0} - \boldsymbol{n}\cdot\boldsymbol{u}_0|_S \\ &= \boldsymbol{n}\cdot\boldsymbol{b}|_{t=0}, \end{aligned}$$

by virtue of the boundary condition imposed on Φ_0 in problem (1.8). Thus, provided that the initial velocity field is modified according to this procedure, the fulfillment of the compatibility condition (1.7) between the boundary and initial data can be ensured, even for problems involving an impulsive motion of the boundaries. It should be noted that in these cases a discontinuity in the tangential components of the velocity on the boundary is usually produced by the introduction of the initial potential flow $\boldsymbol{\nabla}\Phi_0$, namely,

$$\boldsymbol{n}\times\boldsymbol{b}|_{t=0} \neq \boldsymbol{n}\times\boldsymbol{u}_0^\star|_S.$$

The discontinuity of the tangential components of velocity on the boundary at the time of an impulsive start of a body implies that a infinitely thin layer of infinite vorticity is present around the body at that time. This fact is usually referred to as the phenomenon of *vorticity generation* on rigid boundaries. The validity of such a description is limited to the present context of a sudden start of the boundary, with the flow assumed to be inviscid and described by the potential Φ_0.

When passing to the study of viscous flows, it will be shown that the mechanism of vorticity generation on the boundary is no more necessary to account for the effect or influence of the no-slip velocity conditions on the vorticity field. In fact, a completely different picture will emerge, which represent the "interaction" of the vorticity with rigid boundaries in terms of convenient *global constraints* for this variable. The new description presents the advantage of beeing a direct mathematical consequence of the Navier–Stokes equations and of not requiring to consider the inviscid limit with an impulsive time variation of the boundary values of velocity.

The set of equations (1.1)–(1.4) defines a nonlinear initial boundary value problem which, under the assumed conditions (1.5)–(1.7), admits solution(s) $(\boldsymbol{u}(\boldsymbol{x},t), P(\boldsymbol{x},t))$, with the pressure field determined up to an arbitrary additive function of time. In particular, for the linear version of the problem, obtained by dropping the inertial term $(\boldsymbol{u}\cdot\boldsymbol{\nabla})\boldsymbol{u}$, the solution is uniquely defined.

Coming to the numerical solution of the Navier–Stokes equations, a serious difficulty is met in the determination of the pressure field and in the fulfillment of the incompressibility condition. In fact, the continuity equation (1.2) is somewhat peculiar in that it represents a *constraint* for the velocity field. At the same time, the pressure variable, which appears in the momentum equation through the term $\boldsymbol{\nabla}P$, provides the degrees of freedom necessary to accomodate and satisfy such a constraint. Correspondingly, no evolutionary equation exists for the pressure, so that in incompressible problems this variable has not the usual thermodynamical meaning. Here, the rôle of the pressure is that of adjusting itself instantaneously in order for the condition of zero divergence to be satisfied at all times. This behaviour is related to the well known fact that in an incompressible fluid the value of the speed of sound becomes infinite. As a consequence, the pressure field cannot be calculated by an explicit time-advancement procedure but requires instead an implicit determination able to take into account the coupling existing between the pressure and the velocity, as well as the effect of the velocity boundary condition. This aspect can be considered the most distinctive feature of the primitive variable formulation of the incompressible Navier–Stokes equations.

1.3 Organization of the book

Several numerical schemes have been proposed for the calculation of incompressible viscous flows. However, the present study will concentrate almost exclusively on formulations of the Navier–Stokes problem (1.1)–(1.4) based on the reduction of the incompressibility condition to an equation involving the Laplace operator and governing a scalar or a vector unknown. Two quite distinct classes of numerical methods will be considered: the first class allows the use of time discretization of standard (*i.e.*, nonfractional) type, whereas the second class is based on a fractional-step time discretization in which the effect of the viscous term is taken into account separately from the treatment of the incompressibility (fractional-step projection method).

Chapter 2 is devoted to the study of the equations governing two-dimensional flows expressed in terms of the two scalar unknowns vorticity ζ and stream function ψ, namely,

$$\begin{aligned} \frac{\partial \zeta}{\partial t} + J(\zeta, \psi) &= \nu \nabla^2 \zeta, \\ -\nabla^2 \psi &= \zeta, \end{aligned} \tag{1.10}$$

where J denotes the Jacobian. A peculiar difficulty of this system of two equations, one parabolic and the other elliptic, is the lack of boundary conditions for the vorticity variable. After remarking that this obstacle can be circumvented reformulating the system as a biharmonic equation for ψ or considering the equations for the nonprimitive variables ζ and ψ as a system of two coupled equations even in the linear case, we describe a third method which is based on uncoupling the vorticity equation form the stream function, at least for the linear problem. This goal is achieved by introducing conditions of an integral character for the vorticity. The computational ramifications of this approach are described giving the basic algorithmic components which implement the vorticity integral conditions by means of finite differences, finite elements and Chebyshev polynomial approximations.

Chapter 3 discusses nonprimitive variable formulations of the Navier–Stokes equations for general three-dimensional flows. Since a solenoidal velocity field in three dimensions can be represented by means of a vector potential or stream vector nonuniquely, there are several formulations using nonprimitive variables which are the vector generalization of the ζ-ψ formulation for plane problems. For example, one formulation of the incompressible problem is provided by the system of vector equations

$$\begin{aligned} \frac{\partial \boldsymbol{\zeta}}{\partial t} + \boldsymbol{\nabla}\times(\boldsymbol{\zeta}\times\boldsymbol{\nabla}\times\boldsymbol{\psi}) &= \nu \nabla^2 \boldsymbol{\zeta}, \\ -\nabla^2 \boldsymbol{\psi} &= \boldsymbol{\zeta}, \end{aligned} \tag{1.11}$$

where $\boldsymbol{\zeta}$ is the vorticity vector field and $\boldsymbol{\psi}$ is a stream vector. For this representation as well as for other two nonprimitive variable formulations the complete set of boundary conditions is provided which assure the equivalence with the original Navier–Stokes equations. Uncoupled formulations of the governing equations (apart from the nonlinearity) are then obtained by determining the conditions supplementing the vorticity field. These conditions are found always to include conditions of an integral character which represents the generalization to the three-dimensional vector situation of the vorticity integral conditions valid for planar flows. This analysis is limited to simply connected domains but one of the considered representation is characterized by a choice of the boundary conditions for the stream vector which is suitable for dealing with multiply connected three-dimensional regions.

In chapter 4 we investigate another representation of the incompressible Navier–Stokes equations which is intermediate in character between the nonprimitive variable approaches and the primitive variable equations. In this formulation the unknown variables are the vorticity $\boldsymbol{\zeta}$ and the velocity $\boldsymbol{u}$ and the equations for general three-dimensional problems are

$$\begin{aligned} \frac{\partial \boldsymbol{\zeta}}{\partial t} + \boldsymbol{\nabla}\times(\boldsymbol{\zeta}\times\boldsymbol{u}) &= \nu\nabla^2\boldsymbol{\zeta}, \\ -\nabla^2\boldsymbol{u} &= \boldsymbol{\nabla}\times\boldsymbol{\zeta}. \end{aligned} \tag{1.12}$$

The boundary and integral conditions which supplement the vorticity–velocity equations both in two and three dimensions are obtained. Some difficulties which have been met in the application of the formulation for plane two-dimensional problems using finite elements and finite differences on nonstaggered grids are mentioned.

In chapter 5 we come back to the primitive variable formulation of the incompressible problem and consider a time-discretized version of the Navier–Stokes equations obtained by means of a nonfractional-step scheme to give, for instance, the following set of semi-discrete equations (Stokes problem)

$$\begin{aligned} (-\nabla^2 + \gamma)\boldsymbol{u} + \boldsymbol{\nabla}p &= \boldsymbol{g}, \\ \boldsymbol{\nabla}\cdot\boldsymbol{u} &= 0, \end{aligned} \tag{1.13}$$

where $\boldsymbol{u}$ and p are the unknowns at the new time level, $\boldsymbol{g}$ is a known source term which depends on the velocity field at the previous time and $\gamma = 1/(\nu\Delta t)$. The continuity equation is then eliminated by introducing a Poisson equation for the pressure, so that the equations to be solved become

$$\begin{aligned} \nabla^2 p &= \boldsymbol{\nabla}\cdot\boldsymbol{g}, \\ (-\nabla^2 + \gamma)\boldsymbol{u} &= -\boldsymbol{\nabla}p + \boldsymbol{g}. \end{aligned} \tag{1.14}$$

The boundary conditions supplementing this system comprise the original no-slip condition $\boldsymbol{u}|_S = \boldsymbol{u}$ and the derivative boundary condition $\boldsymbol{\nabla}\cdot \boldsymbol{u}|_S = 0$ introduced by Kleiser and Shumann (1980). The presence of these two boundary conditions for the velocity is shown to imply that the (gradient of the) pressure field satisfy conditions of an integral character similar in nature to that for the vorticity. An uncoupled solution method is thereby derived which enforces the integral conditions for the pressure in a computationally efficient way. A related method due to Glowinski and Pironneau for the direct solution of the Stokes problem (Stokes solver) is also described, together with its extension to the equations for compressible flows (generalized Stokes solver).

In chapter 6 the idea of the pressure integral conditions is further investigated in the context of the unsteady Stokes problem with a continuous time dependence, that is, for the linear problem defined by the equations

$$\begin{aligned} \frac{\partial \boldsymbol{u}}{\partial t} &= -\boldsymbol{\nabla} P + \nu \nabla^2 \boldsymbol{u} + \boldsymbol{f}, \\ \boldsymbol{\nabla}\cdot \boldsymbol{u} &= 0, \end{aligned} \tag{1.15}$$

where $\boldsymbol{f}$ denotes some body force acting on the fluid. An uncoupled formulation of the problem is derived which is based on the elimination of the continity equation in favour of the pressure Poisson equation $\nabla^2 P = \boldsymbol{\nabla}\cdot \boldsymbol{f}$ and on the introduction of space-time integral conditions for the pressure field. Such a formulation is helpful in explaining the transition from methods using integral conditions to methods of fractional-step type.

Chapter 7 discusses the celebrated fractional-step projection method introduced by Chorin (1968a-b, 1969) and Temam (1969a-b), which is the most widely used method for solving the primitive-variable Navier–Stokes equations. It relies on a time discretization of fractional-step type in which the effect of viscosity is taken into account separately from the fulfillment of the condition of incompressibility. In a first half-step an intermediate velocity field $\boldsymbol{u}^{n+1/2}$ (not solenoidal) is calculated form a time discretized version of the momentum equations with the pressure term omitted, namely,

$$\frac{\boldsymbol{u}^{n+1/2} - \boldsymbol{u}^n}{\Delta t} = -(\boldsymbol{u}^n \cdot \boldsymbol{\nabla})\boldsymbol{u}^n + \nu \nabla^2 \boldsymbol{u}^n. \tag{1.16}$$

Then, the intermediate field $\boldsymbol{u}^{n+1/2}$ is decomposed into the sum of a solenoidal velocity field $\boldsymbol{u}^{n+1}$ and the gradient of a scalar function proportional to the unknown pressure, namely, $\boldsymbol{\nabla} P^{n+1}$, according to the following equations:

$$\begin{aligned} \frac{\boldsymbol{u}^{n+1} - \boldsymbol{u}^{n+1/2}}{\Delta t} &= -\boldsymbol{\nabla} P^{n+1}, \\ \boldsymbol{\nabla}\cdot \boldsymbol{u}^{n+1} &= 0. \end{aligned} \tag{1.17}$$

The unique character of the projection method is pointed out, giving an interesting abstract geometrical interpretation of the construction underlying the second half-step. The advantages of this method over nonfractional-step approaches are also put in evidence. Special attention is paid to elucidate the form assumed by the boundary conditions in the fractional-step method, according to whether a Poisson equation for the pressure is introduced explicitly or not, and depending on the kind of spatial discretization employed.

The last chapter (8) is devoted to the study of the incompressible equations for an ideal fluid, the so-called incompressible Euler equations, namely,

$$\begin{aligned} \frac{\partial \boldsymbol{u}}{\partial t} + (\boldsymbol{u}\cdot\boldsymbol{\nabla})\boldsymbol{u} &= -\boldsymbol{\nabla}P, \\ \boldsymbol{\nabla}\cdot\boldsymbol{u} &= 0. \end{aligned} \tag{1.18}$$

We will introduce alternative formulations of the Euler equations for incompressible vortical flows using the vorticity variable as unknown, in both two and three dimensions. The complete set of boundary conditions assuring the equivalence of these formulations with the equations (1.18) governing incompressible inviscid flows will be given. In this connection, some new numerical schemes for computing time-dependent solutions of nonlinear advection problems by finite elements will be also described.

The analysis of the various formulations and of the corresponding numerical schemes will be developed mainly in view of spatial approximations based on the finite element method and using unstructured grids. In fact, this kind of spatial representation together with the associated variational formulation of the problem, in addition to allowing to recognize the boundary and nonboundary conditions most appropriate to each situation, leads to numerical schemes of the greatest flexibility and of adequate spatial accuracy.

1.4 Some references

The next several chapters concern the aforementioned formulations of the incompressible Navier–Stokes and Euler equations. For the most part, the exposition is self contained and maintained at a rather elementary level. The reader interested to more specific aspects concerning the equations governing incompressible viscous flows and their approximate solution may wish to consult some of the texts in this area for an alternative presentation or a more detailed discussion.

The standard references for a physical introduction to fluid dynamics in general and incompressible viscous flows in particular are Batchelor [3] and Landau and Lifshitz [26]. A more elementary but nevertheless very excellent treatment of the subject is provided by Currie [11] and by Acheson [1]. A comprehensive account of laminar viscous flows in various sitations and regimes is provided by Churchill [10]. A unique introduction to the equations governing inviscid and viscous incompressible flows is given in the second volume of *The Feynman Lectures* [13], in chapters 40 and 41 entitled respectively "The flow of dry water" and "The flow of wet water."

The mathematical theory of the incompressible Navier–Stokes equations is formulated by Ladyzhensaya in [24] where the basic existence, uniqueness and regularity results for the steady/unsteady and linear/nonlinear problems are discussed in detail. Some important aspects concerning the mathematical structure of the equations for incompressible viscous flows are also examined by Chorin and Marsden [9].

Regarding numerical methods for the Navier–Stokes equations, most texts deals with methods based on a specific spatial discretization technique. For finite differences, Telionis' monograph [34] constitutes the most authoritative presentation of numerical schemes for calculating unsteady viscous flows in the presence of boundaries. Some important results about the convergence of finite difference approximations for the incompressible equations are given by Teman in [35] and by Kreiss and Lorenz in [23].

For finite elements, Temam's work [35] is the classical reference, encompassing both the basic results of functional analysis required to build a modern mathematical theory of the Navier–Stokes equations and detailed error estimates and convergence analyses for several finite-dimensional approximations to the equations. Finite element discretizations of mixed type for the steady-state equations are the subject of the work by Girault and Raviart [16] whereas implementation issues involved by the use of finite elements for solving incompressible problems are discussed by Thomasset [36]. Furthermore, the monography of Glowinski [17] contains a detailed description of direct and iterative methods for solving the Stokes problem and an optimal-control-theory-based method for dealing with the nonlinear terms. Finally, the graduate texts by Pironneau [30] and Carey and Oden [7] on the use of finite elements in computational fluid dynamics contain chapters entirely devoted to the incompressible Navier–Stokes equations.

As far as spectral approximations are concerned, the primitive variable Navier-Stokes equations are briefly discussed by Gottielb and Orszag in [18]. Several spectral methods for different formulations of the incompressible equations are considered by Canuto, Hussaini, Quarteroni and Zang in [6] whereas the spectral approximation of the Stokes problem is thoroughly sudied by Bernardi and Maday in [4].

Finally, for a critical comparison of many of these different numerical techniques the reader can consult Chapter 17 of the second volume of Fletcher [14].

Chapter 2

Nonprimitive variable formulations in 2D

2.1 Introduction

In the study of two-dimensional flows, the incompressible Navier–Stokes equations can be formulated in a convenient alternative manner, by introducing two scalar variables, the vorticity ζ and stream function ψ, in place of the primitive variables, the velocity $\boldsymbol{u}$ and pressure P. The "curl" of ψ is nothing but $\boldsymbol{u}$, whereas the variable ζ itself is the only vector component of $\nabla\times\boldsymbol{u}$ which is different from zero in the particular situation of a plane motion. Such a fully scalar representation of the governing equations is by far the most frequently used for studying both steady and unsteady problems, because it offers the two advantages of reducing the number of the unknowns and of eliminating the incompressibility condition whose fulfillment may be very difficult to be achieved, numerically.

In this chapter, we first derive the equations of the ζ-ψ formulation and demonstrate its equivalence with the primitive variable formulation of the time-dependent Navier–Stokes equations under no-slip boundary conditions. The ζ-ψ representation is however faced with a difficulty: although the unknowns ζ and ψ are governed by two well known equations, the transport (advection–diffusion) equation and the Poisson equation, respectively, the stream function is found to be supplemented by two boundary conditions and no boundary condition is available for the vorticity.

Such a difficulty can be circumvented either by eliminating the vorticity variable from the formulation and introducing a single *biharmonic* equation for the stream function, or by regarding the two second-order partial differential equations for ζ and ψ as constituting a system of two *coupled* equations, the coupling being engendered by the double specification on the boundary for one variable.

After a brief description of these approaches, we will introduce a third method for dealing with the absence of boundary conditions for the vorticity.

The method is based on the idea of supplementing the vorticity transport equation with conditions which are the equivalent substitute of the no-slip conditions for the velocity or the stream function. If definite vorticity conditions can be established for the exact problem in the continuum, this would imply two important consequences: firstly, an independent meaning can be given to the vorticity equation, at least in linear situations; secondly, the study of vorticity dynamics can be made complete by including a description of the influence of solid walls on the vorticity field, aside the mechanisms of vorticity advection and diffusion acting within the fluid.

The appropriate conditions which supplement the vorticity in two-dimensional flows will be derived in the present chapter. They are found to be of an *integral character* and to admit the very simple geometrical interpretation of fixing the *orthogonal projection* of the vorticity field with respect to the linear space of the functions which are harmonic in the domain occupied by the fluid. We will show how this unusual kind of conditions can be implemented in effective numerical schemes for solving the vorticity–stream function equations in an uncoupled manner. The attention is directed mainly to discrete versions of the governing equations obtained by means of the finite element method, but a few numerical results of Chebyshev spectral calculations will also be provided to compare some different ways of enforcing the correct conditions on the vorticity. In this connection, the so-called Glowinski–Pironneau method for solving the ζ-ψ equations in arbitrarily shaped regions will be described and the influence matrix method applied to solve problems in simple domains will be also introduced.

To complete the examination of the computational tools needed to solve the two-dimensional Navier–Stokes equations, the basic schemes for approximating the nonlinear terms within the context of finite differences, finite elements and spectral approximations will be described.

In the last section the vorticity integral conditions for plane two-dimensional flows will be extended to problems with rotational symmetry, and the complete set of equations and conditions in spherical coordinates for this class of fluid motions in three dimensions will be given.

2.2 Vorticity–stream function equations

For fluid motions parallel to the plane xy, the scalar vorticity ζ is the z-component of the vorticity vector $\boldsymbol{\zeta} = \nabla\times\boldsymbol{u}$, normal to that plane, namely,

$$\zeta = \nabla\times\boldsymbol{u}\cdot\boldsymbol{k} \qquad \text{or} \qquad \zeta\,\boldsymbol{k} = \nabla\times\boldsymbol{u}, \tag{2.1}$$

where $\boldsymbol{u} = (u_x, u_y)$, $\boldsymbol{\nabla} = (\partial/\partial x, \partial/\partial y)$ and $\boldsymbol{k}$ is the unit vector normal to the plane xy. The standard notation for vector differential operators is used, see, *e.g.*, Adams [2]. In two dimensions, the condition of incompressibility $\boldsymbol{\nabla}\cdot\boldsymbol{u} = 0$ can be satisfied exactly by expressing $\boldsymbol{u}$ in terms of a stream function ψ according to

$$\boldsymbol{u} = \boldsymbol{\nabla}\psi\times\boldsymbol{k}. \tag{2.2}$$

This equation, once expressed in terms of the vector components, gives $u_x = \partial\psi/\partial y$ and $u_y = -\partial\psi/\partial x$. Thus, one obtains immediately

$$\boldsymbol{\nabla}\cdot\boldsymbol{u} = \boldsymbol{\nabla}\cdot(\boldsymbol{\nabla}\psi\times\boldsymbol{k}) = \frac{\partial}{\partial x}\Big(\frac{\partial\psi}{\partial y}\Big) + \frac{\partial}{\partial y}\Big(-\frac{\partial\psi}{\partial x}\Big) = 0.$$

We now eliminate the pressure from the Navier–Stokes equations taking the curl of the momentum equation. To simplify the derivation, the nonlinear term is first expressed in the so-called Lamb form through the identity

$$(\boldsymbol{u}\cdot\boldsymbol{\nabla})\boldsymbol{u} = (\boldsymbol{\nabla}\times\boldsymbol{u})\times\boldsymbol{u} + \boldsymbol{\nabla}(\tfrac{1}{2}u^2) = \zeta\,\boldsymbol{k}\times\boldsymbol{u} + \boldsymbol{\nabla}(\tfrac{1}{2}u^2).$$

Then, the application of the curl operator to the nonlinear term gives

$$\boldsymbol{\nabla}\times[(\boldsymbol{u}\cdot\boldsymbol{\nabla})\boldsymbol{u}] = \boldsymbol{\nabla}\times[\zeta\,\boldsymbol{k}\times\boldsymbol{u} + \boldsymbol{\nabla}(\tfrac{1}{2}u^2)] = \boldsymbol{\nabla}\times[\zeta\,\boldsymbol{k}\times\boldsymbol{u}],$$

so that the curl of the momentum equation (2.1) gives the equation

$$\frac{\partial}{\partial t}(\zeta\,\boldsymbol{k}) + \boldsymbol{\nabla}\times(\zeta\,\boldsymbol{k}\times\boldsymbol{u}) = \nu\nabla^2(\zeta\,\boldsymbol{k}).$$

Consider now the vector identity

$$\boldsymbol{\nabla}\times(\boldsymbol{a}\times\boldsymbol{b}) = \boldsymbol{a}\,\boldsymbol{\nabla}\cdot\boldsymbol{b} - \boldsymbol{b}\,\boldsymbol{\nabla}\cdot\boldsymbol{a} + (\boldsymbol{b}\cdot\boldsymbol{\nabla})\boldsymbol{a} - (\boldsymbol{a}\cdot\boldsymbol{\nabla})\boldsymbol{b},$$

where $\boldsymbol{\nabla}$ is now the three-dimensional gradient operator, and use it, with $\boldsymbol{a} = \boldsymbol{\nabla}\times\boldsymbol{u}$, $\boldsymbol{b} = \boldsymbol{u}$ and $\boldsymbol{\nabla}\cdot\boldsymbol{u} = 0$, to give

$$\boldsymbol{\nabla}\times[(\boldsymbol{\nabla}\times\boldsymbol{u})\times\boldsymbol{u}] = (\boldsymbol{u}\cdot\boldsymbol{\nabla})\boldsymbol{\nabla}\times\boldsymbol{u} - (\boldsymbol{\nabla}\times\boldsymbol{u}\cdot\boldsymbol{\nabla})\boldsymbol{u}.$$

In the present 2D situation, $\boldsymbol{\nabla}\times\boldsymbol{u} = \zeta\,\boldsymbol{k}$ while the second term on the right-hand side vanishes, since $\boldsymbol{u}$ does not depend on z. It follows that $\boldsymbol{\nabla}\times[\zeta\,\boldsymbol{k}\times\boldsymbol{u}] = (\boldsymbol{u}\cdot\boldsymbol{\nabla})\zeta\,\boldsymbol{k}$, so that the curl of the nonlinear term in two dimensions can be expressed in the following form

$$\boldsymbol{\nabla}\times[(\boldsymbol{u}\cdot\boldsymbol{\nabla})\boldsymbol{u}] = \boldsymbol{\nabla}\times[\zeta\,\boldsymbol{k}\times\boldsymbol{u}] = (\boldsymbol{u}\cdot\boldsymbol{\nabla})\zeta\,\boldsymbol{k}.$$

By virtue of the representation $\boldsymbol{u} = \boldsymbol{\nabla}\psi\times\boldsymbol{k}$, the curl of the nonlinear term can be given the final form

$$\boldsymbol{\nabla}\times[(\boldsymbol{u}\cdot\boldsymbol{\nabla})\boldsymbol{u}] = \frac{\partial(\zeta,\psi)}{\partial(x,y)}\,\boldsymbol{k} = J(\zeta,\psi)\,\boldsymbol{k},$$

where, as usual, J denotes the Jacobian determinant.

In conclusion, taking the curl of the momentum equation leads to the vorticity transport equation:

$$\frac{\partial \zeta}{\partial t} + J(\zeta, \psi) = \nu \nabla^2 \zeta.$$

On the other hand, substituting the expression $\boldsymbol{u} = \boldsymbol{\nabla}\psi \times \boldsymbol{k}$ into the vorticity definition (2.1) gives the following Poisson equation for the stream function

$$-\nabla^2 \psi = \zeta.$$

The boundary conditions supplementing the two equations above are deduced by separating the normal and tangential components of the velocity boundary condition $\boldsymbol{u}|_S = \boldsymbol{b}$. Here S represents the boundary of the two-dimensional domain V, which is always assumed to be simply connected. Let $\boldsymbol{n}$ denote the outward unit vector normal to the boundary S and $\boldsymbol{\tau}$ the unit vector tangential to S with anti-clokwise orientation. Finally, let s be the curvilinear coordinate along the boundary S. Then, the boundary condition $\boldsymbol{u}|_S = \boldsymbol{b}$ yields the condition for the normal component:

$$\boldsymbol{n} \cdot \boldsymbol{\nabla}\psi \times \boldsymbol{k}|_S = \boldsymbol{k} \times \boldsymbol{n} \cdot \boldsymbol{\nabla}\psi|_S = \boldsymbol{\tau} \cdot \boldsymbol{\nabla}\psi|_S = \left.\frac{\partial \psi}{\partial s}\right|_S = \boldsymbol{n} \cdot \boldsymbol{b},$$

and for the tangential component:

$$\boldsymbol{\tau} \cdot \boldsymbol{\nabla}\psi \times \boldsymbol{k}|_S = \boldsymbol{k} \times \boldsymbol{\tau} \cdot \boldsymbol{\nabla}\psi|_S = -\boldsymbol{n} \cdot \boldsymbol{\nabla}\psi|_S = -\left.\frac{\partial \psi}{\partial n}\right|_S = \boldsymbol{\tau} \cdot \boldsymbol{b}.$$

The first boundary condition, after integrating its right-hand side, provides a Dirichlet condition for ψ. By virtue of the global condition $\oint_S \boldsymbol{n} \cdot \boldsymbol{b}\, ds = 0$, such an integration defines a single-valued function up to an arbitrary additive function of time, namely,

$$a(s,t) = \int_{s_1}^{s} \boldsymbol{n}(s') \cdot \boldsymbol{b}(s', t)\, ds' + A(t),$$

where s_1 is the coordinate of any fixed point of S and s' is a dummy variable of integration. To simplify the expression of the boundary conditions for ψ, we drop the term $A(t)$ from the Dirichlet condition and introduce the notation

$$b(s,t) = -\boldsymbol{\tau}(s) \cdot \boldsymbol{b}(s,t),$$

so that the two conditions can be written as follcws

$$\psi|_S = a, \qquad \left.\frac{\partial \psi}{\partial n}\right|_S = b.$$

(Note that b should not be confused with $|\boldsymbol{b}|$, that is, $b \neq |\boldsymbol{b}|$.)

As far as the initial condition for the system of equations governing ζ and ψ is concerned, the initial velocity field $\boldsymbol{u}_0$ provides the following initial condition for the vorticity

$$\zeta|_{t=0} = \boldsymbol{\nabla}\times(\boldsymbol{u}|_{t=0})\cdot\boldsymbol{k} = \boldsymbol{\nabla}\times\boldsymbol{u}_0\cdot\boldsymbol{k}.$$

Collecting the equations and conditions all together, the vorticity–stream function formulation of the Navier–Stokes problem for two-dimensional flows is:

$$\begin{cases} \dfrac{\partial\zeta}{\partial t} - \nu\nabla^2\zeta + J(\zeta,\psi) = 0, \\ -\nabla^2\psi = \zeta, \\ \psi|_S = a, \qquad \left.\dfrac{\partial\psi}{\partial n}\right|_S = b, \\ \zeta|_{t=0} = \boldsymbol{\nabla}\times\boldsymbol{u}_0\cdot\boldsymbol{k}, \end{cases} \tag{2.3}$$

where $a = \int_{s_1}^{s} \boldsymbol{n}\cdot\boldsymbol{b}\, ds'$ and $b = -\boldsymbol{\tau}\cdot\boldsymbol{b}$. The initial datum $\boldsymbol{u}_0(\boldsymbol{x})$ and the boundary datum $a(s,t)$ are assumed to satisfy the conditions

$$\boldsymbol{\nabla}\cdot\boldsymbol{u}_0 = 0, \qquad \frac{\partial a(s,0)}{\partial s} = \boldsymbol{n}\cdot\boldsymbol{u}_0|_S, \tag{2.4}$$

the latter being nothing but the compatibility condition (7) rewritten in terms of the Dirichlet datum $a = a(s,t)$. The global condition $\oint_S \boldsymbol{n}\cdot\boldsymbol{b}\, ds = 0$ does not appear any more because it has been already taken into account in the definition of the single-valued function $a(s,t)$.

Theorem 2.1. *The vorticity–stream function problem (2.3) is equivalent to the original primitive variable Navier–Stokes problem (1.1)–(1.7) in two dimensions, provided that the two conditions (2.4) on the data are satisfied.*

Proof. The implication is evident. Conversely, let us assume that ζ and ψ are solution to the set of equations and conditions (2.3) with the data $\boldsymbol{u}_0$ and a satisfying (2.4). Let us consider, for $t > 0$, the vector field $\boldsymbol{v} = \boldsymbol{\nabla}\psi\times\boldsymbol{k}$. First, $\boldsymbol{v}$ is solenoidal since $\boldsymbol{\nabla}\cdot\boldsymbol{v} = \boldsymbol{\nabla}\cdot(\boldsymbol{\nabla}\psi\times\boldsymbol{k}) = 0$. Furthermore, its curl satisfies $\boldsymbol{\nabla}\times\boldsymbol{v} = \boldsymbol{\nabla}\times(\boldsymbol{\nabla}\psi\times\boldsymbol{k}) = -\nabla^2\psi\,\boldsymbol{k} = \zeta\,\boldsymbol{k}$, since $-\nabla^2\psi = \zeta$. Hence the vorticity equation in (2.3) gives

$$\frac{\partial\boldsymbol{\nabla}\times\boldsymbol{v}}{\partial t} - \nu\nabla^2\boldsymbol{\nabla}\times\boldsymbol{v} + J(\boldsymbol{\nabla}\times\boldsymbol{v},\psi) = 0.$$

By virtue of the vector identity used before and since $\boldsymbol{\nabla}\cdot\boldsymbol{v} = 0$, the nonlinear term can be expressed as follows

$$\begin{aligned} J(\boldsymbol{\nabla}\times\boldsymbol{v},\psi) &= (\boldsymbol{v}\cdot\boldsymbol{\nabla})\boldsymbol{\nabla}\times\boldsymbol{v} = \boldsymbol{\nabla}\times[(\boldsymbol{\nabla}\times\boldsymbol{v})\times\boldsymbol{v}] \\ &= \boldsymbol{\nabla}\times[(\boldsymbol{v}\cdot\boldsymbol{\nabla})\boldsymbol{v} - \boldsymbol{\nabla}(\tfrac{1}{2}v^2)] = \boldsymbol{\nabla}\times[(\boldsymbol{v}\cdot\boldsymbol{\nabla})\boldsymbol{v}]. \end{aligned}$$

The vorticity equation gives

$$\nabla\times\left[\frac{\partial \boldsymbol{v}}{\partial t}-\nu\nabla^2\boldsymbol{v}+(\boldsymbol{v}\cdot\nabla)\boldsymbol{v}\right]=0$$

and therefore

$$\frac{\partial \boldsymbol{v}}{\partial t}-\nu\nabla^2\boldsymbol{v}+(\boldsymbol{v}\cdot\nabla)\boldsymbol{v}=\nabla Q$$

for some scalar function Q. In order that $\boldsymbol{v}$ can be identified with the velocity field $\boldsymbol{u}$ solution of the original problem (and Q with $-P$), it remains to show that $\boldsymbol{v}$ assumes the same boundary and initial values of $\boldsymbol{u}$. For the boundary values it results $\boldsymbol{v}|_S=\nabla\psi\times\boldsymbol{k}|_S$, which, after separating the normal and tangential components, gives

$$\boldsymbol{n}\cdot\boldsymbol{v}|_S=\boldsymbol{n}\cdot\nabla\psi\times\boldsymbol{k}|_S=\boldsymbol{k}\times\boldsymbol{n}\cdot\nabla\psi|_S=\boldsymbol{\tau}\cdot\nabla\psi|_S=\left.\frac{\partial\psi}{\partial s}\right|_S=\frac{\partial a}{\partial s}=\boldsymbol{n}\cdot\boldsymbol{b},$$

$$\boldsymbol{\tau}\cdot\boldsymbol{v}|_S=\boldsymbol{\tau}\cdot\nabla\psi\times\boldsymbol{k}|_S=\boldsymbol{k}\times\boldsymbol{\tau}\cdot\nabla\psi|_S=-\boldsymbol{n}\cdot\nabla\psi|_S=-\left.\frac{\partial\psi}{\partial n}\right|_S=-b=\boldsymbol{\tau}\cdot\boldsymbol{b}.$$

Hence $\boldsymbol{v}|_S=\boldsymbol{b}=\boldsymbol{u}|_S$. Concerning the initial values, one has to determine the values assumed by $\boldsymbol{v}$ as $t\to 0^+$ as a consequence of the imposition of the initial condition for ζ in the ζ-ψ system. One has

$$\boldsymbol{v}|_{t=0}=\lim_{t\to 0^+}\boldsymbol{v}(\boldsymbol{x},t)=\lim_{t\to 0^+}\nabla\psi(\boldsymbol{x},t)\times\boldsymbol{k}=\nabla\Big(\lim_{t\to 0^+}\psi(\boldsymbol{x},t)\Big)\times\boldsymbol{k}.$$

Let ψ_0 denote the solution of the following Dirichlet problem:

$$-\nabla^2\psi_0=\zeta_0=\nabla\times\boldsymbol{u}_0\cdot\boldsymbol{k},\qquad \psi_0|_S=a(s,0).$$

By the assumed continuity of $a(s,t)$ as $t\to 0^+$ and since $\zeta_0=\zeta|_{t=0}$, the well-posedness of the Dirichlet problem implies that $\lim_{t\to 0^+}\psi(\boldsymbol{x},t)=\psi_0(\boldsymbol{x})$, so that

$$\boldsymbol{v}|_{t=0}=\nabla\psi_0(\boldsymbol{x})\times\boldsymbol{k}.$$

Using the identity $\nabla\times(\nabla f\times\boldsymbol{k})=-\nabla^2 f\,\boldsymbol{k}$ in the Poisson equation above, one obtains $\nabla\times(\nabla\psi_0\times\boldsymbol{k})=\nabla\times\boldsymbol{u}_0$ or $\nabla\times(\nabla\psi_0\times\boldsymbol{k}-\boldsymbol{u}_0)=0$. It follows that

$$\nabla\psi_0\times\boldsymbol{k}-\boldsymbol{u}_0=\nabla\alpha$$

for some scalar function α. Now, $\nabla\cdot(\nabla\psi_0\times\boldsymbol{k})=0$ and $\nabla\cdot\boldsymbol{u}_0=0$ by the first condition in (2.4), so that α is harmonic in V. Furthermore, taking the normal component of $\nabla\psi_0\times\boldsymbol{k}-\boldsymbol{u}_0$ on S,

$$\begin{aligned}\boldsymbol{n}\cdot\nabla\psi_0\times\boldsymbol{k}|_S-\boldsymbol{n}\cdot\boldsymbol{u}_0|_S&=\boldsymbol{k}\times\boldsymbol{n}\cdot\nabla\psi_0|_S-\boldsymbol{n}\cdot\boldsymbol{u}_0|_S=\boldsymbol{\tau}\cdot\nabla\psi_0|_S-\boldsymbol{n}\cdot\boldsymbol{u}_0|_S\\&=\left.\frac{\partial\psi_0}{\partial s}\right|_S-\boldsymbol{n}\cdot\boldsymbol{u}_0|_S=\frac{\partial a(s,0)}{\partial s}-\boldsymbol{n}\cdot\boldsymbol{u}_0|_S=0,\end{aligned}$$

by virtue of the second condition in (2.4). It follows that $\boldsymbol{n} \cdot \boldsymbol{\nabla}\alpha|_S = \partial\alpha/\partial n|_S = 0$, so that $\alpha =$ constant. In conclusion, $\boldsymbol{\nabla}\psi_0 \times \boldsymbol{k} = \boldsymbol{u}_0$ everywhere in V and therefore $\boldsymbol{v}|_{t=0} = \boldsymbol{u}_0$. This completes the proof. □

A rigorous demonstration of the theorem has been provided by Guermond and Quartapelle (1993). The interest of this result for the CFD analysts is that the compatibility condition between the initial and boundary data allows for solutions of the Navier–Stokes equations with a *minimal* (reasonable) regularity, *viz.* H^1 for the velocity, which is the regularity occurring in most numerical schemes relying on spatial discretizations of local type.

Remark. The vorticity–stream function problem could be stated also specifying the initial condition for the vorticity directly in terms of a prescribed initial vorticity field ζ_0, namely,

$$\zeta|_{t=0} = \zeta_0,$$

where ζ_0 is an arbitrary function, with no reference to an initial velocity field. Of course, the two conditions (2.4) on the initial and boundary values. In this case, both boundary values $a(s,t)$ and $b(s,t)$ can be specified at the initial time $t = 0$, but they cannot be prescribed independently from the initial vorticity field $\zeta_0(\boldsymbol{x})$. In fact, a Poisson problem for the "initial" stream function ψ_0 with both Dirichlet and Neumann conditions is overdetermined and compatibility conditions between the data $a(s,0)$, $b(s,0)$ and $\zeta_0(\boldsymbol{x})$ must be satisfied in order for the problem to be solvable. This situation is to be compared with that occurring for $t > 0$. At any subsequent time the problem for ψ is still overspecified, the Poisson equation being again supplemented by both Dirichlet and Neumann boundary conditions (cf. problem (2.3)). However, for $t > 0$ the overspecification is only apparent because the compatibility conditions existing between $a(s,t)$, $b(s,t)$ and $\zeta(\boldsymbol{x},t)$ provide the conditions required to supplement the vorticity equation so as to determine the field $\zeta(\boldsymbol{x},t)$ for any $t > 0$. Actually, the very form of these conditions is the main concern of the analysis to be developed in the following.

The solution of the vorticity–stream function problem (2.3) is difficult for three main reasons:

1. The problem is nonlinear due to the presence of the Jacobian term which represents the advection of vorticity by the fluid.

2. The advection term couples (nonlinearly) the two equations for ζ and ψ.

3. The boundary conditions are troublesome, since two boundary conditions are prescribed for ψ and no boundary condition is available for ζ.

The difficulties caused by the nonlinearity can be dealt with by combining explicit or implicit time-stepping algorithm with suitable iterative procedures. On the other hand, the structural difficulty associated with the peculiar nature of the boundary conditions in the ζ-ψ system can be tackled according to one of the following three methods: the biharmonic formulation, the coupled vorticity–stream function approach or the uncoupled (or split) vorticity–stream function method.

2.3 Biharmonic formulation

The simplest and almost obvious way of circumventing the difficulty caused by the lack of a boundary condition for the vorticity consists in eliminating the vorticity variable itself. In other words, ζ is no longer considered as an unknown of the problem but merely as a short-hand notation for the (negative of the) Laplacian of ψ, namely, $\zeta = -\nabla^2\psi$. Substituting this expression into the vorticity transport equation gives the following time-dependent biharmonic problem (see, *e.g.*, Landau and Lifshitz [26, p. 53])

$$\begin{gathered}\frac{\partial\nabla^2\psi}{\partial t} - \nu\nabla^4\psi + J(\nabla^2\psi,\psi) = 0,\\ \psi|_S = a, \qquad \left.\frac{\partial\psi}{\partial n}\right|_S = b,\\ \psi|_{t=0} = \psi_0.\end{gathered} \tag{2.5}$$

Here the "initial" stream function ψ_0 is the solution of the Dirichlet problem $-\nabla^2\psi_0 = \boldsymbol{\nabla}\times\boldsymbol{u}_0\cdot\boldsymbol{k}$, $\psi_0|_S = a(s,0)$, with the data $\boldsymbol{u}_0$ and a satisfying the two conditions (2.4).

In such a formulation, the simultaneous specification of the Dirichlet and Neumann boundary conditions poses no difficulty because they are both required to supplement the fourth-order elliptic operator ∇^4. Therefore, approximate solutions of the biharmonic problem (2.5) either for steady and unsteady problems can be computed using any discretization method, such as finite differences (see, *e.g.*, Schreiber *et al.* 1983), finite elements (see Olson and Tuann 1979) and spectral methods (Orszag 1971). In particular, when finite elements are employed, the fourth-order stream function equation must be recast in a variational form, which reads:

Find $\psi \in H^2(V)$ such that $\psi|_S = a$, $(\partial\psi/\partial n)|_S = b$ and

$$(\boldsymbol{\nabla}\phi, \tfrac{\partial}{\partial t}\boldsymbol{\nabla}\psi) + \nu(\nabla^2\phi,\nabla^2\psi) + (J(\phi,\psi),\nabla^2\psi) = 0, \qquad \forall\phi \in H_0^2(V),$$

where $(\cdot\,,\cdot)$ denotes the L^2 scalar product whereas $H^2(V)$ and $H_0^2(V)$ are the standard notation for Sobolev spaces (see, *e.g.*, Strang and Fix [33]).

Alternatively, a linearized version of problem (2.5) can be solved also by resorting to the classical theory of boundary integral equations (see Jaswon and Symm [21]) and the boundary integral method (Ingham and Kelmanson [19]) for the biharmonic operator ∇^4. In the particular situation of steady flows for which inertial effects can be neglected (Stokes problem), the equation in (2.5) becomes the homogeneous biharmonic equation $\nabla^4\psi = 0$, whose solution is calculated by all these methods in a direct, that is, noniterative, manner.

2.4 Coupled vorticity–stream function formulation

A second method for circumventing the difficulty associated with the double boundary specification for ψ consists in regarding the equations for ζ and ψ as a *coupled system* (Campion-Renson and Crochet 1978).

Of course, the nonlinear term of the vorticity equation implies that the equations for ζ and ψ are already coupled together. However, even in the absence of this term, the equations have still to be considered coupled together through the boundary conditions. In other words, one of the two conditions $\psi|_S = a$ and $(\partial\psi/\partial n)|_S = b$ must be "associated" with the vorticity transport equation which by itself is unable to determine ζ uniquely. In this approach, no numerical boundary condition for ζ is required, but the vorticity equation cannot be solved separately from the stream function equation, not even in the linear case: the two equations must be solved necessarily coupled together.

The basic idea of the method is to express the problem in a form that associates each of the boundary conditions for ψ to each of the governing equations, to give

$$\begin{aligned} -\nu\nabla^2\zeta + \frac{\partial\zeta}{\partial t} + J(\zeta,\psi) &= 0, & \psi|_S &= a, \\ -\nabla^2\psi - \zeta &= 0, & \left.\frac{\partial\psi}{\partial n}\right|_S &= b, \end{aligned} \qquad (2.6)$$

$$\zeta|_{t=0} = \nabla\times\boldsymbol{u}_0\cdot\boldsymbol{k}.$$

Thus, the Dirichlet boundary condition $\psi|_S = a$ is associated with the vorticity equation whereas the Neumann condition $(\partial\psi/\partial n)|_S = b$ supplements the stream function equation. However, it is interesting to note that the equation $-\nabla^2\psi = \zeta$ together with the derivative boundary condition $(\partial\psi/\partial n)|_S = b$ do not constitute a Neumann problem since the variable ζ appearing in the equation is not a given function, but another unknown; correspondingly, the condition $\int \zeta\, dV = -\oint b\, ds$, which should be imposed on the data of a Neumann problem, is not to be considered here: it will be automatically satisfied as a consequence of the fact that ζ and ψ are determined as solution of the entire coupled system (2.6).

When expressed in variational form, the coupled system (2.6) constitutes a so-called *mixed variational formulation* of the vorticity–stream function problem. The most typical characterization of this problem reads as follows:

Find $\zeta \in H^1(V)$ and $\psi \in H^1(V)$ such that $\psi|_S = a$ and

$$\begin{aligned} &\nu(\nabla\phi, \nabla\zeta) + (\phi, \tfrac{\partial\zeta}{\partial t}) + (\phi, J(\zeta, \psi)) = 0, && \forall\phi \in H_0^1(V), \\ &(\nabla\xi, \nabla\psi) - (\xi, \zeta) = \oint \xi b \, ds, && \forall\xi \in H^1(V), \end{aligned} \tag{2.7}$$

where $(\cdot, \cdot)$ denotes the L^2 scalar product whereas $H^1(V)$ and $H_0^1(V)$ are the standard notation for Sobolev spaces (see, *e.g.*, Strang and Fix [33]). Thus, in the mixed formulation one of the two boundary conditions prescribed on ψ is imposed implicitly as a natural condition. In the spatially discrete case, a single large system of algebraic equations is obtained (Campion-Renson and Crochet 1978) which can be solved by Newton–Raphson iteration (Stevens 1982). These ideas have been extended to the equations for three-dimensional flows by Fortin and Thomasset (1979).

The application of the coupled solution method is not limited to spatial discretizations by means of finite elements. For instance, efficient implementations of the method have been proposed for finite differences using ADI techniques (Napolitano 1991) and for a Chebyshev spectral approximation using multigrid acceleration (Heinrichs 1992).

2.5 Vorticity integral conditions

As it has been shown, the set (2.3) containing the equations and conditions for ζ and ψ can be given a mathematically satisfactory interpretation either

i) as a fourth-order differential problem, with ζ being nothing but a shorthand notation for $-\nabla^2\psi$, or

ii) as a coupled system of equations for ζ and ψ, the coupling being brought about by the nature of the boundary conditions, irrespective of the linear or nonlinear character of the vorticity equation.

However, an alternative viewpoint might be introduced if an independent individuality could be attributed to the vorticity equation, at least in the absence of the nonlinear term. In order that this may be possible, conditions for the vorticity are required which are an equivalent substitute of the boundary conditions originally attached to the stream function equation, without involving the values of velocity at internal points of the fluid domain. As a matter of facts, these vorticity conditions do exist, although they are found to be quite different in character from the conditions of the usual boundary-value type. To derive the sought conditions for the vorticity we need one of the well known identities or theorems due to Green (see, *i.e.*, Jackson [20, p. 40]).

2.5.1 Green identities

Green identities are obtained as simple applications of the divergence theorem:

$$\int \boldsymbol{\nabla} \cdot \boldsymbol{A} \, dV = \oint \boldsymbol{n} \cdot \boldsymbol{A} \, ds,$$

where $\boldsymbol{A}$ is any well-behaved vector field defined in the plane domain V and ds is the infinitesimal element of the boundary S of V. Let $\boldsymbol{A} = \psi \boldsymbol{\nabla}\phi$, where ψ and ϕ are arbitrary scalar fields. Now

$$\boldsymbol{\nabla} \cdot (\psi \boldsymbol{\nabla}\phi) = \psi \nabla^2 \phi + \boldsymbol{\nabla}\psi \cdot \boldsymbol{\nabla}\phi$$

and

$$\psi \boldsymbol{n} \cdot \boldsymbol{\nabla}\phi = \psi \frac{\partial \phi}{\partial n},$$

where $\partial/\partial n$ is the outwards normal derivative at the boundary S. When these two relations are substituted into the divergence theorem, there results *Green first identity:*

$$\int (\psi \nabla^2 \phi + \boldsymbol{\nabla}\psi \cdot \boldsymbol{\nabla}\phi) \, dV = \oint \psi \frac{\partial \phi}{\partial n} \, ds.$$

If we write this relation again with ψ and ϕ interchanged, and then subtract the new relation from the old one, the terms $\boldsymbol{\nabla}\psi \cdot \boldsymbol{\nabla}\phi$ cancel, and we obtain *Green second identity:*

$$\int (\psi \nabla^2 \phi - \phi \nabla^2 \psi) \, dV = \oint \left(\psi \frac{\partial \phi}{\partial n} - \phi \frac{\partial \psi}{\partial n} \right) ds. \tag{2.8}$$

In the following, identity (2.8) will be simply referred to as *Green identity.*

2.5.2 Vorticity integral conditions

The conditions satisfied by, and to be imposed on, the vorticity are a consequence of the following simple theorem (Quartapelle and Valz-Gris 1981).

Theorem 2.2. *A function ζ defined in V is such that $\zeta = -\nabla^2 \psi$, with $\psi|_S = a$ and $(\partial \psi / \partial n)|_S = b$, if and only if*

$$\int \zeta \eta \, dV = \oint \left(a \frac{\partial \eta}{\partial n} - b \eta \right) ds \tag{2.9}$$

for any function η harmonic in V, i.e., any function η such that $\nabla^2 \eta = 0$ in V.

Proof. To prove this theorem, first let $\zeta = -\nabla^2 \psi$ for some function with $\psi|_S = a$ and $(\partial \psi / \partial n)|_S = b$. By Green identity, it results, for any function ϕ,

$$\begin{aligned}
\int \zeta \phi \, dV &= \int (-\nabla^2 \psi) \phi \, dV \\
&= \int \psi (-\nabla^2 \phi) \, dV + \oint \left(\psi \frac{\partial \phi}{\partial n} - \frac{\partial \psi}{\partial n} \phi \right) ds \\
&= \int \psi (-\nabla^2 \phi) \, dV + \oint \left(a \frac{\partial \phi}{\partial n} - b \phi \right) ds.
\end{aligned}$$

If now the function ϕ is taken to be harmonic in V, namely, one takes $\phi = \eta$, with $-\nabla^2\eta = 0$, the relation becomes

$$\int \zeta\eta \, dV = \oint \left(a\frac{\partial\eta}{\partial n} - b\eta \right) ds.$$

Conversely, let ω be a function satisfying

$$\int \omega\eta \, dV = \oint \left(a\frac{\partial\eta}{\partial n} - b\eta \right) ds$$

for any function η harmonic in V. Then, let ψ denote the unique solution to the Poisson equation $-\nabla^2\psi = \omega$ supplemented by the Dirichlet condition $\psi|_S = a$. By Green identity, it results, for any harmonic function η,

$$\begin{aligned} \int \omega\eta \, dV &= \int (-\nabla^2\psi)\eta \, dV \\ &= \int \psi(-\nabla^2\eta) \, dV + \oint \left(\psi\frac{\partial\eta}{\partial n} - \frac{\partial\psi}{\partial n}\eta \right) ds \\ &= \oint \left(a\frac{\partial\eta}{\partial n} - \frac{\partial\psi}{\partial n}\eta \right) ds. \end{aligned}$$

Hence, by the assumption,

$$\oint b\eta \, ds = \oint \frac{\partial\psi}{\partial n}\eta \, ds.$$

From the arbitrariness of the boundary values of η, it follows that $(\partial\psi/\partial n)|_S = b$. Thus $\omega = -\nabla^2\psi$ with $\psi|_S = a$ and $(\partial\psi/\partial n)|_S = b$, which means $\omega \equiv \zeta$. □

Integral conditions of the same kind as (2.9) have been considered for the first time by Lanczos [25] in his discussion of overdetermined problems (p. 205). He obtained the integral conditions as conditions of *compatibility* between the data ζ, a and b of the overdetermined boundary value problem for the stream function consisting in a Poisson equation supplemented by both Dirichlet and Neumann boundary conditions.

Theorem 2.2 enlarges the category of conditions which are legitimate candidates for supplementing second-order parabolic or elliptic equations beyond the set of classical conditions of boundary value type, that is, the Dirichlet, Neumann and Newton (also called Robin) boundary conditions. As the subsequent analysis will show, the integral conditions are of a primitive character and must be considered as conditions of an entirely new kind, because it is impossible to interpret them in terms of the aforementioned conditions of local boundary-value type. After all, the fact that the vorticity is subject to integral conditions should not be a great surprise considering that the vorticity is the derivative of a variable—the velocity—whose boundary values are specified. This point had been anticipated by Sin-I Cheng (1975) when observing that the boundary condition for the normal derivative of ψ

... represents some integrated condition on the vorticity field rather than the local values of the vorticity.

From a functional analytic standpoint, the need for integral conditions for ζ is almost evident: the vorticity variable has regularity properties lower than those of the velocity and therefore it must be subject to stronger conditions than those of boundary value type supplementing the velocity to guarantee the same regularity of this variable.

2.5.3 What the integral conditions are not

It is important to clarify the relationship between the use of Green identity (2.8) for deriving the integral conditions for ζ and the use of the same identity for deriving a *boundary integral formulation* of the Poisson equation. In this kind of formulations one introduces fundamental solutions $G(\boldsymbol{x}, \boldsymbol{x}')$ defined by the equation

$$-\nabla'^2 G(\boldsymbol{x}, \boldsymbol{x}') = 4\pi\delta^{(2)}(\boldsymbol{x} - \boldsymbol{x}'),$$

where $\delta^{(2)}(\boldsymbol{x} - \boldsymbol{x}')$ is Dirac distribution in two dimensions, $\boldsymbol{x}$ is the so-called observation point and $\boldsymbol{x}'$ is the integration variable. Let us now assume to substitute $\phi = G(\boldsymbol{x}, \boldsymbol{x}')$ in Green identity (2.8) in which ψ satisfies the equation $-\nabla^2\psi = \zeta$ and the two boundary conditions $\psi|_S = a$ and $(\partial\psi/\partial n)|_S = b$. We obtain

$$\int \Big[- 4\pi\delta^{(2)}(\boldsymbol{x} - \boldsymbol{x}')\psi(\boldsymbol{x}') + G(\boldsymbol{x}, \boldsymbol{x}')\zeta(\boldsymbol{x}')\Big]\, dV' = \oint \left[a(s')\frac{\partial G(\boldsymbol{x}, \boldsymbol{x}'_s)}{\partial n'} - b(s')G(\boldsymbol{x}, \boldsymbol{x}'_s)\right] ds'.$$

Evaluating the integral of the first term we obtain

$$- 4\pi\psi(\boldsymbol{x})K_V(\boldsymbol{x}) + \int G(\boldsymbol{x}, \boldsymbol{x}')\zeta(\boldsymbol{x}')\, dV' = \oint \left[a(s')\frac{\partial G(\boldsymbol{x}, \boldsymbol{x}'_s)}{\partial n'} - b(s')G(\boldsymbol{x}, \boldsymbol{x}'_s)\right] ds',$$

where $K_V(\boldsymbol{x}) = 1, 0, \frac{1}{2}$ according to whether $\boldsymbol{x}$ is inside V, outside V or a (regular) boundary point of V. If we now evaluate the integral relation above on S and use again the Dirichlet boundary condition we obtain

$$- 2\pi a(s) + \int G(\boldsymbol{x}_s, \boldsymbol{x}')\zeta(\boldsymbol{x}')\, dV' = \oint \left[a(s')\frac{\partial G(\boldsymbol{x}_s, \boldsymbol{x}'_s)}{\partial n'} - b(s')G(\boldsymbol{x}_s, \boldsymbol{x}'_s)\right] ds'.$$

This result does not represent a boundary integral equation for the vorticity since the entire vorticity field is involved by the volume integral. The relation so obtained is therefore a different form of expressing the volume integral conditions (2.9) for the vorticity and like the latter need to be combined with the vorticity equation to have a well defined problem. In this respect, we can observe that to arrive at a physical interpretation of the vorticity integral conditions their effect must be analyzed together with the dynamical consequences implied by the vorticity transport equation. The two forms of the integral conditions have their respective advantages and drawbacks. The vorticity conditions (2.9) involve the harmonic functions η which depend on the considered domain V and are in general not known analitically for region of an arbitrary shape. On the contrary, the analytic expression of the Green function $G(\boldsymbol{x}, \boldsymbol{x}')$ is well known and does not depend on the particular domain under examination. Instead, from the viewpoint of the evaluation of the integrals, those occurring in (2.9) appear simpler to be evaluated while those required by the Green function form of the vorticity conditions are weakly singular.

An awkward application of the integral conditions, which can give the illusion of getting a boundary integral method for the vorticity, consists in taking the time derivative of the vorticity condition, namely,

$$\int \frac{\partial \zeta}{\partial t} \eta \, dV = \oint \left(\frac{\partial a}{\partial t} \frac{\partial \eta}{\partial n} - \frac{\partial b}{\partial t} \eta \right) ds$$

in order to substitute $\partial\zeta/\partial t$ from the vorticity transport equation, to give

$$\int \left(-J(\zeta, \psi) + \nu \nabla^2 \zeta \right) \eta \, dV = \oint \left(\frac{\partial a}{\partial t} \frac{\partial \eta}{\partial n} - \frac{\partial b}{\partial t} \eta \right) ds.$$

Green identity can now be applied to the term containing $\nabla^2\zeta$ to transform it into a boundary integral, as follows,

$$-\int J(\zeta, \psi) \eta \, dV + \nu \oint \left(\frac{\partial \zeta}{\partial n} \eta - \zeta \frac{\partial \eta}{\partial n} \right) ds = \oint \left(\frac{\partial a}{\partial t} \frac{\partial \eta}{\partial n} - \frac{\partial b}{\partial t} \eta \right) ds.$$

Unfortunately, the integral relationship so obtained is nonlinear and couples the variables ζ and ψ together, whereas the original integral conditions were linear and involved only one unknown. It is also to be remarked that, even assuming to disregard the nonlinear term, the integral relationship above does not permit to obtain a boundary integral formulation for ζ. In fact, in the linear situation the integral relationship simplifies to

$$\nu \oint \left(\frac{\partial \zeta}{\partial n} \eta - \zeta \frac{\partial \eta}{\partial n} \right) ds = \oint \left(\frac{\partial a}{\partial t} \frac{\partial \eta}{\partial n} - \frac{\partial b}{\partial t} \eta \right) ds,$$

but neither ζ nor $\partial\zeta/\partial n$ is specified on the boundary. Therefore, even assuming the linearity, the integral relationship above represent a boundary integral *condition* which still need to be combined with the vorticity equation to obtain a well defined problem.

The distinction between the vorticity integral conditions and a boundary integral formulation must be emphasized. Perhaps, such a difference can be better appreciated by examining several available boundary integral formulations of the Navier–Stokes equations expressed in terms of the nonprimitive variables ψ and ζ. All methods indicated in the following list are in fact essentially different form the standard boundary integral formulation of second-order elliptic or parabolic problems.

- The boundary integral formulation introduced by Ingham and Kelmanson [19] solves directly the stream function biharmonic equation.
- The boundary integral method recently proposed by Rodriguez-Prada *et al.* (1990) solves the system of vorticity and stream function equations. This method is however characterized by the use of fundamental solutions which take into account the coupling existing between the two unknowns, by virtue of the presence of the relationship $-\nabla^2\psi = \zeta$.
- A boundary integral formulation for determining only the vorticity boundary values has been proposed very recently by Ruas (1991). This method is nonclassical in that it relies on the use of a basis of special functions—the space of the harmonic functions. Thus, the method represents an elaboration of the idea of the vorticity integral conditions.
- Finally, a mixed integro-differential formulation of the Navier–Stokes equations governing the vorticity and the velocity potentials in three dimensions has been developed by Achard and Canot (1988). In this approach, the equations for the scalar and vector potentials of velocity are formulated as classical boundary integral equations, whereas the vorticity equation is retained in its differential form and is solved supplemented by boundary and integral conditions.

2.6 Split vorticity–stream function formulation

The introduction of the integral conditions (2.9) allows to express the vorticity–stream function problem (2.3) as a system of two second-order problems, in the following form:

$$\begin{aligned} \left(-\nu\nabla^2 + \tfrac{\partial}{\partial t}\right)\zeta &= -J(\zeta,\psi), &\qquad \int \zeta\eta\, dV &= \oint\left(a\frac{\partial\eta}{\partial n} - b\eta\right) ds, \\ -\nabla^2\psi &= \zeta, & \psi|_S &= a, \end{aligned} \tag{2.10}$$

where η denotes any function harmonic in V. Note that, in the absence of the nonlinear term, a complete splitting of the two equations is achieved thanks to the fact that the integral conditions do not involve the values of ψ at any internal points of V: only the values of ψ and $\partial\psi/\partial n$ prescribed on the boundary enter the vorticity integral conditions. Thus, at least in the linear case, the two equations can be solved in sequence, in the order indicated above.

In the split or uncoupled formulation (2.10), while the stream function equation is provided with standard Dirichlet boundary conditions, the vorticity equation is supplemented by uncommon conditions of an integral character. The important point about these vorticity conditions is that they admit a very simple geometrical interpretation. In fact, the left-hand side of the integral conditions is the scalar product of ζ and η in the Hilbert space $L^2(V)$ of square integrable functions. In particular, when $a = b = 0$, ζ is orthogonal in the L^2-sense to the linear space of the harmonic functions. For general nonhomogeneous boundary conditions for ψ, the orthogonal projection of ζ on this space is determined by the functions a and b, defined on S. The "number" of integral conditions to be satisfied is equal to the number of linearly independent functions η which are harmonic in the considered domain V. This number is equal to the "number" of boundary points, so that the projection integral conditions in (2.10) provide the correct number of conditions needed to supplement the parabolic equation for ζ so as to give a completely determined problem for such a variable.

It is worth pointing out that the equation $-\nabla^2\psi = \zeta$ can be solved also using the Neumann boubdary condition, $(\partial\psi/\partial n)|_S = b$, in place of the Dirichlet condition, provided that the vorticity field satisfies the integral conditions. If the Neumann problem is considered to determine ψ, the solvability condition for this problem would read

$$\int \zeta(\boldsymbol{x}, t)\, dV = -\oint b(s, t)\, ds, \tag{2.11}$$

which is simply the Stokes theorem for plane flows. In the split formulation (2.10), the solvability condition (2.11) is automatically satisfied since it is nothing but the vorticity integral condition with respect to the trivial harmonic function $\eta(\boldsymbol{x}) \equiv 1$. Therefore, the use of the projection conditions for the vorticity allows a complete freedom in the choice of the Dirichlet condition or the Neumann condition for supplementing the stream function equation.

It is also worth noticing that the vorticity integral conditions are invariant with respect to the change in ψ associated with the arbitrariness of the function $A(t)$ occurring in the definition of a. In fact, adding the function $A(t)$ to a implies that the right-hand side of the integral conditions contains the following additional term

$$\oint A(t) \frac{\partial \eta}{\partial n}\, ds,$$

and this term is zero since, using the divergence theorem, one has

$$\oint A(t)\frac{\partial \eta}{\partial n}\, ds = A(t)\oint \frac{\partial \eta}{\partial n}\, ds = A(t)\oint \boldsymbol{n}\cdot\boldsymbol{\nabla}\eta\, ds$$
$$= A(t)\int \boldsymbol{\nabla}\cdot\boldsymbol{\nabla}\eta\, dV = A(t)\int \nabla^2\eta\, dV = 0,$$

where the last equality results from the harmonic character of η.

An important aspect of the vorticity conditions is their *nonlocal* character: the vorticity distribution in the interior of the domain and on its boundary is affected at each time by the instantaneous values of the tangential and normal components of the velocity along the entire boundary. In other words, the distribution of vorticity in the whole domain is constrained by the velocity boundary values.

An evident consequence of the integral character of the vorticity conditions is that it is impossible to obtain the values assumed by the vorticity on the boundary by means of purely local operations performed near the boundary and using only the boundary data for the stream function. Admitting this limitation can be essential for a correct understanding of several results to be presented in this study. Unfortunately, the fact that second-order elliptic and parabolic operators are usually supplemented by boundary conditions of Dirichlet, Neumann or possibly Newton type makes the prejudice in favour of the existence of conditions of boundary value type also for the vorticity very difficult to be defeated. In particular, the formal equivalence of the vorticity transport equation for two dimensional flows with the equation governing the evolution of the temperature in a heat-conducting incompressible fluid is sometimes considered as an indication that boundary conditions for the vorticity ζ should exist similar to those usually imposed on the temperature T. However, this is not the case and the analogy existing between the equations does not carry over to the conditions to be imposed on each of the two variables: temperature and vorticity remain basically different variables at the level of their respective conditioning.

2.7 One-dimensional vorticity integral conditions

The first application of the vorticity integral conditions of the kind discussed so far were in fact made by Dennis and co-workers to problems involving the flow past geometrically simple bodies, such as flat plates of finite length aligned with the stream (Dennis and Dunwoody 1966), circular cylinders (Dennis and Chang 1970) and spheres (Dennis and Walker 1971). In this kind of application the method of the series truncation was used to expand the unknowns ζ and ψ in a series of convenient orthogonal functions. For instance, for the flow past a circular cylinder:

$$\zeta(\xi,\theta,t)=\sum_k \zeta_k(\xi,t)\sin(k\theta), \tag{2.12}$$

with a similar expression for ψ. Here, $\xi=\ln(r/a)$, (r,θ) are the polar coordinates and a is the radius of the cylinder. After substituting these expansions into the equations (2.3) for ζ and ψ, Galerkin method provides the following set of partial differential equations for the expansion coefficients $\zeta_k(\xi,t)$ and $\psi_k(\xi,t)$:

$$\begin{gathered}
-\nu\Big(\frac{\partial^2}{\partial\xi^2}-k^2\Big)\zeta_k+e^{2\xi}\frac{\partial\zeta_k}{\partial t}=f_k,\\
-\Big(\frac{\partial^2}{\partial\xi^2}-k^2\Big)\psi_k=e^{2\xi}\zeta_k,\\
\psi_k(\xi_i,t)=a_k(i,t),\qquad \frac{\partial\psi_k(\xi_i,t)}{\partial\xi}=b_k(i,t),\qquad i=1,2,
\end{gathered}\tag{2.13}$$

where f_k is the Galerkin projection of the nonlinear term $-J(\zeta,\psi)$ on the k-th function of the basis whereas a_k and b_k are the expansion coefficients of the boundary data a and b; it has been assumed that $\xi_1\le\xi\le\xi_2$. Now, by expanding the harmonic functions η on the same basis, the two-dimensional integral conditions (2.9) transform themselves into a set *one-dimensional* integral conditions the coefficients of vorticity: there are two integral conditions for each coefficient ζ_k, of the form

$$\int_{\xi_1}^{\xi_2}\zeta_k(\xi,t)e^{(2\pm k)\xi}\,d\xi=\Big[\big(\pm k a_k(i,t)-b_k(i,t)\big)e^{\pm k\xi_i}\Big]\Big|_{i=1}^{i=2}, \tag{2.14}$$

where $e^{\pm k\xi}=\eta_k(\xi)$ are two linearly independent solutions of the harmonic modal equation $(d^2/d\xi^2-k^2)\eta_k(\xi)=0$.

A more direct method for obtaining these one-dimensional integral conditions is to start from problem (2.13) for the expansion coefficients and to resort to Green identity for the ordinary differential operator $d^2/d\xi^2$, which reads

$$\int_{\xi_1}^{\xi_2}\Big(\psi\frac{d^2\phi}{d\xi^2}-\phi\frac{d^2\psi}{d\xi^2}\Big)d\xi=\Big(\psi\frac{d\phi}{d\xi}-\phi\frac{d\psi}{d\xi}\Big)\Big|_{\xi_1}^{\xi_2},$$

where $\psi(\xi)$ and $\phi(\xi)$ are arbitrary functions. By taking $\phi=\eta_k$ solution to the equation $(d^2/d\xi^2-k^2)\eta_k(\xi)=0$ so that $\eta_k(\xi)=e^{\pm k\xi}$, and by taking $\psi=\psi_k$ satisfying (2.13), one obtains again the integral conditions (2.14).

By means of these conditions, the equations governing the expansion coefficients can be written in the split form, apart from the nonlinearities,

$$-\nu\Big(\frac{\partial^2}{\partial\xi^2} - k^2\Big)\zeta_k + e^{2\xi}\frac{\partial\zeta_k}{\partial t} = f_k,$$

$$\int_{\xi_1}^{\xi_2} \zeta_k(\xi,t)e^{(2\pm k)\xi}\,d\xi = \Big[\Big(\pm k a_k(i,t) - b_k(i,t)\Big)e^{\pm k\xi_i}\Big]\Big|_{i=1}^{i=2}; \tag{2.15}$$

$$-\Big(\frac{\partial^2}{\partial\xi^2} - k^2\Big)\psi_k = e^{2\xi}\zeta_k, \qquad \psi_k(\xi_i,t) = a_k(i,t),\ i = 1,2.$$

Numerical methods of this kind using one-dimensional vorticity integral conditions have been applied to solve several steady and unsteady flow problems of basic fluid dynamic interest, see Anwar and Dennis (1988) and the references therein.

Application of this kind should not convey the idea that the validity of integral conditions for the vorticity is limited to problems involving particularly simple geometrical regions or that they pertain only to spatial discretizations obtained by means of series truncation method. On the contrary, the vorticity integral conditioning is a structural component of the vorticity–stream function system, that will be shown to underlie the mathematical theory of the Navier–Stokes equations, whenever these equations are expressed as a system of uncoupled second-order parabolic and elliptic equations.

We end this section by noting that, whereas the splitting of the vorticity from the stream function is incomplete for the full (nonlinear) Navier–Stokes equations, it becomes complete whenever a linearized version of the vorticity equation can be considered. For instance, let us consider the unsteady Oseen equations, namely,

$$\begin{aligned} -\nu\nabla^2\zeta + \frac{\partial\zeta}{\partial t} + J(\zeta,\psi_\infty) &= 0,\\ -\nabla^2\psi &= \zeta, \end{aligned} \tag{2.16}$$

where ψ_∞ is the stream function of the uniform flow at a large distance from a (cylindrical) body. In this case, the vorticity integral conditions allow to determine the vorticity field without considering the stream function ψ at all. In particular, the fully uncoupled vorticity problem for the steady version of Oseen equations gives

$$-\nu\nabla^2\zeta + J(\zeta,\psi_\infty) = 0, \qquad \int \zeta\eta\,dV = \oint\Big(a\frac{\partial\eta}{\partial n} - b\eta\Big)\,ds, \tag{2.17}$$

where η is any solution to $\nabla^2\eta = 0$ in V. This approach has been considered by Dennis and Kocabiyik (1990) (see also Kocabiyik 1987) to study the flow past cylinders (stationary elliptic cylinders and possibly rotating circular cylinders) at low Reynolds numbers. For instance, in the case of symmetrical flow past a nonrotating circular cylinder of radius a, the series expansion (2.12) can be used in the Oseen problem (2.17) for the uncoupled vorticity. Performing an asymptotic analysis in the Reynolds number $R = 2aU/\nu$ as $R \to 0$, where U represents the velocity of the uniform flow, the integral conditions allow to obtain the following result for the drag coefficient (Dennis and Kocabiyik 1990):

$$C_{\text{drag}} = -\frac{8\pi}{R}\left[\ln(\tfrac{1}{8}R) + \gamma - \tfrac{1}{2}\right]^{-1}\left[1 + \mathrm{O}(R^2)\right].$$

where $\gamma = 0.5772156\ldots$ is Euler constant. This is the well known formula first obtained by Lamb in 1911, a result showing that the independent determination of the vorticity field in the Oseen approximation represents an important application of the concept of the vorticity integral conditioning.

2.8 Orthogonal projection operator

To investigate the structure of the operation of orthogonal projection of the vorticity field we now examine a time-discretized version of the vorticity and stream function equations. For definiteness, let us consider a two-level time-integration scheme with the viscous term evaluated implicitly and the nonlinear term taken into account explicitly. The equations determining the unknowns $\zeta \equiv \zeta^{n+1}$ and $\psi \equiv \psi^{n+1}$ at the new time level t^{n+1} assume the form

$$\begin{aligned} (-\nabla^2 + \gamma)\zeta &= f, \qquad & \int \zeta\eta\, dV &= \oint \left(a\frac{\partial\eta}{\partial n} - b\eta\right) ds; \\ -\nabla^2\psi &= \zeta, \qquad & \psi|_S &= a; \end{aligned} \tag{2.18}$$

where $\gamma = 1/(\nu\Delta t)$, $f = \gamma\zeta^n - \nu^{-1}J(\zeta^n, \psi^n)$, whereas now $a = a(s) = a(s, t^{n+1})$ and $b = b(s) = b(s, t^{n+1})$. We rederive the split formulation (2.18) by means of an abstract argument and thereby introduce a geometrical interpretation of the vorticity problem which is convenient for imposing the integral conditions in the numerical schemes. In fact, the idea of projecting the vorticity field orthogonally onto the harmonic functions allows to devise a numerical scheme for solving the vorticity–stream function problem in terms of elliptic operators supplemented only by standard conditions of boundary value type. As it will be shown in the subsequent chapters, the geometrical construction underlying this scheme can be generalized to three-dimensional situations and extended to different representations of the Navier–Stokes equations.

Consider any linear operator A the range of which is a subspace of $L^2(V)$ and let $\mathcal{R}(A)$ and $\mathcal{N}(A)$ denote the range (or image) and null space (or kernel) of A, respectively. A fundamental orthogonality theorem of functional analysis (see, *e.g.*, Oden [29, p. 312]) and of linear algebra (see, *e.g.*, Strang [32, p. 87]) states that

$$\overline{\mathcal{R}(A)} = \mathcal{N}(A^\dagger)^\perp \tag{2.19}$$

where the bar denotes closure, $A^\dagger$ is the adjoint operator of A and $\perp$ indicates the orthogonal complement in $L^2(V)$.

The application of this theorem to the vorticity and stream function problem is as follows. Consider first the situation of homogeneous boundary conditions for ψ, that is, $a = b = 0$. Then, the vorticity field must be such that $\zeta = -\nabla^2\psi$, with $\psi|_S = 0$ and $(\partial\psi/\partial n)|_S = 0$. Let ∇^2_{DN} denote the Laplace operator supplemented by both Dirichlet and Neumann boundary conditions. Such a characterization of ζ can also be expressed in the form $\zeta \in \mathcal{R}(-\nabla^2_{\text{DN}})$, and, by the orthogonality theorem, the latter means that ζ must be orthogonal to the linear space (or subspace) $\mathcal{N}((-\nabla^2_{\text{DN}})^\dagger)$, namely $\zeta \perp \mathcal{N}((-\nabla^2_{\text{DN}})^\dagger)$. To identify the adjoint operator $(-\nabla^2_{\text{DN}})^\dagger$ one has to determine both the differential operator and the boundary conditions associated with it. This is done by inspecting the two members of Green identity (2.8): the left-hand side shows that the adjoint operator is still the Laplacian, whereas the right-hand side indicates that the adjoint operator is supplemented by no boundary condition because the boundary integral is already zero by virtue of the two boundary conditions associated with $-\nabla^2_{\text{DN}}$. Therefore it is $(-\nabla^2_{\text{DN}})^\dagger = -\widetilde{\nabla}^2$, where $\widetilde{\nabla}^2$ denotes the Laplace operator with *no* boundary condition. Now, $\mathcal{N}(-\widetilde{\nabla}^2)$ is the linear space of functions η that satisfy the equation $-\nabla^2\eta = 0$ (without boundary conditions). Thus ζ must be orthogonal to the space of the harmonic functions, namely, $\zeta \perp \{\eta \,|\, -\nabla^2\eta = 0\}$.

Let us now assume that ζ is to be found as the solution of the elliptic equation $(-\nabla^2 + \gamma)\zeta = f$, with $\gamma \geq 0$ fixed, and let us introduce the linear space Z_γ defined as follows: $Z_\gamma = \{z \,|\, (-\nabla^2 + \gamma)z = g, \text{for all } g\}$. Since the space of the harmonic functions $\mathcal{H} = \{\eta \,|\, -\nabla^2\eta = 0\}$ is a closed subspace of Z_γ, there exists a unique orthogonal projection operator from Z_γ onto $\mathcal{H}$, which will be denoted by $\mathbb{P}_{\mathcal{H}}$ or, more explicitly, by $\mathbb{P}_{\mathcal{N}(-\widetilde{\nabla}^2)}$. We now consider the solutions of the homogeneous version of the vorticity equation, that is, the functions ζ' such that $(-\nabla^2 + \gamma)\zeta' = 0$, which are called *metaharmonic* functions, and introduce the linear space $\mathcal{M}_\gamma = \{\zeta' \,|\, (-\nabla^2 + \gamma)\zeta' = 0\}$ of the metaharmonic functions. If the projection operator $\mathbb{P}_{\mathcal{N}(-\widetilde{\nabla}^2)}$ is made to act only on the metaharmonic functions, a linear operator is obtained which maps the elements $\zeta' \in \mathcal{M}_\gamma$ onto $\mathcal{H}$. This operator is called the *restriction* of the orthogonal projection operator $\mathbb{P}_{\mathcal{N}(-\widetilde{\nabla}^2)}$ to the linear space $\mathcal{M}_\gamma$ and will be denoted by P^γ, according to the following definition:

$$P^\gamma = \mathbb{P}_{\mathcal{H}}\Big|_{\mathcal{M}_\gamma} = \mathbb{P}_{\mathcal{N}(-\widetilde{\nabla}^2)}\Big|_{\mathcal{N}(-\widetilde{\nabla}^2+\gamma)}. \tag{2.20}$$

The restricted projection operator P^γ establishes a mapping between the two spaces $\mathcal{M}_\gamma$ and $\mathcal{H}$, which is made definite by the following theorem.

Theorem 2.3. *Let $\widetilde{\nabla}^2$ denote the Laplacian operator without any boundary condition and let $\mathbb{P}_{\mathcal{N}(-\widetilde{\nabla}^2)}$ be the operator of orthogonal projection onto the space of the harmonic functions in V, $\mathcal{H} = \{\eta \,|\, -\nabla^2\eta = 0\} = \mathcal{N}(-\widetilde{\nabla}^2)$. Then,*

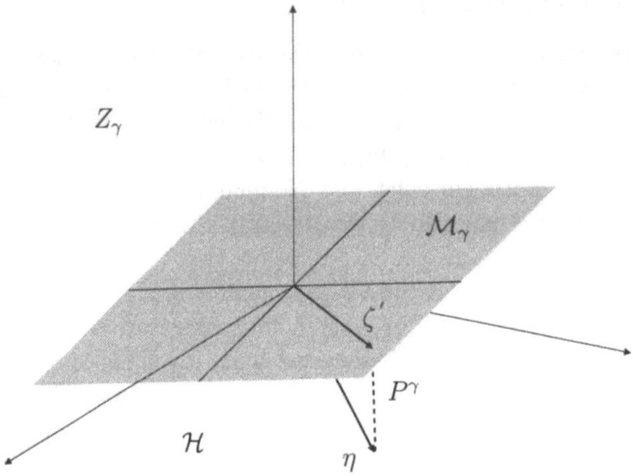

Figure 2.1: Invertibility of the restricted projection operator P^γ.

the operator P^γ, representing the restriction of $\mathbb{P}_{\mathcal{N}(-\widetilde{\nabla}^2)}$ to the linear space of the metaharmonic functions in V, $\mathcal{M}_\gamma = \{\zeta' \,|\, (-\nabla^2+\gamma)\zeta' = 0\} = \mathcal{N}(-\widetilde{\nabla}^2+\gamma)$, is one-to-one and therefore invertible.

Proof. Roughly speaking, the theorem asserts that the relationship existing between the metaharmonic and harmonic linear spaces is similar to that between two nonparallel planes passing through the origin of the three-dimensional Euclidean space (see Fig. 2.1).

The (nonhomogeneous) boundary conditions that can be associated with the operators $-\nabla^2$ and $(-\nabla^2+\gamma)$ for constructing the spaces $\mathcal{H}$ and $\mathcal{M}_\gamma$ are equal, so these two spaces have the same dimensionality. Then, P^γ will be one-to-one if and only if $\mathcal{N}(P^\gamma) = \{0\}$. This condition is equivalent to the assertion that in $\mathcal{M}_\gamma$ there is no function $\zeta' \neq 0$ orthogonal to the harmonic space $\mathcal{H}$. (The null space $\mathcal{N}(P^\gamma)$ should not be confused with the null space $\mathcal{N}(-\widetilde{\nabla}^2+\gamma) \equiv \mathcal{M}_\gamma$, which is the entire *domain* of the restricted projection operator P^γ.)

For $\gamma = 0$, $\mathcal{M}_{\gamma=0} \equiv \mathcal{H}$ and $P^{\gamma=0} \equiv I$, where I denotes the identity operator acting on $\mathcal{M}_\gamma$, and therefore $P^{\gamma=0}$ is one-to-one, trivially.

Consider now the nontrivial case $\gamma > 0$. We prove that for any given $\zeta' \in \mathcal{M}_\gamma$, with $\zeta' \neq 0$, it is not possible to have $(\zeta', \eta) = 0$ for all $\eta \in \mathcal{H}$, and that there is at least one harmonic function η' such that

$$(\zeta', \eta') \neq 0.$$

Let ζ' be a nontrivial solution to the equation $(-\nabla^2+\gamma)\zeta' = 0$ and let η'

denote the harmonic function which satisfies the same boundary conditions of ζ', namely, such that $\eta'|_S = \zeta'|_S$. Since $\zeta' = \gamma^{-1}\nabla^2\zeta'$, one has, using Green identity (2.8), the harmonic character of η' and the divergence theorem,

$$\begin{aligned}(\zeta',\eta') &= \gamma^{-1}(\nabla^2\zeta',\eta')\\ &= \gamma^{-1}\left[(\zeta',\nabla^2\eta') + \oint\left(\frac{\partial\zeta'}{\partial n}\eta' - \zeta'\frac{\partial\eta'}{\partial n}\right)ds\right]\\ &= \gamma^{-1}\oint\left(\frac{\partial\zeta'}{\partial n}\zeta' - \eta'\frac{\partial\eta'}{\partial n}\right)ds\\ &= \gamma^{-1}\oint\left(\zeta'\boldsymbol{n}\cdot\nabla\zeta' - \eta'\boldsymbol{n}\cdot\nabla\eta'\right)ds\\ &= \gamma^{-1}\int\left(\nabla\cdot(\zeta'\nabla\zeta') - \nabla\cdot(\eta'\nabla\eta')\right)dV.\end{aligned}$$

Then, exploiting again the harmonic character of η' and using the equation $(-\nabla^2+\gamma)\zeta' = 0$,

$$\begin{aligned}(\zeta',\eta') &= \gamma^{-1}\int\left(\nabla\cdot(\zeta'\nabla\zeta') - \nabla\cdot(\eta'\nabla\eta')\right)dV\\ &= \gamma^{-1}\int\left(|\nabla\zeta'|^2 + \zeta'\nabla^2\zeta' - |\nabla\eta'|^2 - \eta'\nabla^2\eta'\right)dV\\ &= \gamma^{-1}\int\left(|\nabla\zeta'|^2 + \gamma(\zeta')^2 - |\nabla\eta'|^2\right)dV.\end{aligned}$$

Since η' and ζ' are coincident on S, the Dirichlet principle for the Laplace equation supplemented by Dirichlet boundary condition (see Brezis [5, p. 177]) gives

$$\int|\nabla\eta'|^2\,dV \le \int|\nabla\zeta'|^2\,dV.$$

Using this inequality in the previous equation one obtains

$$(\zeta',\eta') \ge \int(\zeta')^2\,dV > 0,$$

since it has been assumed $\zeta' \neq 0$. Thus $(\zeta',\eta') > 0$ and there is no nontrivial subspace of $\mathcal{M}_\gamma$ orthogonal to $\mathcal{H}$. □

By a similar argument, it is readily found that the scalar products of two arbitrary functions ζ_1' and ζ_2' of $\mathcal{M}_\gamma$ with the functions η_1' and η_2' of $\mathcal{H}$, which satisfy identical boundary conditions as ζ_1' and ζ_2', are such that $(\zeta_1',\eta_2') = (\zeta_2',\eta_1')$. In the discrete case this result means that the matrix corresponding to (ζ',η) is symmetric.

Coming back to the solution of the time-discretized vorticity and stream function equations (2.18), the invertibility of the restricted projection operator P^γ allows to determine the solution ζ of $(-\nabla^2+\gamma)\zeta = f$, subject to the orthogonality condition $\zeta \perp \mathcal{H}$, as follows: first compute a "wrong" vorticity field ζ^0 solution of the equation $(-\nabla^2+\gamma)\zeta^0 = f$ and that assumes arbitrary values on

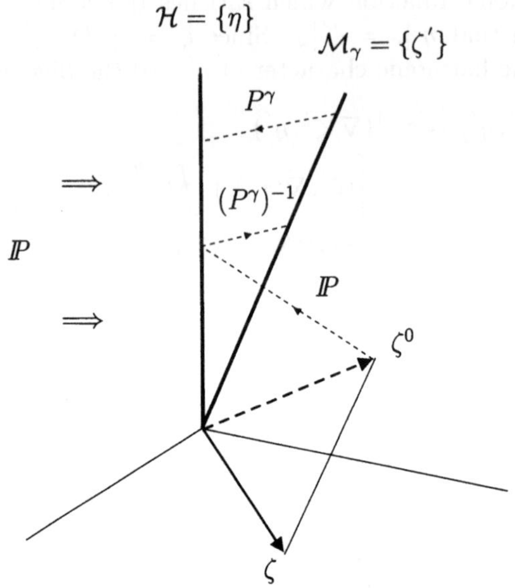

Figure 2.2: Construction of the vorticity field for homogeneous boundary conditions of velocity.

the boundary S, for instance, $\zeta^0|_S = 0$; the sought solution ζ is then obtained by subtracting from ζ^0 its component in $\mathcal{M}_\gamma$ whose projection on $\mathcal{H}$ is nonzero, according to the relationship:

$$\zeta = \zeta^0 - (P^\gamma)^{-1} I\!P_{\mathcal{N}(-\widetilde{\nabla}^2)}\zeta^0 = \left[I\!I - (P^\gamma)^{-1} I\!P_{\mathcal{N}(-\widetilde{\nabla}^2)}\right]\zeta^0, \tag{2.21}$$

where $I\!I$ denotes the identity operator acting on the space Z_γ. The geometrical construction necessary to satisfy the orthogonality is depicted in Figure 2.2. Note that the linear spaces $\mathcal{H}$ and $\mathcal{M}_\gamma$ are now represented by straight lines instead of the planes shown in the previous figure, and that the space $\mathcal{H}$ is plotted vertically instead of horizontally.

This construction has to be modified for nonhomogeneous boundary conditions. In this case, ζ is no more required to be orthogonal to $\mathcal{H}$, but only to have a fixed projection on $\mathcal{H}$. Let $\eta[a,b]$ denote the prescribed projection of ζ on $\mathcal{H}$, which, of course, depends on the boundary data a and b. It is not difficult to verify that the vorticity field with the correct projection on the space of the harmonic functions is given by the expression:

$$\begin{aligned}\zeta &= \zeta^0 - (P^\gamma)^{-1}\left[I\!P_{\mathcal{N}(-\widetilde{\nabla}^2)}\zeta^0 - \eta[a,b]\right] \\ &= \left[I\!I - (P^\gamma)^{-1} I\!P_{\mathcal{N}(-\widetilde{\nabla}^2)}\right]\zeta^0 - (P^\gamma)^{-1}\eta[a,b],\end{aligned} \tag{2.22}$$

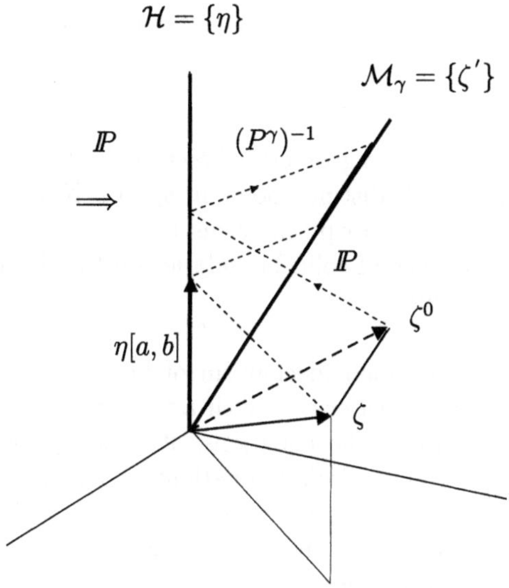

Figure 2.3: Construction of the vorticity field for nonhomogeneous boundary conditions of velocity.

whose geometrical meaning is illustrated in Figure 2.3.

Some final comments regarding these abstract geometrical considerations. When a time dependent problem is studied, the operation of projecting the vorticity field onto $\mathcal{H}$ must be repeated at each time step, in order that the no-slip velocity condition be satisfied exactly at every time. According to this geometrical picture, the vorticity field in viscous flows in the presence of solid boundaries must satisfy a global constraint represented by the orthogonal projection. The latter assures the "coherence" of the entire vorticity distribution with the velocity prescribed on the boundary, thus keeping the vorticity field within the restricted class of kinematically admissible fields, as clearly pointed out by Lighthill (1963, p. 57):

> *... there is only a restricted class of vorticity distributions that correspond to real flows satisfying also the no-slip condition on the tangential velocity.*

As time advances, the vorticity field evolves but still under the global control of the orthogonal projection, so that the field can be kinematically admissible at any time. Always in Lighthill's words:

> *For a step-by-step computation of vorticity development, we need a*

method of progressing from one vorticity distribution within this restricted class to the vorticity distribution a short time later.

Admittedly, the proposed geometrical picture does not give any insight about what is going on at each distinct point in space. Nevertheless, being mathematically sound, such a description seems the appropriate theoretical tool for investigating the influence of solid boundaries on the dynamics of the vorticity in viscous incompressible flows.

Resorting to the so-called mechanism of boundary vorticity generation in viscous problems can be criticized. Such a view requires in fact to proceed to a *double singularization* of the equations: it is necessary first to consider the inviscid limit of the Navier–Stokes equations and then to replace the smooth time dependence of the boundary values of velocity with a discontinous stepwise dependence. In this way, one obtains sheets of infinite vorticity, localized on the moving walls; only subsequently the physical mechanisms inherent to the viscous fluid are left to come into play so as to allow the diffusion of the vorticity and its transport in the interior of the fluid.

By contrast, the picture based on the orthogonal projection of the vorticity field applies directly to the Navier–Stokes equations without requiring any limiting process. The picture is therefore perfectly adequate to study the interaction of vorticity with rigid walls under all circumstance, including the special but interesting case of fixed boundaries. As an example, consider the well known phenomenon of a vortex ring impinging normally on a plane wall, which is parallel to the plane of the ring. Coming closer to the wall, the radius of the ring becomes larger and in the meanwhile the inner vortical structure of the ring is affected simultaneously by viscous diffusion, vortex stretching and the presence of the no-slip boundary. Thus, the rigid wall can affect the vorticity distribution even in the absence of a direct contact of the ring with the wall. Such an interaction process can be more easily understood in terms of the nonlocal influence of the rigid boundary through the projection conditions than by means of any model assuming the generation of vorticity on the boundary.

Perhaps, one should have to accept the abstract character of the description of wall/vorticity interaction *via* an orthogonal projection for the same reason put forward by Feynman in supporting the abstract field idea in the study of electromagnetism:

That it is abstract is unfortunate, but necessary.

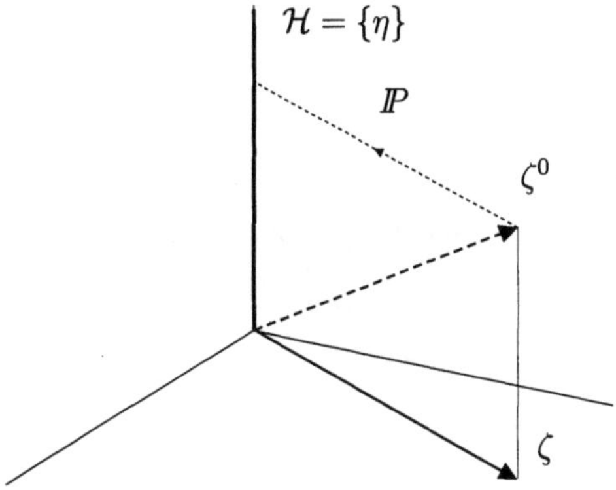

Figure 2.4: Construction of the vorticity field for the steady-state Stokes problem with homogeneous velocity conditions.

2.9 Factorized vorticity–stream function problem

Using the projective construction for the vorticity indicated by (2.21), it is possible to represent the solution ψ of the vorticity–stream function problem (2.18) in the following *factorized* form:

$$\psi = (-\nabla^2)^{-1}\big[I\!I - (P^\gamma)^{-1} I\!P_{\mathcal{N}(-\widetilde{\nabla}^2)}\big](-\nabla^2 + \gamma)^{-1} f. \tag{2.23}$$

Here, ∇^2 denotes the Laplace operator supplemented by Dirichlet boundary conditions, so that $(-\nabla^2)^{-1}$ represents formally the solution of a Dirichlet problem, and similarly for $(-\nabla^2+\gamma)^{-1}$; on the other hand, $\widetilde{\nabla}^2$ denotes the Laplace operator with no boundary conditions. (For simplicity, homogeneous boundary conditions have been considered in the factorization above.)

The operator within the square brackets in relation (2.23) provides the mathematical description of the effect of the no-slip conditions on the vorticity in two-dimensional flows. In fact, in the absence of solid boundaries or with periodic boundary conditions, this operator disappears and the factorization simplifies to

$$\psi = (-\nabla^2)^{-1}(-\nabla^2 + \gamma)^{-1} f. \tag{2.24}$$

We can also consider the simple situation of the steady-state linear equations

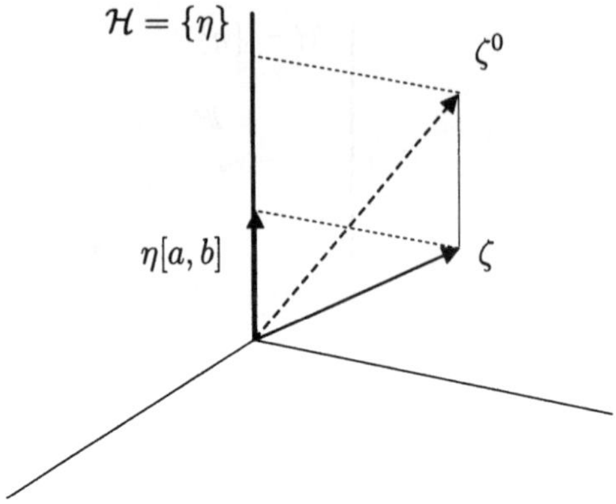

Figure 2.5: Construction of the vorticity field for the steady-state Stokes problem with nonhomogeneous velocity conditions.

which give the Stokes problem, namely,

$$\begin{aligned} -\nu\nabla^2\zeta &= f, & \int \zeta\eta \, dV &= \oint \left(a\frac{\partial\eta}{\partial n} - b\eta\right) ds; \\ -\nabla^2\psi &= \zeta, & \psi|_S &= a; \end{aligned} \tag{2.25}$$

where $a = a(s)$ and $b = b(s)$. The previous analysis applies in a simplified form since $\gamma = 0$ and $\mathcal{M}_{\gamma=0} \equiv \mathcal{H}$. It follows that the factorization of the Stokes problem is, always assuming homogeneous boundary conditions,

$$\psi = (-\nabla^2)^{-1}\left[\mathbb{I} - \mathbb{P}_{\mathcal{N}(-\widetilde{\nabla}^2)}\right](-\nabla^2)^{-1} f. \tag{2.26}$$

Thus the projection operation is still present, but the geometrical construction is simplified thanks to the coincidence of the metaharmonic and harmonic linear spaces. The picture of the vorticity construction for the Stokes problem is given in Figures 2.4 and 2.5, for homogeneous and nonhomogeneous boundary conditions, respectively.

2.10 Numerical schemes: local discretizations

If we turn now to the question of finding discretized versions of the problem (2.18), there are several ways for satisfying the vorticity integral conditions and performing the underlying orthogonal projection. Before discussing the various methods which belong to this class, we describe the method first introduced in finite difference methods to circumvent the difficulty caused by the absence of boundary conditions for the vorticity (Woods 1954).

2.10.1 Boundary vorticity formula methods

Such a method consists in defining the boundary values of vorticity in terms of the stream function by means of some approximate formula—hence the name of vorticity boundary formula method. The various formula are derived by expanding the stream function in a Taylor series along the *inward* normal to the boundary, $\overline{\boldsymbol{n}} = -\boldsymbol{n}$,

$$\psi_1 = \psi|_S + h\frac{\partial\psi}{\partial\overline{n}}\bigg|_S + \frac{h^2}{2}\frac{\partial^2\psi}{\partial\overline{n}^2}\bigg|_S + \frac{h^3}{6}\frac{\partial^3\psi}{\partial\overline{n}^3}\bigg|_S + \cdots$$

where subscript "1" indicates the interior point nearest to S. Then, the equation $-\nabla^2\psi = \zeta$ is written on S, namely,

$$\zeta|_S = (-\nabla^2\psi)|_{S},= \left(-\frac{\partial^2\psi}{\partial\overline{n}^2} - \frac{\partial^2\psi}{\partial s^2}\right)\bigg|_S$$

in order to obtain the missing boundary condition for ζ. We notice that both steps (the Taylor expansion and evaluating the Poisson equation on the boundary) represent a functional "crime" because involve the evaluation of the second (and also higher) derivative of ψ on S, which is meaningless insofar as the solution is sought in the space $H^1(V)$. Using the Dirichlet condition $\psi|_S = a$ gives

$$\zeta|_S = -\frac{\partial^2\psi}{\partial\overline{n}^2}\bigg|_S - \frac{d^2a}{ds^2}.$$

The value of the normal second derivative is obtained from the previous Taylor series, taking into account the two boundary conditions for ψ, namely, $\psi|_S = \psi_0 = a$ and $(\partial\psi/\partial\overline{n})|_S = -b$. Substituting the value so obtained into the expression for $\zeta|_S$ gives

$$\zeta|_S = \frac{2}{h^2}(-\psi_1 + a - hb) - \frac{d^2a}{ds^2} + \frac{h}{3}\frac{\partial^3\psi}{\partial\overline{n}^3}\bigg|_S + \cdots$$

The boundary vorticity formula of Woods is obtained by truncating this expression after the second term on the righ-hand side, to give

$$\zeta|_S = \frac{2}{h^2}(-\psi_1 + a - hb) - \frac{d^2a}{ds^2}.$$

This approach is iterative in nature, since the value $\zeta|_S$ is given in terms of the value ψ_1 at an internal point, which must be known from a previous iteration. As it could have been anticipated, the iteration is found to diverge in general, even for the linear problem and even when the elliptic equations for ζ and ψ are solved by direct methods, see, among others, Quartapelle and Valz-Gris (1981). The divergence can be avoided by introducing some sort of relaxation in the vorticity boundary formula. If the value of the relaxation parameter is chosen within a proper range, the iteration can be made to converge (McLaurin 1974). This kind of iterative method can be effective for calculating the solution of the steady-state equations

$$-\nu\nabla^2\zeta + J(\zeta,\psi) = 0, \qquad -\nabla^2\psi = \zeta.$$

The problem being nonlinear, an iterative method is mandatory and the iteration for converging to the correct boundary values of vorticity can be exploited to deal also with the nonlinear term of the first equation. For unsteady solutions, the iteration for vorticity boundary relaxation ought to be used at each time step, to have an accurate representation of the evolution and also not to loose the correct asymptotic value of physically important quantities.

Concerning these iterative methods for the solution of the ζ-ψ equations, it may be interesting to note that they are expected to satisfy the vorticity integral conditions at convergence, in spite of the fact that these conditions are not even mentioned in this kind of methods. As a matter of facts, here the effect of nonlocality brought about by the integral conditions is achieved indirectly through the iteration process. It may also be noted that, according to this interpretation, the observed convergence superiority of boundary vorticity formulas of lower accuracy with respect to higher-order approximations can be explained. In fact, the effectiveness of a boundary vorticity formula must not be measured on the basis of its accuracy but only on its capability of achieving a high rate of convergence for the iterative process, toward the eventual and ultimate fulfillment of the vorticity integral conditions.

We notice incidentally that even in the coupled formulation (2.6) the integral conditions for the vorticity are satisfied. In this case the fulfillment is achieved by imposing the two boundary conditions for ψ through the direct coupling existing between the equations for ζ and ψ.

2.10.2 Decomposition scheme

Coming now to direct, *i.e.*, noniterative, methods for imposing the vorticity conditions, we first note that the number of integral conditions to be satisfied corresponds to the number of linearly independent harmonic functions $\eta(\boldsymbol{x})$ in the given domain V. Since the latter is equal to the number of boundary points associated with the considered spatial discretization, the vorticity equation is always supplemented with the correct number of independent conditions required to give a well determined problem. Obviously, as the degree of the approximation is improved, more boundary points and hence harmonic functions are employed.

The most immediate method for satisfying the vorticity conditions is to use the integral conditions in (2.18) directly to give a linear equation for every boundary point. Each equation relates the values of the unknown ζ at all grid points, with coefficients equal to the point values of the harmonic functions η. The set of such linear equations corresponding to the space of the discrete set of harmonic functions then closes the system of algebraic equations resulting from the spatial discretization of the vorticity equation. Unfortunately, this full system of equations has a rather cumbersome profile since the equations expressing the integral conditions have almost all coefficients different from zero. Therefore, for arbitrarily shaped two-dimensional regions the method requires the handling of matrices which are much less sparse than those associated with the Dirichlet problem on the same spatial discretization.

On the contrary, for simple domains that allow the separation of variables and for boundary conditions allowing to express the integral conditions in one-dimensional form, the direct imposition of the vorticity conditions can be convenient. In fact, for this kind of problems, the unknowns can be expanded in a truncated series of trigonometric functions and a set of ordinary differential equations is obtained for the expansion coefficients, which can be solved by means of any approximation method. In particular, the linear system of discrete equations for each expansion coefficient of vorticity contains the two equations representing the integral conditions. These equations have all coefficients different from zero and efficient Gauss elimination schemes can be developed for the resulting bordered banded matrices. An example of the direct imposition of the vorticity integral conditions in one-dimensional form will be given later in this chapter.

For plane two-dimensional problems in arbitrarily shaped regions, the difficulty of the cumbersome matrix profile can be circumvented resorting to a method that makes use of the superposition principle and requires to solve Dirichlet problems and a supplementary small linear problem, with only the vorticity boundary value as unknown. The vorticity field $\zeta(\boldsymbol{x})$ is decomposed in the following form:

$$\zeta(\boldsymbol{x}) = \zeta^0(\boldsymbol{x}) + \oint \zeta'(\boldsymbol{x};\sigma')\lambda(\sigma')\,d\sigma', \tag{2.27}$$

where the fields $\zeta^0(\boldsymbol{x})$ and $\zeta'(\boldsymbol{x};\sigma')$, for any $\sigma' \in S$, are the components already considered in Figure 2.2. They are defined as solution to the following problems

$$(-\nabla^2 + \gamma)\zeta^0 = f, \qquad \zeta^0|_S = 0; \tag{2.28a}$$

$$(-\nabla^2 + \gamma)\zeta' = 0, \qquad \zeta'(\boldsymbol{x};\sigma')|_S = \delta(s-\sigma'); \tag{2.28b}$$

with δ denoting the Dirac delta distribution over the boundary curve S. The boundary unknown $\lambda(\sigma)$, with $\sigma \in S$, is determined by imposing ζ to satisfy the integral conditions in (2.18) with respect to all of the harmonic functions $\eta(\boldsymbol{x};\sigma)$

defined by the problem

$$-\nabla^2\eta = 0, \qquad \eta(\boldsymbol{x};\sigma)|_S = \delta(s-\sigma), \tag{2.29}$$

for any $\sigma \in S$. It is important to note that η is not a Green function since $\delta(s-\sigma)$ does not represent a source term of the equation satisfied by η, but only indicates that the boundary condition for obtaining the harmonic functions must be nonhomogeneous. Substituting (2.27) into the integral conditions gives the linear problem

$$\oint A(\sigma,\sigma')\lambda(\sigma')\,d\sigma' = \beta(\sigma), \tag{2.30}$$

where

$$A(\sigma,\sigma') = \int \zeta'(\boldsymbol{x};\sigma')\eta(\boldsymbol{x};\sigma)\,dV, \tag{2.31}$$

and

$$\beta(\sigma) = -\int \zeta^0(\boldsymbol{x})\eta(\boldsymbol{x};\sigma)\,dV + \oint \left[a(s)\frac{\partial\eta(\boldsymbol{x}_s;\sigma)}{\partial n} - b(s)\eta(\boldsymbol{x}_s;\sigma)\right]ds. \tag{2.32}$$

Such a computational scheme is therefore based on the solution of only Dirichlet problems for the operators $(-\nabla^2+\gamma)$ and $-\nabla^2$, together with the additional linear problem (2.30) to determine the unknown λ. In the spatially discrete case, the matrix corresponding to the linear operator $A(\sigma,\sigma')$ is symmetric, as previously noticed. It is important to note that the operator $A(\sigma,\sigma')$ depends only on the geometry of the domain through the functions ζ' and η. Therefore, it can be determined once and for all at the beginning of the calculations and used subsequently to solve as many different right-hand sides as requested.

Equations (2.31) and (2.32) have been used in this form in earlier applications of the two-dimensional integral conditions using finite differences (Quartapelle and Valz-Gris 1981, see also Quartapelle 1981) and in a Chebyshev pseudo-spectral method recently proposed for the solution of problems with nonperiodic boundary conditions (Nguyen *et al.* 1991). However, the decomposition method presents the disadvantage that the harmonic functions η must be stored in the computer memory and that the evaluation of the volume integrals can become very time-consuming for large grids. Fortunately, both inconveniences can be eliminated by applying to the equations (2.18) the method proposed by Glowinski and Pironneau (1979a-c) for the solution of the two-dimensional Stokes problem, as described in the next subsection.

2.10.3 Glowinski–Pironneau method

In the Glowinski–Pironneau method, the determination of the harmonic functions η and the evaluation of the volume integrals appearing in expressions (2.31) and (2.32) are avoided at the cost of doubling the number of elliptic equations to be solved. After the vorticity problems (2.28), one has to solve the stream function problems

$$-\nabla^2\psi^0 = \zeta^0, \qquad \psi^0|_S = a; \tag{2.33a}$$

$$-\nabla^2\psi' = \zeta', \qquad \psi'|_S = 0. \tag{2.33b}$$

Then, by introducing for each $\sigma \in S$ the simple function

$$w(\boldsymbol{x};\sigma) \text{ arbitrary in } V, \qquad w(\boldsymbol{x};\sigma)|_S = \delta(s-\sigma), \tag{2.34}$$

Green identity (2.8) can be used to characterize A and β in an equivalent form, that does not involve the harmonic functions η. In fact, using (2.33b) and Green identity, and since η is harmonic, one obtains

$$\begin{aligned} A(\sigma,\sigma') &= \int \zeta'\eta\, dV = \int (-\nabla^2\psi')\eta\, dV \\ &= \int \psi'(-\nabla^2\eta)\, dV + \oint \left(\psi'\frac{\partial\eta}{\partial n} - \eta\frac{\partial\psi'}{\partial n}\right) ds \\ &= \oint \left(0\cdot\frac{\partial\eta}{\partial n} - \eta\frac{\partial\psi'}{\partial n}\right) ds = \oint \left(-\eta\frac{\partial\psi'}{\partial n}\right) ds. \end{aligned}$$

Since the boundary conditions of η and w have been taken identical (compare (2.29) with (2.34)), the divergence theorem gives

$$\begin{aligned} A(\sigma,\sigma') &= \oint (-\eta\, \boldsymbol{n}\cdot\boldsymbol{\nabla}\psi')\, ds = \oint (-w\, \boldsymbol{n}\cdot\boldsymbol{\nabla}\psi')\, ds \\ &= \int \left(-\boldsymbol{\nabla}\cdot(w\boldsymbol{\nabla}\psi')\right) dV \\ &= \int (-\boldsymbol{\nabla}w\cdot\boldsymbol{\nabla}\psi' - w\nabla^2\psi')\, dV \\ &= \int (\zeta' w - \boldsymbol{\nabla}\psi'\cdot\boldsymbol{\nabla}w)\, dV, \end{aligned}$$

using again equation (2.33b). A similar calculation can be performed to express β without η, using the solution ψ^0 of problem (2.33a). The application of Green identity in expression (2.32) gives

$$\begin{aligned} \beta(\sigma) &= -\int \zeta^0\eta\, dV + \oint \left(a\frac{\partial\eta}{\partial n} - b\eta\right) ds \\ &= -\int (-\nabla^2\psi^0)\eta\, dV + \oint \left(a\frac{\partial\eta}{\partial n} - b\eta\right) ds \\ &= -\int \psi^0(-\nabla^2\eta)\, dV - \oint \left(\psi^0\frac{\partial\eta}{\partial n} - \eta\frac{\partial\psi^0}{\partial n}\right) ds + \oint \left(a\frac{\partial\eta}{\partial n} - b\eta\right) ds \\ &= -\oint \left(-\eta\frac{\partial\psi^0}{\partial n}\right) dV - \oint b\eta\, ds, \end{aligned}$$

and, since the boundary values of η and w are coincident,

$$\begin{aligned}\beta(\sigma) &= -\oint(-w\,\boldsymbol{n}\cdot\boldsymbol{\nabla}\psi^0)\,ds - \oint bw\,ds\\ &= -\int\big(-\boldsymbol{\nabla}\cdot(w\boldsymbol{\nabla}\psi^0)\big)\,dV - \oint bw\,ds\\ &= -\int(-\boldsymbol{\nabla}w\cdot\boldsymbol{\nabla}\psi^0 - w\nabla^2\psi^0)\,dV - \oint bw\,ds\\ &= -\int(\zeta^0 w - \boldsymbol{\nabla}\psi^0\cdot\boldsymbol{\nabla}w)\,dV - \oint bw\,ds.\end{aligned}$$

In conclusion, the quantity A and β can be calculated through the following expressions:

$$\begin{aligned}A(\sigma,\sigma') &= \int(\zeta' w - \boldsymbol{\nabla}\psi'\cdot\boldsymbol{\nabla}w)\,dV,\\ \beta(\sigma) &= -\int(\zeta^0 w - \boldsymbol{\nabla}\psi^0\cdot\boldsymbol{\nabla}w)\,dV - \oint bw\,ds.\end{aligned}$$

The arbitrariness of the functions w at all internal point of V is exploited by choosing $w = 0$ inside V so that the integration domain in the two previous relations becomes a narrow strip along the boundary (Glowinski and Pironneau 1979c). Once the linear problem $A\lambda = \beta$ has been solved, the sought solution (ζ,ψ) is determined by solving the two elliptic problems:

$$\begin{aligned}(-\nabla^2+\gamma)\zeta &= f, \qquad & \zeta|_S &= \lambda;\\ -\nabla^2\psi &= \zeta, \qquad & \psi|_S &= a.\end{aligned} \tag{2.35}$$

Finite element implementations of this method have been described by Glowinski and Pironneau (1979c) for the Stokes problem and by Quartapelle and Napolitano (1984) for general boundary conditions. In the discrete case, the linear operator A becomes a symmetric full matrix of order equal to the number of boundary points where the no-slip condition is prescribed. For large scale applications, this number can become so large to make the storage of the matrix in the computer memory prohibitive and thus preventing the use of direct solution methods. In these situations, it is possible to resort to iterative solution methods which avoid the explicit determination of the matrix. Glowinski and Pironneau have proposed several iterative methods of this kind for the solution of the symmetric system $A_h\lambda_h = \beta_h$, which is the discrete counterpart of linear problem $A\lambda = \beta$ (1979c).

Of course, to be solved by means of the finite element method, the various elliptic problems must be expressed in variational form. The typical pair of vorticity and stream function equations to be solved will be written as follows:

Find $\zeta \in H^1(V)$ with $\zeta|_S = \lambda$ such that

$$\nu(\boldsymbol{\nabla}\xi,\boldsymbol{\nabla}\zeta) + \gamma(\xi,\zeta) = (\xi,f), \qquad \forall\xi \in H_0^1(V); \tag{2.36}$$

then find $\psi \in H^1(V)$ with $\psi|_S = a$ such that

$$(\nabla\phi, \nabla\psi) = (\phi, \zeta), \qquad \forall \phi \in H_0^1(V). \tag{2.37}$$

As it has been remarked, the fact that both the boundary data a and b enter the right-hand side of the vorticity integral conditions gives a complete freedom in choosing the Dirichlet or the Neumann boundary condition to supplement the Poisson equation for the stream function. The equivalence of the two conditions has been verified in numerical implementations using finite elements. However, in computational schemes which avoid the explicit determination of the harmonic functions η, such as the Glowinski–Pironneau method just described, it is necessary to make a distinction between the two boundary data a and b. We note in fact that in this method only the datum b appears explicitly in the term β, whereas the Dirichlet datum a affects β through the presence of the function ψ^0 which satisfies the boundary condition $\psi^0|_S = a$. To distinguish these two different situations the boundary datum b is said to give an *essential* contribution to the boundary term of the integral conditions (*i.e.*, to the right-hand side of the linear problem $A\lambda = \beta$), whereas the datum a is said to give a *natural* contribution. The harmonic function η multiplying the corresponding datum in the boundary integral appears in a nondifferentiated or differentiated form, respectively. In other words, if the function η enters the boundary integral in nondifferentiated form, then the corresponding boundary datum gives an essential contribution to the integral conditions; on the contrary, if the derivative of the function is present, then the corresponding boundary datum gives a natural contribution to the condition. In the first case, the boundary value of the projection function can be replaced in the boundary integral itself by the auxiliary function, such as w, which is introduced to avoid the explicit determination of the harmonic function η. When such a replacement is possible we will say that the integral conditions can be reduced in Glowinski–Pironneau form. To ease the recognition of the datum giving an essential contribution, the corresponding term will always be written as the last one of the boundary integral of the integral conditions.

2.10.4 Discretization of the nonlinear terms

The analysis of the numerical schemes for the ζ-ψ equations would be incomplete without saying some words about the approximation of the nonlinear term $J(\zeta, \psi)$. We describe the most typical methods for discretizing this term by means of central finite differences and finite elements. The standard second-order accurate centred approximation of $J(\zeta, \psi)$ over a uniform finite-difference grid is given by

$$J(\zeta, \psi)_{\mathrm{FD}} = \frac{1}{4h^2}\Big[(\zeta_{i+1,j} - \zeta_{i-1,j})(\psi_{i,j+1} - \psi_{i,j-1}) - (\zeta_{i,j+1} - \zeta_{i,j-1})(\psi_{i+1,j} - \psi_{i-1,j})\Big].$$

This discretization presents some drawbacks concerning the numerical stability of the scheme constructed to solve the inviscid version of the vorticity equation. The issue has been addressed in a fundamental paper by Arakawa (1966) who has shown that the instability can be explained as a result of the strong anisotropy of this discrete representation of the Jacobian. To overcome this problem, he has proposed to combine this most simple discrete form of the Jacobian with two other centred approximations of $J(\zeta,\psi)$ written in the conservative form $\boldsymbol{\nabla}\cdot(\zeta\boldsymbol{u})$, as follows,

$$J(\zeta,\psi)^{+\times}_{\mathrm{FD}} = \frac{1}{4h^2}\Big[\zeta_{i+1,j}(\psi_{i+1,j+1}-\psi_{i+1,j-1}) - \zeta_{i-1,j}(\psi_{i-1,j+1}-\psi_{i-1,j-1}) - \zeta_{i,j+1}(\psi_{i+1,j+1}-\psi_{i-1,j+1}) + \zeta_{i,j-1}(\psi_{i+1,j-1}-\psi_{i-1,j-1})\Big]$$

and

$$J(\zeta,\psi)^{\times+}_{\mathrm{FD}} = \frac{1}{4h^2}\Big[\zeta_{i+1,j+1}(\psi_{i,j+1}-\psi_{i+1,j}) - \zeta_{i-1,j-1}(\psi_{i-1,j}-\psi_{i,j-1}) - \zeta_{i-1,j+1}(\psi_{i,j+1}-\psi_{i-1,j}) + \zeta_{i+1,j-1}(\psi_{i+1,j}-\psi_{i,j-1})\Big].$$

Arakawa demonstrated that the two conservative quadratic quantities kinetic energy $\int \frac{1}{2}|\boldsymbol{\nabla}\psi|^2\,dV$ and *enstrophy* $\int \zeta^2\,dV$ are conserved also in the spatially discretized problem provided the Jacobian is approximated by the following cocktails' golden rule

$$J(\zeta,\psi)_{\mathrm{Arakawa}} = \frac{1}{3}\Big[J(\zeta,\psi)_{\mathrm{FD}} + J(\zeta,\psi)^{+\times}_{\mathrm{FD}} + J(\zeta,\psi)^{\times+}_{\mathrm{FD}}\Big].$$

This is the best possible approximation with a second-order spatial accuracy. Arakawa's paper contains also a fourth-order discretization of the Jacobian, endowed with similar conservation properties as well as a good isotropic response. A simpler, but less isotropic, fourth-order accurate approximation of $J(\zeta,\psi)$ can be obtained by approximating all the first derivatives of the Jacobian by means of the five-point centred formula

$$\frac{\partial u}{\partial x}\bigg|_i = \frac{1}{12h}(-u_{i+2}+8u_{i+1}-8u_{i-1}+u_{i-2}),$$

for all internal points, excluding the first internal point $x=x_1$ located near to the boundary. At this special location, the five-point uncentred fourth-order approximation

$$\frac{\partial u}{\partial x}\bigg|_1 = \frac{1}{12h}(u_4-6u_3+18u_2-10u_1-3u_0)$$

can be used.

Coming to finite elements, we first notice that the Galerkin projection method applied to the a nonlinear model term of the type $\zeta_x\psi_x$ gives a more satisfactory representation of the nonlinear interaction than finite differences. In fact, while standard central differences give

$$\left.\frac{\partial\zeta}{\partial x}\frac{\partial\psi}{\partial x}\right|_i = \frac{1}{4h^2}(\zeta_{i+1}-\zeta_{i-1})(\psi_{i+1}-\psi_{i-1})$$

unless some special and clever strategy is adopted, the Galerkin method with linear finite elements on the same uniform grid gives

$$\left\langle \phi_i, \frac{\partial\zeta}{\partial x}\frac{\partial\psi}{\partial x}\right\rangle = \frac{1}{2h^2}\Big[(\zeta_{i+1}-\zeta_i)(\psi_{i+1}-\psi_i)+(\zeta_i-\zeta_{i-1})(\psi_i-\psi_{i-1})\Big].$$

Therefore the finite element Galerkin method is legitimately expected to representing the two-dimensional nonlinear term $J(\zeta,\psi)$ better than the approximation $J(\zeta,\psi)_{\mathrm{FD}}$. As a matter of fact, the Galerkin approximation using a uniform grid of bilinear elements is found to be coincident with second-order Arakawa differencing, as shown by Jespersen (1974). Thus, for a uniform rectangular mesh, we have the important result

$$\langle\phi,\, J(\zeta,\psi)\rangle_{\substack{\text{bilinear}\\ \text{elements}}} = J(\zeta,\psi)_{\text{Arakawa}}.$$

Considering a uniform mesh of equilateral triangles with a linear interpolation of the variables, the Galerkin method provides the approximation (Saiac 1985):

$$\begin{aligned}\langle\phi_0,\, J(\zeta,\psi)\rangle_{\substack{\text{linear}\\ \text{elements}}} = \frac{1}{6h^2}\Big[&\zeta_1(\psi_6-\psi_2)+\zeta_2(\psi_1-\psi_3)+\zeta_3(\psi_2-\psi_4)\\ &+\zeta_4(\psi_3-\psi_5)+\zeta_5(\psi_4-\psi_6)+\zeta_6(\psi_5-\psi_1)\Big],\end{aligned}$$

where the subscripts $1,2,\ldots,6$ indicate the six nodal points placed in the hexagonal and anti-clockwise pattern around the central node 0. Both finite element approximations satisfy the conservation properties of the Arakawa scheme, and on any mesh. Thus, the Galerkin method adds the intrinsic flexibility characteristic of any finite-element-based spatial discretization and the fulfillment of basic conservation properties inherent to the inviscid equations.

Of course, such conclusions are drawn without considering the rôle played by the time discretization of the dynamical equation, an issue which will be briefly touched upon later when studying the inviscid equations.

2.11 Numerical schemes: spectral method

In order to apply spectral methods [18], [6], to the solution of the Navier–Stokes equations under no-slip conditions special attention must be paid to the determination of the lacking boundary values for the vorticity or the pressure. The method most frequently used for this purpose is the influence matrix method, that has been proposed by Kleiser and Schumann (1980) to solve the primitive variable equations for the spectral simulation of plane channel flows. A very similar method has also been considered independently by Gadzag, Takao and Fromm (1982) as a means for computing the solution of the Fourier modes of the vorticity–stream function equations. The influence matrix method for the primitive variable equations has been subsequently extended to problems with more complicated boundaries by Le Quéré and Alziary de Roquefort (1982), Marcus (1984) and Le Quéré and Pécheux (1989). Influence matrix methods for the spectral solution of the vorticity-stream function equations have been employed by Ehrenstein and Peyret (1989).

An alternative to the influence matrix method is to impose the vorticity integral conditions so that the equations of vorticity and stream function can be solved in an uncoupled manner, as already described in this chapter (cf. section 2.7). A technique of this kind has been implemented by Dennis and Quartapelle (1983) for solving the spectral equations in one-dimensional form by adapting the finite element method of Glowinski and Pironneau discussed in section 2.10.3 to the Galerkin equations provided by a Chebyshev polynomial approximation. Two-dimensional vorticity integral conditions have been considered recently by Ng, Paik and Chung (1991) to calculate pseudospectral solutions of the classical driven-cavity problem. Integral conditions still of one-dimensional form but suitably extended to the vector character of the vorticity variable in three-dimensional problems in spherical coordinates have also been considered to solve the Navier–Stokes equations in a spherical gap by means of Chebyshev polynomials and spherical harmonics (Dennis and Quartapelle 1985).

In this section we intend to compare a few methods for determining the laking vorticity boundary values under no-slip conditions, by confronting their respective numerical accuracy and computational efficiency. We will consider the equations governing a single Fourier mode of the (linearized) ζ-ψ equations.

First, the nonprimitive variable counterpart of the Kleiser–Schumann method will be described. Then, three different implementations of the method using integral conditions will be considered.

In the first two methods of this kind, the vorticity integral contitions are used either to supplement directly the second-order differential equation for the vorticity, or to determine a decomposition of the vorticity into elementary solutions to the second-order differential problems supplemented by conditions of standard boundary value type (Dirichlet conditions). Both these variants rely on the explicit use of the functions which constitute the kernel of the integral conditions. For problems in geometrically simple domains which allow the separation of variables and the reduction of the integral conditions to a one-dimensional form, such as in the Chebyshev spectral methods of interest here, the analytical expression of the kernel functions is obtained very easily and their use does not cause any computational difficulty. This is to be contrasted with the situation of regions of arbitrary shape, where the determination and the storage of the discrete counterpart of the kernel functions represent a very heavy computational burden.

The third method based on the integral conditions is very similar to the influence matrix method in that it employs an auxiliary matrix whose construction requires to solve the same differential equations of the influence matrix method. In this technique the vorticity integral conditions for the Chebyshev approximation are expressed as in the Glowinski and Pironneau method (1979c) and do not require the kernels occurring in the integral.

2.11.1 Modal equations

To analyze the differences between the classical influence matrix method and the three methods for imposing the vorticity integral conditions either through an auxiliary matrix or directly, plane problems admitting a Fourier analysis are considered. The system of ordinary differential equations governing the Fourier coefficients of the vorticity ζ and the stream function ψ assumes the typical form

$$\left(-\mathcal{D}^2 + \gamma\epsilon(x)\right)\zeta = f(x), \tag{2.38a}$$

$$-\mathcal{D}^2\psi = \epsilon(x)\zeta. \tag{2.38b}$$

Here, $\mathcal{D}^2$ represents the second-order linear differential operator

$$\mathcal{D}^2 = t\frac{d^2}{dx^2} + u\frac{d}{dx} + v = tD^2 + uD + v, \tag{2.39}$$

where t, u and v are constant coefficients and D denotes the differentiation with respect to x. In equations (2.38a,b), ζ and ψ are the unknowns, $\epsilon(x)$ is a given function and $f(x)$ represents a known source term, both defined on the integration interval $[x_1, x_2]$. Equations of this type are encountered when the method of the truncated series expansion is employed in the solution of the vorticity–stream function equations in domains which admit the separation of variables and a Fourier representation of the dependence on one of the two spatial variables. Actually, a system of coupled nonlinear equations governing the expansion coefficients is obtained, as already shown in section 2.7. For steady problems, the differential equations are ordinary and, after a suitable linearization, one obtains equation pairs like (2.38) with $\gamma = 0$. For unsteady problems, the truncated expansion method leads to a system of partial differential equations governing the evolution of the Fourier coefficients. Therefore, the equations must be linearized and also discretized in time to obtain systems of ordinary differential equations like (2.38), with typically $\gamma \propto 1/(\nu\Delta t)$, Δt being the time step.

The boundary conditions attached to this ζ-ψ system are usually

$$\psi(x_i) = a_i, \qquad \psi'(x_i) = b_i, \qquad i = 1, 2,$$

where the values a_i and b_i are obtained from the (Fourier coefficients of the) normal and tangential components of the velocity specified on the boundary.

2.11.2 Influence matrix method

The simultaneous specification of both ψ and ψ' at the two boundary points makes the solution of the equations (2.38) difficult since too many boundary conditions are prescribed on one unknown (ψ) whereas no boundary condition is available for the other (ζ). A method for circumventing such a difficulty consists in introducing a 2×2 matrix of influence between the redundant boundary conditions $\psi'(x_i) = b_i$, $i = 1, 2$, and the missing boundary values $\zeta(x_i)$, $i = 1, 2$. To this aim, the solution (ζ, ψ) is expanded in the form

$$\begin{bmatrix} \zeta \\ \psi \end{bmatrix} = \begin{bmatrix} \zeta_0 \\ \psi_0 \end{bmatrix} + \lambda_1 \begin{bmatrix} \zeta_1 \\ \psi_1 \end{bmatrix} + \lambda_2 \begin{bmatrix} \zeta_2 \\ \psi_2 \end{bmatrix}, \tag{2.40}$$

where λ_1 and λ_2 are two unknowns to be determined so as to satisfy the redundant boundary conditions for ψ and the functions (ζ_l, ψ_l), $l = 0, 1, 2$, are the solutions to the following problems: for $l = 0$

$$\left(-\mathcal{D}^2 + \gamma\epsilon(x)\right)\zeta_0 = f(x), \qquad \zeta_0(x_i) = 0, \quad i = 1, 2; \tag{2.41a}$$

$$-\mathcal{D}^2\psi_0 = \epsilon(x)\zeta_0, \qquad \psi_0(x_i) = a_i, \quad i = 1, 2; \tag{2.41b}$$

whereas, for $l = k = 1, 2$,

$$\left(-\mathcal{D}^2 + \gamma\epsilon(x)\right)\zeta_k = 0, \qquad \zeta_k(x_i) = \delta_{ki}, \quad i = 1, 2; \tag{2.42a}$$

$$-\mathcal{D}^2\psi_k = \epsilon(x)\zeta_k, \qquad \psi_k(x_i) = 0, \quad i = 1, 2; \tag{2.42b}$$

where δ_{ki} is the Kronecker delta. Imposing the fulfillment of the derivative boundary condition $\psi'(x_i) = b_i$, $i = 1, 2$, gives the linear system

$$\widetilde{A}\boldsymbol{\lambda} = \widetilde{\boldsymbol{\beta}}, \tag{2.43}$$

with the influence matrix $\widetilde{A}$ and the right-hand side $\widetilde{\boldsymbol{\beta}}$ defined as follows

$$\widetilde{A} = \begin{bmatrix} \psi_1'(x_1) & \psi_2'(x_1) \\ \psi_1'(x_2) & \psi_2'(x_2) \end{bmatrix}, \qquad \widetilde{\boldsymbol{\beta}} = -\begin{bmatrix} \psi_0'(x_1) \\ \psi_0'(x_2) \end{bmatrix} + \begin{bmatrix} b_1 \\ b_2 \end{bmatrix}. \tag{2.44}$$

Due to the boundary conditions imposed on the vorticity in problems (2.41a) and (2.42a), the solution $\boldsymbol{\lambda}$ to the influence matrix system (2.43) gives the requested vorticity boundary values. They can be used to evaluate the linear combination (2.40), thus providing the solution (ζ, ψ) to the original problem at the cost of solving a total of 6 second-order equations.

2.11.3 Integral conditions

The specification of both ψ and ψ' at the two boundary points introduces a coupling between the linear equations for ζ and ψ, so that, under the considered set of boundary conditions for ψ, they have to be solved coupled together. As a matter of facts, in the discrete case the influence matrix method just described can be regarded as a Gaussian elimination taylored to the specific profile of the matrix of the system comprising the full set of coupled equations. The coupling is eliminated by deriving independent conditions for the vorticity by means of Green identity for the (ordinary differential) operator $\mathcal{D}^2$, namely,

$$\int_{x_1}^{x_2} \left[\phi(-\mathcal{D}^2\psi) - \psi(-\mathcal{D}^{2\dagger}\phi)\right] dx = \left[t(\psi\phi' - \psi'\phi) - u\psi\phi\right]_{x_1}^{x_2}, \tag{2.45}$$

where $\mathcal{D}^{2\dagger}$ denotes the formal adjoint operator to $\mathcal{D}^2$, which, by the definition of $\mathcal{D}^2$ in (2.39), is given by

$$\mathcal{D}^{2\dagger} = tD^2 - uD + v. \tag{2.46}$$

Note that the operator $\mathcal{D}^2$ is self-adjoint only when $u = 0$. By this Green identity, it follows that $\epsilon(x)\zeta = -\mathcal{D}^2\psi$, with $\psi(x_i) = a_i$ and $\psi'(x_i) = b_i$, $i = 1, 2$, if and only if ζ satisfies the two integral conditions

$$\int_{x_1}^{x_2} \epsilon\zeta\eta_k \, dx = \left[ta_i\eta_k'(x_i) - (ua_i + tb_i)\eta_k(x_i)\right]_{i=1}^{i=2}, \qquad k = 1, 2, \tag{2.47}$$

in which the "kernel functions" η_k, $k = 1, 2$ are any two linearly independent solutions η to the adjoint equation

$$\mathcal{D}^{2\dagger}\eta = 0.$$

For definitness, the two solutions can be characterized as the solutions to the problems

$$\mathcal{D}^{2\dagger}\eta_k = 0, \qquad \eta_k(x_i) = \delta_{ki}, \quad i = 1, 2,$$

for $k = 1, 2$. These solutions are found immediately to be

$$\eta_1(x) = \frac{e^{\lambda_+(x-x_2)} - e^{\lambda_-(x-x_2)}}{e^{\lambda_+(x_1-x_2)} - e^{\lambda_-(x_1-x_2)}},$$

$$\eta_2(x) = \frac{e^{\lambda_+(x-x_1)} - e^{\lambda_-(x-x_1)}}{e^{\lambda_+(x_2-x_1)} - e^{\lambda_-(x_2-x_1)}},$$

where

$$\lambda_\pm = \frac{+u \pm \sqrt{u^2 - 4tv}}{2t}.$$

By means of the integral conditions (2.47) the original problem can be re-stated in the following uncoupled or split form

$$\left(-\mathcal{D}^2 + \gamma\epsilon(x)\right)\zeta = f(x), \qquad \int_{x_1}^{x_2} \epsilon\zeta\eta_k \, dx = c_k, \ k = 1, 2; \tag{2.48a}$$

$$-\mathcal{D}^2\psi = \epsilon(x)\zeta, \qquad \psi(x_i) = a_i, \ i = 1, 2; \tag{2.48b}$$

where

$$c_k = \Big[ta_i\eta_k'(x_i) - (ua_i + tb_i)\eta_k(x_i)\Big]_{i=1}^{i=2}. \tag{2.49}$$

Each equation is therefore supplemented by its own conditions, so that they can be solved in sequence one after the other. As a consequence, the solution method based on the integrally conditioned vorticity equation requires to solve only two scond-order problems.

If the vorticity (or its derivative) is specified at one end point, such as, for instance, $\zeta(x_1) = z$, only one integral condition has to be satisfied by ζ. In this case the function η to be used as the kernel of the single integral condition satisfies a condition at $x = x_1$ which depends on whether the boundary value of ψ or the derivative of ψ is imposed at $x = x_1$ where the vorticity boundary condition is specified. If, for example, the condition $\psi(x_1) = a_1$ is prescribed, then η is any solution to the problem

$$\mathcal{D}^{2\dagger}\eta = 0, \qquad \eta(x_1) = 0, \quad \eta(x_2) \neq 0,$$

and the two conditions for ζ are

$$\int_{x_1}^{x_2} \epsilon\zeta\eta\, dx = c \qquad \text{and} \qquad \zeta(x_1) = z,$$

where

$$c = ta_2\eta'(x_2) - (ua_2 + tb_2)\eta(x_2) - ta_1\eta'(x_1).$$

On the other hand, if the derivative boundary condition $\psi'(x_1) = b_1$ is specified, then η is any solution to the problem

$$\mathcal{D}^{2\dagger}\eta = 0, \qquad t\eta'(x_1) - u\eta(x_1) = 0, \quad \eta(x_2) \neq 0,$$

and the two vorticity conditions are again

$$\int_{x_1}^{x_2} \epsilon\zeta\eta\, dx = c \qquad \text{and} \qquad \zeta(x_1) = z,$$

but now

$$c = ta_2\eta'(x_2) - (ua_2 + tb_2)\eta(x_2) + tb_1\eta(x_1).$$

Decomposed integral conditions

There are other ways of satifying the integral conditions for the vorticity. In the method above they are used to complete the second-order differential operator of the vorticity equation, and to obtain, in the discrete case, a (nonsingular) linear system of equations whose solution satisfies automatically the two global conditions. If instead one prefers to deal only with standard boundary value problems for such an operator, it is possible to express the solution ζ as the linear combination

$$\zeta = \zeta_0 + \lambda_1\zeta_1 + \lambda_2\zeta_2, \tag{2.50}$$

where ζ_0, ζ_1 and ζ_2 are the solutions to the "one-dimensional Dirichlet problems"

$$\left(-\mathcal{D}^2 + \gamma\epsilon(x)\right)\zeta_0 = f(x), \qquad \zeta_0(x_i) = 0, \quad i = 1,2; \tag{2.51a}$$

$$\left(-\mathcal{D}^2 + \gamma\epsilon(x)\right)\zeta_k = 0, \qquad \zeta_k(x_i) = \delta_{ki}, \quad i = 1,2; \tag{2.51b}$$

and then to require that the linear combination satify the integral conditions. In this way one obtains the linear system

$$A\boldsymbol{\lambda} = \boldsymbol{\beta}, \tag{2.52}$$

with the matrix A and the right-hand side $\boldsymbol{\beta}$ defined by

$$A = \begin{bmatrix} \int\epsilon\zeta_1\eta_1\,dx & \int\epsilon\zeta_2\eta_1\,dx \\ \int\epsilon\zeta_1\eta_2\,dx & \int\epsilon\zeta_2\eta_2\,dx \end{bmatrix}, \qquad \boldsymbol{\beta} = -\begin{bmatrix} \int\epsilon\zeta_0\eta_1\,dx \\ \int\epsilon\zeta_0\eta_2\,dx \end{bmatrix} + \begin{bmatrix} c_1 \\ c_2 \end{bmatrix}. \tag{2.53}$$

Once the unknown $\boldsymbol{\lambda}$ has been determined, it remains to solve the problem $-\mathcal{D}^2\psi = \epsilon(x)\zeta$, $\psi(x_i) = a_i$, $i = 1,2$. This method has some resemblance with the influence matrix method, but is different in that the correct vorticity is determined without any reference to the stream function equation. It imposes the vorticity integral conditions in a *decomposed* form and requires to solve four second-order problems, three for the vorticity and only one for the stream function. A two-dimensional version of this method has been implemented by Nguyen, Paik and Chung (1991) using numerically evaluated kernel functions $\eta_k(x,y)$.

Glowinski–Pironneau method

Both integral condition methods described so far employ the analytical expressions of the kernel appearing in the vorticity condition. The same integral conditions can be expressed in a form which does not require an explicit knowledge of the functions η_k by an adaptation of the method introduced by Glowinski and Pironneau (1979c) for finite element solution of the two-dimensional equations, which has been described in section 2.10.3. For the present spectral situation of one-dimensional equations, the Glowinski–Pironneau procedure requires to solve exactly the same set of second-order systems (2.41) and (2.42) considered by the influence matrix method. Then, the vorticity is written as the linear combination $\zeta = \zeta_0 + \lambda_1\zeta_1 + \lambda_2\zeta_2$, where the unknown $\boldsymbol{\lambda}$ represents as always the boundary values of vorticity. The latter are now determined by imposing the fulfillment of the integral conditions, which give the same system of the decomposed method, namely,

$$A\boldsymbol{\lambda} = \boldsymbol{\beta}. \tag{2.54}$$

However, the integrals which define A and $\boldsymbol{\beta}$ are transformed using Green identity (2.45) and expressing the functions $\epsilon(x)\zeta_l(x)$, $l = 0, 1, 2$, in terms of $-\mathcal{D}^2\psi_l$, so that only the boundary values of η_k are involved. At this stage, one introduces the two auxiliary functions

$$w_k(x) = \text{arbitrary in }]x_1, x_2[, \qquad w_k(x_i) = \delta_{ik},$$

to replace the boundary values of η_k occurring in the boundary term of Green identity. Then, using Green identity in the other direction and employing again the equations $-\mathcal{D}^2\psi_l = \epsilon(x)\zeta_l$, $l = 0, 1, 2$, one obtains

$$A_{ki} = \int_{x_1}^{x_2} \Big[(\epsilon\zeta_i + u\psi_i' + v\psi_i)w_k - t\psi_i' w_k'\Big]dx,$$

$$\beta_k = -\int_{x_1}^{x_2} \Big[(\epsilon\zeta_0 + u\psi_0' + v\psi_0)w_k - t\psi_0' w_k'\Big]dx - t\Big[b_i w_k(x_i)\Big]_{i=1}^{i=2}.$$

This method enforces the integral conditions without employing the functions η_k. On the other hand, it requires to solve the same second-order equations of the influence matrix method and, like the latter, provides the requested boundary values of vorticity passing through the solution of stream function equations.

The number of second-order differential equations which are solved by the four different methods is summarized in the following Table.

Number of second-order equations	
Influence matrix method	6
Integral conditions à la Glowinski–Pironneau	6
Decomposed integral conditions	4
Direct integral conditions	2

2.11.4 Chebyshev spectral approximation

Tau equations

Assuming that $x_1 = -1$ and $x_2 = 1$, let the functions ζ and ψ be expanded in a truncated series of Chebyshev polynomials $T_n(x) = \cos[n\cos^{-1}(x)]$, namely,

$$\zeta(x) = \sum_{n=0}^{N} \zeta_n T_n(x) \qquad \text{and} \qquad \psi(x) = \sum_{n=0}^{N} \psi_n T_n(x).$$

The *tau* projection method (see, e.g., [6, p. 335]) applied to the vorticity equation (2.38a) leads to the following Chebyshev equations

$$-t\left(\zeta_n^{[2]} + u\zeta_n^{[1]} + v\zeta_n^{[0]}\right) + \gamma \sum_{m=0}^{N} E_{nm}\zeta_m = f_n^{[0]},$$

where the notation $\zeta_n^{[j]}$ indicates the coefficients of the jth derivative of a Chebyshev series with coefficients ζ_n, multiplied by the normalization constant Π_n, with $\Pi_0 = \pi$ and $\Pi_n = \pi/2$, for $n \geq 1$. For instance, $\zeta_n^{[0]} = \Pi_n\zeta_n$. In other words, the discrete equations are left unnormalized. The matrix of coefficients E_{nm} is defined by

$$E_{nm} = \int_{-1}^{1} \frac{T_n(x)T_m(x)\epsilon(x)}{\sqrt{1-x^2}}\, dx.$$

The two integral conditions supplementing the *tau* equations for vorticity are written in the form

$$\sum_{n=0}^{N} H_n[\eta_k]\zeta_n = c_k, \qquad k = 1, 2,$$

where

$$H_n[\eta] = \int_{-1}^{1} \epsilon(x)T_n(x)\eta(x)\, dx.$$

The integrals defining the matrix coefficients E_{nm} are evaluated numerically by means of the Chebyshev–Gauss–Legendre quadrature formula with $N+1$ internal points, namely,

$$\int_{-1}^{1} \frac{u(x)}{\sqrt{1-x^2}}\, dx = \frac{\pi}{N+1} \sum_{r=1}^{N+1} u(x_r),$$

where $x_r = \cos\theta_r = \cos\frac{(2r-1)\pi}{2(N+1)}$, $r = 1, 2, \ldots, N+1$. The integrals which define the coefficients $H_n[\eta]$ appearing in the integral conditions are evaluated using the Gauss–Legendre quadrature formula with $N+1$ points.

Replacing the last two *tau* equations for ζ by the integral conditions gives the complete linear system

$$\left(t\zeta_n^{[2]} + u\zeta_n^{[1]} + v\zeta_n^{[0]}\right) + \gamma \sum_{m=0}^{N} E_{nm}\zeta_m = f_n^{[0]}, \qquad 0 \leq n \leq N-2,$$

$$\sum_{n=0}^{N} H_n[\eta_k]\zeta_n = c_k, \qquad k = 1, 2.$$

Similarly, the discrete equations for the stream function subject to the boundary conditions $\psi(\mp 1) = a_{1,2}$ are

$$\left(t\psi_n^{[2]} + u\psi_n^{[1]} + v\psi_n^{[0]}\right) = \sum_{m=0}^{N} E_{nm}\zeta_m,$$

$$\sum_{n=0}^{N}(-1)^n\psi_n = a_1, \qquad \sum_{n=0}^{N}\psi_n = a_2.$$

The values on the boundary of the derivative ψ', required by the influence matrix method, are calculated by means of

$$\psi'(\pm 1) = \sum_{n=0}^{N}(\pm 1)^{n+1} n^2 \psi_n.$$

To reduce possible cancellation errors in this summation, the negative and positive contributions are accumulated separately and added only at the end of the summation.

Glowinski–Pironneau method

The Chebyshev approximation to the integrals involving the auxiliary functions $w_k(x)$ required to implement the Glowinski–Pironneau procedure is obtained as follows. The two auxiliary functions $w_k(x)$, $k = 1, 2$, are taken in the subspace of the first two Chebyshev polynomials, namely,

$$w_1(x) = \tfrac{1}{2}(1 - x), \qquad w_2(x) = \tfrac{1}{2}(1 + x),$$

and are expressed for convenience in the following form

$$w_k(x) = W + W_k x, \; k = 1, 2, \qquad W = \tfrac{1}{2}, \quad W_1 = -\tfrac{1}{2}, \quad W_2 = \tfrac{1}{2}.$$

Then, the aforementioned integrals can be written as

$$\begin{aligned} I_k &= \int_{-1}^{1} [(\epsilon\zeta + u\psi' + v\psi)w_k - t\psi' w_k']dx \\ &= W \int_{-1}^{1} \xi \, dx + W_k \int_{-1}^{1} x\xi \, dx + (uW - tW_k) \int_{-1}^{1} \psi' \, dx + uW_k \int_{-1}^{1} x\psi' \, dx, \end{aligned}$$

where the variable $\xi = \epsilon\zeta + v\psi$ has been introduced. The various integrals of this expressions can be evaluated in terms of the integrals of Chebyshev polynomials and of the Chebyshev coefficients of ξ and ψ. By means of relationships (A.11), (A.9) and (A.18) of [18], one obtains

$$x\xi(x) = \sum_{n=0}^{N} \rho_n T_n(x), \qquad \begin{aligned} \rho_0 &= \tfrac{1}{2}\xi_1 \\ \rho_n &= \tfrac{1}{2}(c_{n-1}\xi_{n-1} + \xi_{n+1}), \quad 1 \le n \le N-1, \\ \rho_N &= \tfrac{1}{2} c_{N-1}\xi_{N-1} \end{aligned}$$

$$\psi'(x) = \sum_{n=0}^{N} \frac{2}{c_n} \Bigg(\sum_{\substack{p=n+1 \\ p+n \text{ odd}}}^{N} p\psi_p \Bigg) T_n(x),$$

and

$$x\psi'(x) = \sum_{n=0}^{N} \frac{1}{c_n}\left(n\psi_n + 2 \sum_{\substack{p=n+2 \\ p+n \text{ even}}}^{N} p\psi_p \right) T_n(x),$$

respectively. Here the constant c_n is such that $c_0 = 2$ and $c_n = 1$ for $n \geq 1$. Evaluating the integral of $T_n(x)$ and taking into account that it vanishes for n odd, whereas $\int_{-1}^{1} T_n(x)\,dx = 2/(1-n^2)$, for n even, the final expression is obtained in the form

$$I_k = \sum_{\substack{n=0 \\ n \text{ even}}}^{N} \left[W\xi_n + W_k\rho_n + (uW - tW_k)\frac{2}{c_n} \sum_{\substack{p=n+1 \\ p+n \text{ odd}}}^{N} p\psi_p \right.$$

$$\left. + uW_k \left(n\psi_n + \frac{2}{c_n} \sum_{\substack{p=n+2 \\ p+n \text{ even}}}^{N} p\psi_p \right) \right] \frac{2}{1-n^2}.$$

Such an expression can be evaluated efficiently in only $O(N)$ operations using recurrence relations, see, for instance, [18, p. 117].

Bordered quasi-pentadiagonal linear system

For the constant coefficient second-order operator $\mathcal{D}^2$, the *tau* projection method provides Chebyshev equations whose matrix can be recast in a banded form by a suitable linear combination of the equations. Consider the Chebyshev equations for the unknown ψ, with the right-hand side rewritten as a single term, namely, $\sum_{m=0}^{N} E_{nm}\zeta_m = \Pi_n r_n$, with $\Pi_0 = \pi$ and $\Pi_n = \pi/2$, for $n \geq 1$. After normalizing the discrete equations by dividing by $\pi/2$, they read

$$t \sum_{\substack{m=n+2 \\ m+n \text{ even}}}^{N} m(m^2 - n^2)\psi_m + 2u \sum_{\substack{m=n+1 \\ m+n \text{ odd}}}^{N} m\psi_m + vc_n\psi_n = c_n r_n,$$

where $c_0 = 2$ and $c_n = 1$, for $n \geq 1$ ($\Pi_n = \frac{\pi}{2}c_n$). Of course, only the first $N-1$ equations for $0 \leq n \leq N-2$ are considered, since there are two equations resulting from the boundary (or possibly integral) conditions. Now, assuming in any case $N \geq 6$, for $2 \leq n \leq N-4$ we can consider the following linear combination of the equations above

$$\frac{\text{Eq}(n-2)}{4n(n-1)} - \frac{\text{Eq}(n)}{2(n^2-1)} + \frac{\text{Eq}(n+2)}{4n(n+1)},$$

where $\text{Eq}(n')$ means the equation above for $n = n'$. By means of simple calculations we obtain

$$\frac{vc_{n-2}}{4n(n-1)}\psi_{n-2} + \frac{u}{2n}\psi_{n-1} + \left[t - \frac{v}{2(n^2-1)} \right] \psi_n - \frac{u}{2n}\psi_{n+1} + \frac{v}{4n(n+1)}\psi_{n+2}$$

$$= \frac{c_{n-2}r_{n-2}}{4n(n-1)} - \frac{r_n}{2(n^2-1)} + \frac{r_{n+2}}{4n(n+1)}, \qquad 2 \leq n \leq N-4.$$

The final linear system for the unknowns $\{\psi_0, \psi_1, \ldots, \psi_N\}$ so obtained is composed by the first two equations for imposing the boundary conditions, the new $N-5$ trasformed equations for $2 \le n \le N-4$ and 4 original equations for $N-5 \le n \le N-2$. In total there are $2+(N-5)+4 = N+1$ equations for the $N+1$ unknowns $\{\psi_n\}$. The profile of the resulting matrix is bordered pentadiagonal, except for a single coefficient different from zero in the last entry of the fourth row from the bottom. Such a bordered quasi-pentadiagonal profile is here shown.

$$\begin{bmatrix}
g_1 & g_2 & g_3 & g_4 & g_5 & g_6 & \cdot & \cdot & \cdot & g_{n-2} & g_{n-1} & g_n \\
h_1 & h_2 & h_3 & h_4 & h_5 & h_6 & \cdot & \cdot & \cdot & h_{n-2} & h_{n-1} & h_n \\
a_1 & b_2 & c_3 & d_4 & e_5 & & & & & & & \\
 & a_2 & b_3 & c_4 & d_5 & e_6 & & & & & & \\
 & & a_3 & b_4 & c_5 & d_6 & e_7 & & & & & \\
 & & & \cdot & \cdot & \cdot & \cdot & \cdot & & & & \\
 & & & & \cdot & \cdot & \cdot & \cdot & \cdot & & & \\
 & & & & & a_{n-6} & b_{n-5} & c_{n-4} & d_{n-3} & e_{n-2} & & \\
 & & & & & & a_{n-5} & b_{n-4} & c_{n-3} & d_{n-2} & e_{n-1} & f \\
 & & & & & & & a_{n-4} & b_{n-3} & c_{n-2} & d_{n-1} & e_n \\
 & & & & & & & & a_{n-3} & b_{n-2} & c_{n-1} & d_n \\
 & & & & & & & & & a_{n-2} & b_{n-1} & c_n
\end{bmatrix}$$

It is important to take the equations for the boundary conditions as the first two equations at the top of the linear system. In fact, with this ordering the system can be solved by UL factorization which proceedes from the bottom toward the top avoiding the fill-in and which is, at the same time, numerically stable so that no pivoting is requested. This is not the case if the equations for the boundary conditions are placed at the bottom of the system and the LU factorization is employed to avoid the fill-in.

The algorithm which performs the UL factorization as well as the corresponding backward and forward substitutions has been given by Dennis and Quartapelle (1985) and is reported here for completeness. This algorithm can be used to solve the vorticity equations only if $\gamma = 0$. Otherwise, the matrix of the linear system for the vorticity equations is full since it contains the matrix $\{E_{nm}\}$. The standard LU factorization for a full matrix (without pivoting) will be used in this case.

UL Factorization and Backward and Forward Substitutions for a Bordered Quasi-Pentadiagonal Matrix

$a_{n-1}, a_n, b_n \leftarrow 0$
do i **from** n **step** -1 **until** 3
  $c_i \leftarrow c_i - d_{i+1}b_i - e_{i+2}a_i$
  if $i > 3$ **then** $d_i \leftarrow d_i - e_{i+1}b_i$
  if $i = n-1$ **then** $e_i \leftarrow e_i - fb_i$
  if $i = n-2$ **then** $d_i \leftarrow d_i - fa_i$
  $h_i \leftarrow h_i - h_{i+1}b_i - h_{i+2}a_i$
  $g_i \leftarrow g_i - g_{i+1}b_i - g_{i+2}a_i$
  $b_{i-1} \leftarrow (b_{i-1} - d_{i+1}a_{i-1})/c_i$
  $a_{i-2} \leftarrow a_{i-2}/c_i$
end

$i \leftarrow 2$
$h_i \leftarrow h_i - h_{i+1}b_i - h_{i+2}a_i$
$g_i \leftarrow g_i - g_{i+1}b_i - g_{i+2}a_i$
$h_{i-1} \leftarrow (h_{i-1} - h_{i+1}a_{i-1})/h_i$

$i \leftarrow 1$
$g_i \leftarrow g_i - g_{i+1}h_i - g_{i+2}a_i$

$x_n \leftarrow x_n/c_n$
$x_{n-1} \leftarrow (x_{n-1} - d_n x_n)/c_{n-1}$
do i **from** $n-2$ **step** -1 **until** 3
  if $i = n-3$ **then** $x_i \leftarrow x_i - fx_{i+3}$
  $x_i \leftarrow (x_i - d_{i+1}x_{i+1} - e_{i+2}x_{i+2})/c_i$
end

$x \leftarrow x_2$
do j **from** 3 **step** 1 **until** n, $x \leftarrow x - h_j x_j$ **end**
$x_2 \leftarrow x/h_2$

$x \leftarrow x_1$
do j **from** 2 **step** 1 **until** n, $x \leftarrow x - g_j x_j$ **end**
$x_1 \leftarrow x/g_1$

$x_2 \leftarrow x_2 - h_1 x_1$
do i **from** 3 **step** 1 **until** n
  $x_i \leftarrow x_i - a_{i-2}x_{i-2} - b_{i-1}x_{i-1}$
end

Nonlinear terms

The presence of nonlinear terms in the vorticity transport equation produces corresponding nonlinear terms in the modal equations governing the Fourier coefficients of vorticity. Of course, these terms do couple all the Fourier coefficients of the vorticity and the stream function, so that some iterative method is necessary to determine the solution. We describe a possible iteration scheme which takes into account these nonlinear terms as variable coefficients terms, with the dependence on the Chebyshev expansion coefficients ζ_n always accounted for implicitly. It should be remembered that the most popular method for dealing with the nonlinear terms in spectral codes is based on a fully explicit evaluation with the aid of the FFT.

Suppose first that the vorticity modal equation contains (on the left-hand side) an additional term of the form

$$J[\alpha, \beta, \gamma]\zeta = \alpha(x)\zeta' + \beta(x)\zeta + \gamma'(x)\zeta,$$

where α, β and γ are given functions, whose first $N+1$ Chebyshev coefficients are assumed to be known. Applying the *tau* projection method, this term originates the following term to be included in the left-hand side of the (not normalized) Chebyshev equations

$$\sum_{m=0}^{N} J_{nm}[\alpha, \beta, \gamma]\zeta_m,$$

where

$$J_{nm}[\alpha, \beta, \gamma] = \sum_{p=0}^{N}(L_{npm}\alpha_p + M_{npm}\beta_p + L_{nmp}\gamma_p).$$

The L and M coefficients in this expression are the integrals of the three-fold products of Chebyshev polynomials, called also interaction coefficients, namely,

$$M_{npq} = \int_{-1}^{1} \frac{T_n(x)T_p(x)T_q(x)}{\sqrt{1-x^2}}\, dx,$$

and

$$L_{npq} = \int_{-1}^{1} \frac{T_n(x)T_p(x)T_q'(x)}{\sqrt{1-x^2}}\, dx.$$

Using well known results [15, p. 52], the value of the M integrals are found to be

$$M_{npq} = \tfrac{\pi}{2}c_n\tfrac{1}{2}(\delta_{n,p+q} + \delta_{n,|p-q|}).$$

The value of the L integrals can be expressed in terms of the function $U_n(\cos\theta) = [\sin(n+1)\theta]/\sin\theta$ [15, p. 53] to give

$$L_{npq} = \frac{q}{4}\Big\{\int_{-1}^{1} \frac{U_{q+p+n-1}(x)}{\sqrt{1-x^2}}\, dx + \int_{-1}^{1} \frac{U_{q+p-n-1}(x)}{\sqrt{1-x^2}}\, dx$$
$$+ \int_{-1}^{1} \frac{U_{q-p+n-1}(x)}{\sqrt{1-x^2}}\, dx + \int_{-1}^{1} \frac{U_{q-p-n-1}(x)}{\sqrt{1-x^2}}\, dx\Big\}.$$

The component integrals can be evaluated using the result

$$\int_{-1}^{1} \frac{U_n(x)}{\sqrt{1-x^2}}\,dx = \begin{cases} 0 & \text{if } n \text{ is odd (positive or negative)} \\ \pi & \text{if } n \text{ is even and } \geq 0 \\ -\pi & \text{if } n \text{ is even and } < 0 \end{cases}$$

The complete discrete vorticity equations including the considered variable coefficient term and comprising the integral conditions would read as

$$-\left(t\zeta_n^{[2]} + u\zeta_n^{[1]} + v\zeta_n^{[0]}\right) + \sum_{m=0}^{N} \left\{\gamma E_{nm} + J_{nm}[\alpha,\beta,\gamma]\right\}\zeta_m = f_n^{[0]},$$

$$\sum_{n=0}^{N} H_n[\eta_k]\zeta_n = c_k, \qquad k = 1, 2.$$

We remark that such a system of equations embodies the most implicit treatment possible for the vorticity, accounting simultaneously for the global conditions on ζ and the influence on ζ coming from the nonlinear advection terms.

If now an actually nonlinear ζ-ψ problem of the form

$$\left(-\mathcal{D}^2 + \gamma\epsilon(x)\right)\zeta + \psi\zeta' + \psi\zeta + \psi'\zeta = f(x),$$
$$-\mathcal{D}^2\psi = \epsilon(x)\zeta,$$

is considered, then a possible simple iterative scheme could be

$$\begin{aligned}
&\boldsymbol{\psi}^0 \leftarrow \mathbf{0}, \quad \boldsymbol{\zeta}^0 \leftarrow \mathbf{0}, \\
&\quad \left\{-\mathbf{D}^2 + \gamma\mathbf{E} + \mathbf{J}[\boldsymbol{\psi}^0]\right\}\boldsymbol{\zeta} = \mathbf{f}, \\
&\quad -\mathbf{D}^2\boldsymbol{\psi} = \mathbf{E}\boldsymbol{\zeta}, \\
&\quad \text{if } \|\boldsymbol{\zeta} - \boldsymbol{\zeta}^0\| \text{ and } \|\boldsymbol{\psi} - \boldsymbol{\psi}^0\| < \text{tolerance, then stop} \\
&\quad \boldsymbol{\psi}^0 \leftarrow \boldsymbol{\psi}, \quad \boldsymbol{\zeta}^0 \leftarrow \boldsymbol{\zeta},
\end{aligned}$$

with an obvious meaning of the notation.

2.11.5 Numerical comparisons

To compare the different methods for solving the vorticity–stream function equations we consider some simple problems whose analytical solution is known. Furthermore, to have a better appreciation of the numerical errors caused by that part of the procedures which is involved in the determination of the missing boundary values of ζ, for each test problem we compute its numerical solution using Dirichlet boundary conditions also for the vorticity.

	Influence matrix method		Integral conditions à la Glowinski–Pironneau	
N	$\text{rerr}(\zeta)$	$\text{rerr}(\psi)$	$\text{rerr}(\zeta)$	$\text{rerr}(\psi)$
6	0.54 (−04)	0.78 (−05)	0.20 (−04)	0.65 (−05)
8	0.13 (−06)	0.17 (−07)	0.54 (−07)	0.15 (−07)
10	0.26 (−09)	0.32 (−10)	0.11 (−09)	0.29 (−10)
12	0.38 (−12)	0.47 (−13)	0.16 (−12)	0.43 (−13)
14	0.59 (−14)	0.14 (−14)	0.43 (−14)	0.12 (−14)
16	0.37 (−14)	0.14 (−14)	0.42 (−14)	0.12 (−14)
	Decomposed integral conditions		Direct integral conditions	
N	$\text{rerr}(\zeta)$	$\text{rerr}(\psi)$	$\text{rerr}(\zeta)$	$\text{rerr}(\psi)$
6	0.58 (−05)	0.58 (−05)	0.58 (−05)	0.58 (−05)
8	0.14 (−07)	0.15 (−07)	0.14 (−07)	0.15 (−07)
10	0.28 (−10)	0.28 (−10)	0.28 (−10)	0.28 (−10)
12	0.42 (−13)	0.41 (−13)	0.41 (−13)	0.41 (−13)
14	0.48 (−14)	0.72 (−15)	0.39 (−14)	0.10 (−14)
16	0.40 (−14)	0.88 (−15)	0.32 (−14)	0.98 (−15)
	$-\zeta'' + \zeta' + \zeta = f, \quad -\psi'' + \psi' + \psi = \zeta,$ $f = \zeta = \psi = e^{x-1}$			

Table 2.1:

The accuracy of a computed Chebyshev solution $u^{\mathrm{c}}(x)$ is expressed by the relative error defined by

$$\text{rerr}(u) = \frac{||u^{\mathrm{c}} - u^{\mathrm{t}}||}{||u^{\mathrm{t}}||},$$

where $u^{\mathrm{t}}(x)$ represent the truncated Chebyshev approximation to the exact anayltical solution and $||u||$ denotes the L^2 norm of u, namely,

$$||u||^2 = \frac{\pi}{N+1} \sum_{r=1}^{N+1} [u(x_r)]^2,$$

where $x_r = \cos[(r-1)\pi/N]$, $r = 1, 2, \ldots, N+1$.

The first test example is given by the system of equations

$$-\zeta'' + \zeta' + \zeta = f,$$
$$-\psi'' + \psi' + \psi = \zeta.$$

Assuming the function $\psi(x) = e^{x-1}$ as solution, one has immediately $f(x) = \zeta(x) = \psi(x) = e^{x-1}$. The numerical errors of the solutions calculated by means of the four methods are given in Table 2.1.

	Vorticity boundary conditions imposed	
N	rerr(ζ)	rerr(ψ)
6	0.56 (−05)	0.59 (−05)
8	0.14 (−07)	0.15 (−07)
10	0.28 (−10)	0.28 (−10)
12	0.41 (−13)	0.41 (−13)
14	0.24 (−15)	0.12 (−14)
16	0.22 (−15)	0.12 (−14)
	$-\zeta'' + \zeta' + \zeta = f,$ $-\psi'' + \psi' + \psi = \zeta,$ $f = \zeta = \psi = e^{x-1}$	

Table 2.2:

First, we observe that all methods have the spectral convergence. Then, we note that the stream function solution is calculated with almost the same accuracy by the four methods, the most precise results being provided by the two methods imposing integral conditions and using the kernel functions. These methods provide exactly the same ψ solution obtained imposing boundary conditions for the vorticity, whose errors are shown for comparison in Table 2.2.

Finally, the ψ results given by the Glowinski–Pironneau method are marginally more accurate than those of the influence matrix method. The comparison of the ζ results is more interesting. Here, again the two methods imposing integral conditions without solving stream function equations compute essentially the same solution provided by the boundary value solver, except when the round-off limit of the floating-point arithmetic is reached

The vorticity provided by the other two methods are instead less accurate, in some cases by an order of magnitude. Similarly to the ψ results, the Glowinski–Pironneau method is always found to be more accurate than the influence matrix method, this time by a factor of 2 in the relative error, as already noted by Dennis and Quartapelle (1983). In conclusion, the influence matrix method and the Glowinski–Pironneau method of imposing the integral conditions are found to be less accurate than the other two methods.

In order to elucidate these findings, we have considered another problem for the same equation system, in which the solution ψ contains some component belonging to the null space of the operator $\mathcal{D}^2$. For instance, we take

$$\psi(x) = e^{x-1} + 100e^{\mu_1(x-1)} + 100e^{\mu_2(x-1)}.$$

The corresponding numerical errors are shown in Tables 2.3 and 2.4.

The loss of accuracy in the vorticity becomes now dramatic: from four to five orders of magnitude. We notice, however, that in any case, the ψ solution is always calculated correctly, with an accuracy which only marginally better in methods using the integral conditions. It should be remarked that, irrespectively of such a deterioration of the accuracy, the spectral convergence is always verified. In this respect, we note that also the two integral condition methods, which do not suffer from the weakness discussed above, in this problems face with convergence difficulties before the round-off limit is reached.

It is however not difficult to remedy to such a situation, since in these two methods the loss of convergence can be caused only by an inaccurate evaluation of the right-hand side of the integral conditions, namely, of the quantity

$$c_k = \Big[ta_i\eta_k'(x_i) - (ua_i + tb_i)\eta_k(x_i)\Big]_{i=1}^{i=2}.$$

By calculating the expression on the right (and only this quantity) in extended precision, we obtain the results shown in the last part of Table 2.3. The improvement in the convergence near the round-off limit is such that the higher accuracy by four orders of magnitude w.r.t. the other two methods covers now the entire range.

To conclude this section on the numerical results for the vorticity–stream function equations, we consider the solution of the nonlinear problem with a vorticity modal equation of the form

$$-\zeta'' + \zeta' + \zeta + \psi\zeta' + (\psi + \psi' + 2)\zeta = f,$$

and with a solution

$$\psi(x) = e^{x-1} + 2e^{\mu_1(x-1)} + 2e^{\mu_2(x-1)}.$$

The iterative method described in the previous section is used with the tolerance error set to 10^{-13}. Each iteration requires to solve first the integrally conditioned vorticity equation with nonconstant coefficents and then the stream function equation. The linear system of equations for the vorticity problem in discrete form is solved by means of the standard LU factorization for a full matrix, whereas the bordered quasi-pentadiagonal linear system associated to the stream function problem is solved, as previously, by means of the algorithm described above. The numerical errors of the computed solutions with respect to the exact solution are given in Table 2.5.

	Influence matrix method		Integral conditions à la Glowinski–Pironneau	
N	rerr(ζ)	rerr(ψ)	rerr(ζ)	rerr(ψ)
6	0.88 (−01)	0.29 (−04)	0.35 (−01)	0.24 (−04)
8	0.59 (−03)	0.17 (−06)	0.23 (−03)	0.17 (−06)
10	0.30 (−05)	0.83 (−09)	0.12 (−05)	0.76 (−09)
12	0.11 (−07)	0.31 (−11)	0.47 (−08)	0.28 (−11)
14	0.34 (−10)	0.97 (−14)	0.15 (−10)	0.87 (−14)
16	0.11 (−11)	0.19 (−14)	0.13 (−11)	0.19 (−14)
	Decomposed integral conditions		Direct integral conditions	
N	rerr(ζ)	rerr(ψ)	rerr(ζ)	rerr(ψ)
6	0.58 (−05)	0.21 (−04)	0.58 (−05)	0.21 (−04)
8	0.14 (−07)	0.14 (−06)	0.14 (−07)	0.14 (−06)
10	0.28 (−10)	0.74 (−09)	0.28 (−10)	0.74 (−09)
12	0.28 (−12)	0.27 (−11)	0.28 (−12)	0.27 (−11)
14	0.28 (−12)	0.82 (−14)	0.28 (−12)	0.82 (−14)
16	0.28 (−12)	0.19 (−14)	0.28 (−12)	0.19 (−14)
	integral condition rhs in extended precision		*integral condition rhs in extended precision*	
N	rerr(ζ)	rerr(ψ)	rerr(ζ)	rerr(ψ)
6	0.58 (−05)	0.21 (−04)	0.58 (−05)	0.21 (−04)
8	0.14 (−07)	0.14 (−06)	0.14 (−07)	0.14 (−06)
10	0.28 (−10)	0.74 (−09)	0.28 (−10)	0.74 (−09)
12	0.42 (−13)	0.27 (−11)	0.41 (−13)	0.27 (−11)
14	0.50 (−14)	0.82 (−14)	0.40 (−14)	0.82 (−14)
16	0.40 (−14)	0.17 (−14)	0.32 (−14)	0.17 (−14)

$$-\zeta'' + \zeta' + \zeta = f, \quad -\psi'' + \psi' + \psi = \zeta,$$
$$\psi = e^{x-1} + C_1 e^{\mu_1(x-1)} + C_2 e^{\mu_2(x-1)},$$
$$\mu_{1,2} = (1 \pm \sqrt{5})/2, \qquad C_1 = C_2 = 100$$

Table 2.3:

	Vorticity boundary conditions imposed	
N	$\mathrm{rerr}(\zeta)$	$\mathrm{rerr}(\psi)$
6	0.56 (−05)	0.21 (−04)
8	0.14 (−07)	0.14 (−06)
10	0.28 (−10)	0.74 (−09)
12	0.41 (−13)	0.27 (−11)
14	0.24 (−15)	0.82 (−14)
16	0.22 (−15)	0.17 (−14)

$$-\zeta'' + \zeta' + \zeta = f,$$
$$-\psi'' + \psi' + \psi = \zeta,$$
$$\psi = e^{x-1} + C_1 e^{\mu_1(x-1)} + C_2 e^{\mu_2(x-1)},$$
$$\mu_{1,2} = (1 \pm \sqrt{5})/2$$
$$C_1 = C_2 = 100$$

Table 2.4:

Nonlinear equations Direct integral conditions			
N	$\mathrm{rerr}(\zeta)$	$\mathrm{rerr}(\psi)$	iterations
6	0.85 (−05)	0.20 (−04)	9
8	0.26 (−07)	0.13 (−06)	9
10	0.39 (−09)	0.68 (−09)	9
12	0.32 (−13)	0.25 (−11)	9
14	0.22 (−13)	0.74 (−14)	9
16	0.10 (−13)	0.33 (−15)	9
$-\zeta'' + \zeta' + \zeta + \psi\zeta' + (\psi + \psi' + 2)\zeta = f,$ $-\psi'' + \psi' + \psi = \zeta,$ $\psi = e^{x-1} + 2e^{\mu_1(x-1)} + 2e^{\mu_2(x-1)},$ $\mu_{1,2} = (1 \pm \sqrt{5})/2$ tolerance 10^{-13}			

Table 2.5:

2.12 Higher-order time discretization

In this section we want to mention the possibility of adopting time-integration schemes different from the two-level scheme of first-order time accuracy considered to discretize the vorticity equation. In particular, a method which is employed very frequently to solve the incompressible Navier–Stokes equations is the second-order accurate three-level scheme, with the nonlinear term taken into account explicitly using the three-level Adams–Bashforth method, and the linear viscous term evaluated implicitly by means of the Crank–Nicolson method. The resulting vorticity–stream function equations for this scheme, usually called ABCN method, are

$$\begin{aligned}(-\nabla^2+2\gamma)\zeta^{n+1} &= (\nabla^2+2\gamma)\zeta^n-\nu^{-1}\Big[3J(\zeta^n,\psi^n)-J(\zeta^{n-1},\psi^{n-1})\Big];\\ -\nabla^2\psi^{n+1} &= \zeta^{n+1},\end{aligned}$$

where $\gamma=1/(\nu\Delta t)$. Of course, to ensure an actual second-order accuracy at any time level, a second-order accurate solution (ζ^1,ψ^1) has be calculated at the first time step. This solution can be obtained by the Crank–Nicolson scheme

$$\begin{aligned}&(-\nabla^2+2\gamma)\zeta^1+\nu^{-1}J(\zeta^1,\psi^1)=(\nabla^2+2\gamma)\zeta^0-\nu^{-1}J(\zeta^0,\psi^0),\\ &-\nabla^2\psi^1-\zeta^1=0,\end{aligned}$$

where $\zeta^0=\zeta|_{t=0}=\boldsymbol{\nabla}\times\boldsymbol{u}_0\cdot\boldsymbol{k}$, and ψ^0 is the solution to the Dirichlet problem $-\nabla^2\psi^0=\zeta^0$, $\psi^0|_S=a(s,0)$. The problem for ζ^1 and ψ^1 is a nonlinear system of two coupled equations that can be solved in terms of uncoupled equations by means of an iterative method of the type

$$\begin{aligned}(-\nabla^2+2\gamma)\zeta^1_r &= -\nu^{-1}J(\zeta^1_{r-1},\psi^1_{r-1})+f^0;\\ -\nabla^2\psi^1_r &= \zeta^1_r;\end{aligned}$$

where $f^0=(\nabla^2+2\gamma)\zeta^0-\nu^{-1}J(\zeta^0,\psi^0)$. Of course, the vorticity equations appearing in both the nonlinear and linear problems above are supplemented by integral conditions.

Finally, it is worth remembering that, once the vorticity and the stream function have been determined, the pressure field P can be calculated by evaluating the *line integral* of the momentum equation written in the form:

$$-\boldsymbol{\nabla}\Big(P+\tfrac{1}{2}|\boldsymbol{\nabla}\psi|^2\Big)=\Big(\frac{\partial\boldsymbol{\nabla}\psi}{\partial t}+\nu\boldsymbol{\nabla}\zeta\Big)\times\boldsymbol{k}+\zeta\boldsymbol{\nabla}\psi.$$

This procedure is the extension to the time-dependent equations of the method introduced by Burggraf (1966) for steady problems.

2.13 Rotationally symmetric equations

The last section of the chapter is devoted to the analysis of the equations governing rotationally symmetric flows, expressed in spherical coordinates. The analysis of this particular class of three-dimensional fluid motions can be preparatory for the study of nonprimitive variable Navier–Stokes equations for general flows in three dimensions to be undertaken in the next chapter.

Let us consider the flow of a viscous incompressible fluid occupying an annular region and let us introduce a system of spherical coordinates (r, θ, ϕ), with the axis $\theta = 0$ coincident with the axis of the annulus. The cross section of the region with an axial plane will be denoted by S and it is given by the parametric representation

$$r = r(s), \qquad \theta = \theta(s),$$

where $r(s)$ and $\theta(s)$ are two given functions of the arc length parameter s of S.

Assuming that the flow is symmetrical about the axis $\theta = 0$, all quantities are independent of the azimuthal angle ϕ. It is then possible to express the vectors velocity $\boldsymbol{u}$ and vorticity $\boldsymbol{\zeta}$ in terms of their azimuthal components $u = u_\phi$ and $\zeta = \zeta_\phi$ together with a "stream" function ψ introduced to represent the fluid motion in the azimuthal plane. This scalar unknown is actually the azimuthal component of the stream vector potential $\boldsymbol{\psi}$, to be considered later for the study of genuinely three-dimensional flows.

In terms of these three scalar variables the vector fields $\boldsymbol{u}$ and $\boldsymbol{\zeta}$ are given by

$$\boldsymbol{u} = u\,\widehat{\boldsymbol{\phi}} + \boldsymbol{\nabla}\times(\psi\boldsymbol{\phi}) = u\,\widehat{\boldsymbol{\phi}} + \frac{\boldsymbol{\nabla}(\psi r\sin\theta)}{r\sin\theta} \times \widehat{\boldsymbol{\phi}},$$

$$\boldsymbol{\zeta} = \zeta\,\widehat{\boldsymbol{\phi}} + \boldsymbol{\nabla}\times(u\boldsymbol{\phi}) = \zeta\,\widehat{\boldsymbol{\phi}} + \frac{\boldsymbol{\nabla}(ur\sin\theta)}{r\sin\theta} \times \widehat{\boldsymbol{\phi}},$$

where $\widehat{\boldsymbol{\phi}} = \widehat{\boldsymbol{\phi}}(\theta, \phi)$ is the unit vector normal to the axial plane $\phi =$ constant and $\boldsymbol{\nabla}$ is the two-dimensional gradient operator in the same plane, *i.e.*, $\boldsymbol{\nabla} = (\partial/\partial r, r^{-1}\partial/\partial\theta)$. The Navier-Stokes equations for the dependent variables ζ, ψ and u assume the form

$$- \nu I\!D^2\zeta + \frac{\partial\zeta}{\partial t} + F(\zeta, \psi, u) = 0,$$

$$- I\!D^2\psi = \zeta,$$

$$- \nu I\!D^2 u + \frac{\partial u}{\partial t} + G(u, \psi) = 0.$$

The elliptic operator $I\!D^2$ for the rotationally symmetric problem in spherical coordinates is defined by

$$\mathbb{D}^2 = \nabla^2 - \frac{1}{r^2 \sin^2\theta},$$

$$\nabla^2 = \frac{\partial^2}{\partial r^2} + \frac{2}{r}\frac{\partial}{\partial r} + \frac{1}{r^2}\frac{\partial^2}{\partial \theta^2} + \frac{\cot\theta}{r^2}\frac{\partial}{\partial \theta}.$$

The nonlinear terms F and G are given by the following relationships

$$F(\zeta, \psi, u) = \frac{1}{r} J(\zeta, \psi) + \frac{1}{r^2 \sin\theta}[\psi J_1(\zeta) + \zeta J_1(\psi) - 2u J_1(u)],$$

$$G(u, \psi) = \frac{1}{r} J(u, \psi) + \frac{1}{r^2 \sin\theta}[\psi J_1(u) - u J_1(\psi)],$$

where $J(\alpha, \beta)$ represents the standard Jacobian, namely,

$$J(\alpha, \beta) = \frac{\partial \alpha}{\partial r}\frac{\partial \beta}{\partial \theta} - \frac{\partial \alpha}{\partial \theta}\frac{\partial \beta}{\partial r},$$

whereas $J_1(\alpha) = J(\alpha, r\sin\theta)$.

Assuming that the velocity is prescribed on the entire boundary S, one has

$$\boldsymbol{u}|_S = \boldsymbol{b}(s,t) = \boldsymbol{b}(\boldsymbol{x}_s, t) = \boldsymbol{b}(r(s), \theta(s), t).$$

By virtue of the assumed velocity representation, such a boundary condition can be shown to reduce itself to the following conditions for the two scalar unknowns ψ and u

$$\psi|_S = a(s,t),$$

$$\boldsymbol{n} \cdot \boldsymbol{\nabla}\psi|_S = b(s,t),$$

$$u|_S = b_\phi(s,t),$$

where the boundary data $a(s,t)$ and $b(s,t)$ are defined according to

$$a(s,t) = \frac{1}{r(s)\sin(\theta(s))} \int_{s_1}^{s} \boldsymbol{n}(s') \cdot \boldsymbol{b}(s', t)\, r(s') \sin(\theta(s'))\, ds'$$

and

$$b(s,t) = -\boldsymbol{\tau}(s) \cdot \boldsymbol{b}(s,t) - \frac{\boldsymbol{n}(s) \cdot \boldsymbol{\nabla}[r(s)\sin(\theta(s))]}{r(s)\sin(\theta(s))}\, a(s,t).$$

Here $\boldsymbol{\tau}(s)$ represents the unit vector tangential to the boundary S in the axial plane.

The pair of boundary conditions for ψ is then converted into a condition of integral type for ζ by resorting to the Green identity for the spherical coordinate elliptic operator $\mathbb{D}^2$, which reads

$$\int\!\!\int (\psi\, \mathbb{D}^2\phi - \phi\, \mathbb{D}^2\psi)\, r^2 \sin\theta\, dr\, d\theta = \oint \boldsymbol{n} \cdot (\psi \boldsymbol{\nabla}\phi - \phi \boldsymbol{\nabla}\psi)\, r \sin\theta\, ds.$$

By introducing the linear space of functions η solution to the elliptic equation $\mathbb{D}^2\eta = 0$, Green identity can be used with $\phi = \eta$, $-\mathbb{D}^2\psi = \zeta$, and ψ satisfying the two boundary conditions, to give the following integral conditions for the vorticity in rotationally symmetric flows

$$\iint \zeta\eta\, r^2 \sin\theta \, dr \, d\theta = \oint (a\, \boldsymbol{n} \cdot \boldsymbol{\nabla}\eta - b\,\eta)\, r(s) \sin(\theta(s))\, ds.$$

A finite difference implementation of this set of equations imposing the integral conditions by means of the Glowinski–Pironneau method and using fast elliptic solvers has been employed by Dennis and Quartapelle (1984) to compute steady and unsteady flows in a spherical gap.

To conclude this chapter on the nonprimitive variable formulation for two-dimensional flows, we want to mention that the concept of the vorticity integral conditions has been applied also to the calculation of natural convection flows by means of finite elements by Quartapelle and Napolitano (1983).

Chapter 3

Nonprimitive variable formulations in 3D

3.1 Introduction

The analysis developed in the previous chapter, regarding the vorticity and stream function equations under no-slip conditions, has revealed that an uncoupled formulation of the linear or linearized problem is possible, provided that the vorticity variable is supplemented by conditions of an integral character. These conditions have been found to admit a very simple geometrical interpretation: they impose the orthogonality of the vorticity field with respect to the linear space of the functions which are harmonic in the considered domain, or, at least, fix the orthogonal projection of the vorticity with respect to such a linear space. This geometrical result allows to recast the Navier–Stokes equations for two-dimensional flows in terms of two second-order scalar equations, one parabolic and the other elliptic, the vorticity equation being completely independent in linear situations. It seems therefore worthwhile to investigate whether these abstract geometrical ideas can be generalized to three dimensions.

An investigation of this kind, however, is rather difficult for the following reasons:

i) In three-dimensional flows the vorticity is a vector variable and two tangential components of the vorticity have to be determined on solid boundaries.

ii) The vorticity field must be guaranteed to be solenoidal.

iii) Different choices are possible for representing the (solenoidal) velocity field: either by means of the combination of a scalar and a vector potential, or by means of a vector potential alone, called stream vector.

iv) In order to define the vector potential or the stream vector uniquely, it is necessary to introduce additional conditions, *e.g.*, a gauge condition which typically requires that the vector field be solenoidal.

v) Last but not least, different choices are possible for the boundary conditions to be imposed on the stream vector, the choice depending on the topology of the domain occupied by the fluid.

Previous studies concerning the representation of the three-dimensional Navier–Stokes equations in terms of the vorticity and of the velocity potentials have been focussed mainly on the difficulties associated with the specification of the boundary conditions for the potentials, a particular attention being paid to the situation of multiply connected regions (see, *e.g.*, Hirasaki and Hellums 1968, 1970 and Richardson and Cornish 1977).

By contrast, the problem of establishing the conditions to be imposed on, or satisfied by, the vorticity field has not received a comparable attention. This is rather disappointing, when one considers the great importance of this variable in many forms of fluid motions containing vortical structures, from both the theoretical and the experimental viewpoints, see, *e.g.*, Saffman [31] and Lugt [28].

While the vorticity dynamics in unbounded spaces is described by the well known Helmholtz vortex theorems and Kelvin circulation theorem complemented by the viscous diffusion mechanism (see, *e.g.*, Acheson [1, p. 157] and the monograph by Ting and Klein [37]), the influence of solid walls on the vorticity field in three-dimensional flows is not yet clearly understood. Only two works have attempted to extend to three dimensions the thoretical results established for plane motions and which have been the subject of the previous chapter. In the work by Quartapelle and Valz-Gris (1981) investigating the vorticity–velocity potentials formulation, the existence of integral conditions for the vorticity vector in three dimensions has been demonstrated. In another article, Achard and Canot (1988) have investigated the nonprimitive variable Navier–Stokes equations and have proposed two solution methods, which combine a boundary integral formulation of the Poisson problem for the potentials with the standard differential representation of the vorticity equation, the latter being supplemented by conditions of both boundary and integral type.

In this chapter a number of nonprimitive variable formulations of the equations governing incompressible viscous flows in three dimensions will be discussed. We will attempt to address all the aforementioned issues, aiming at establishing a comprehensive and unitary framework for their analysis. Complete statements of the Navier–Stokes problem expressed as a system of parabolic and elliptic equations will be presented.

The first formulation to be discussed is the classical one based on representing the solenoidal velocity field $\boldsymbol{u}$ in terms of both a scalar and a vector potential, according to the definition $\boldsymbol{u} = \boldsymbol{\nabla}\varphi + \boldsymbol{\nabla}\times\boldsymbol{A}$. The second formulation is based on the introduction of a stream vector $\boldsymbol{\psi}$ together with a scalar potential q_S, defined only on the surface of the (simply connected) domain. In this formulation the velocity is expressed only in terms of $\boldsymbol{\psi}$ through the relation $\boldsymbol{u} = \boldsymbol{\nabla}\times\boldsymbol{\psi}$, the stream vector $\boldsymbol{\psi}$ being the generalization to three dimensions of the stream function ψ used in plane problems. Finally, a third representation of the velocity field will be considered, still relying on a single stream vector $\boldsymbol{\phi}$, $\boldsymbol{u} = \boldsymbol{\nabla}\times\boldsymbol{\phi}$, but different from the previous one by the kind of the boundary conditions imposed on $\boldsymbol{\phi}$ in order to deal with multiply connected domains.

For all these representations we will provide the complete set of boundary conditions which allow to establish the equivalence of the nonprimitive variable formulation with the original statement of the primitive variable Navier–Stokes equations. In each case an uncoupled version of the governing equations is derived by introducing the conditions which supplement the respective unknown variables. In particular, the vorticity transport equation will be shown to be supplemented by a set of conditions comprising both integral and boundary conditions. Surprisingly enough, the vorticity conditions are found to assume a different form depending on the considered formulation: in other words, there are different ways of expressing the same set of linearly independent conditions for the vorticity field, each way being specifically appropriate for each distinct nonprimitive variable formulation. A crucial point in performing such an analysis is the transition from the formulation using the stream vector $\boldsymbol{\psi}$ to the formulation using the stream vector $\boldsymbol{\phi}$. Such a transition will be shown to be accompanied by a reduction of the vorticity integral conditions, half of which can be replaced by a standard condition of boundary value type for the normal component of the vorticity.

3.2 Vorticity vector equation

For three-dimensional flows, consider the vorticity vector variable $\boldsymbol{\zeta}$ defined by

$$\boldsymbol{\zeta} = \boldsymbol{\nabla}\times\boldsymbol{u}. \tag{3.1}$$

Taking the curl of the momentum equation (1.1) and recalling that $(\boldsymbol{u}\cdot\boldsymbol{\nabla})\boldsymbol{u} = (\boldsymbol{\nabla}\times\boldsymbol{u})\times\boldsymbol{u} + \boldsymbol{\nabla}(\frac{1}{2}u^2)$ gives the vorticity equation

$$\frac{\partial\boldsymbol{\zeta}}{\partial t} + \boldsymbol{\nabla}\times(\boldsymbol{\zeta}\times\boldsymbol{u}) = \nu\nabla^2\boldsymbol{\zeta}. \tag{3.2}$$

In all of the formulations of the 3D equations using the vorticity as unknown, the initial condition and one (scalar) boundary condition to be imposed on $\boldsymbol{\zeta}$ result from the vorticity definition (3.1) and the vorticity equation (3.2). The initial condition is an immediate consequence of the correspondent condition for $\boldsymbol{u}$, through the definition of vorticity (3.1)

$$\zeta|_{t=0} = \nabla\times \boldsymbol{u}_0. \tag{3.3}$$

Concerning the scalar boundary condition, by the definition of $\boldsymbol{\zeta}$, $\nabla\cdot\boldsymbol{\zeta} = 0$ for any $t > 0$. Now, taking the divergence of the vorticity equation (3.2) gives the homogeneous heat equation for the scalar variable $\nabla\cdot\boldsymbol{\zeta}$, namely,

$$\frac{\partial \nabla\cdot\zeta}{\partial t} = \nu\nabla^2\nabla\cdot\zeta,$$

supplemented by the initial condition $\nabla\cdot\zeta|_{t=0} = \nabla\cdot\nabla\times\boldsymbol{u}_0 = 0$, by (3.3). If the heat equation is supplemented also by the homogeneous boundary condition

$$\nabla\cdot\zeta|_S = 0, \tag{3.4}$$

for any $t > 0$, its solution will vanish identically, *i.e.*, $\nabla\cdot\boldsymbol{\zeta} = 0$, in V for all $t > 0$. Therefore, the two conditions (3.3) and (3.4) are necessary and sufficient conditions for the vorticity field to be solenoidal at any time, as required by its definition (3.1). Boundary condition (3.4) has been considered for the first time by Lighthill (1963, p. 55).

3.3 φ-ζ-A formulation

In this section we introduce a formulation of the Navier–Stokes equations for three-dimensional flows which is based on a representation of the velocity field in terms of a scalar and a vector potential. Firstly, the equations for these potentials will be derived together with their respective boundary conditions. Then, the equivalence of the complete set of equations and conditions with the original incompressible problem will be demonstrated. Finally, an uncoupled formulation of the equations for the vorticity and the velocity potentials will be derived, introducing a three-dimensional and vectorial generalization of the integral conditions for vorticity discussed in the previous chapter.

3.3.1 Equations and boundary conditions for the velocity potentials

Assume to represent the velocity field in terms of a scalar potential φ and a vector potential $\boldsymbol{A}$, namely,

$$\boldsymbol{u} = \nabla\varphi + \nabla\times\boldsymbol{A}. \tag{3.5}$$

To satisfy the incompressibility condition, the scalar potential must satisfy the Laplace equation

$$-\nabla^2\varphi = 0, \tag{3.6}$$

so that φ is harmonic in V. The boundary conditions to be imposed on φ and $\boldsymbol{A}$ are derived from that of $\boldsymbol{u}$ by separating the normal and tangential components of $\boldsymbol{u}|_S = \boldsymbol{b}$. Following Hirasaki and Hellums (1970), let us assume that the fulfillment of the velocity boundary condition for the normal component on S is imposed through the following two (separate) boundary conditions

$$\boldsymbol{n}\cdot\boldsymbol{\nabla}\varphi|_S = \boldsymbol{n}\cdot\boldsymbol{b} \tag{3.7}$$

and

$$\boldsymbol{n}\cdot\boldsymbol{\nabla}\times\boldsymbol{A}|_S = 0. \tag{3.8}$$

(We notice in passing that the representation (3.5), with $\boldsymbol{A}$ satisying boundary condition (3.8), constitutes an orthogonal decomposition of the velocity field, in the sense that

$$\int \boldsymbol{\nabla}\varphi\cdot\boldsymbol{\nabla}\times\boldsymbol{A}\, dV = 0.$$

In fact, by the divergence theorem,

$$\int \boldsymbol{\nabla}\varphi\cdot\boldsymbol{\nabla}\times\boldsymbol{A}\, dV = \int \boldsymbol{\nabla}\cdot(\varphi\boldsymbol{\nabla}\times\boldsymbol{A})\, dV = \oint \varphi\boldsymbol{n}\cdot\boldsymbol{\nabla}\times\boldsymbol{A}\, dS = 0,$$

due to the boundary condition (3.8).)

Thus, the scalar potential φ is defined by the following Neumann problem:

$$-\nabla^2\varphi = 0, \qquad \boldsymbol{n}\cdot\boldsymbol{\nabla}\varphi|_S = \boldsymbol{n}\cdot\boldsymbol{b}, \tag{3.9}$$

which, in view of the global condition $\oint \boldsymbol{n}\cdot\boldsymbol{b}\, dS = 0$, can be solved uniquely apart from an arbitrary additive function of time, $\varphi(\boldsymbol{x},t) \to \varphi(\boldsymbol{x},t) + \Phi(t)$.

Assuming that the Neumann problem (3.9) has been solved for any $t > 0$, the tangential part of the velocity boundary condition, namely, $\boldsymbol{n}\times\boldsymbol{u}|_S = \boldsymbol{n}\times\boldsymbol{b}$, becomes, by virtue of the velocity representation (3.5),

$$\boldsymbol{n}\times\boldsymbol{\nabla}\times\boldsymbol{A}|_S = \boldsymbol{n}\times(-\boldsymbol{\nabla}\varphi|_S + \boldsymbol{b}). \tag{3.10}$$

Taking the curl of the velocity definition (3.5) in terms of the potentials and using the vorticity definition (3.1), gives the equation $\boldsymbol{\nabla}\times\boldsymbol{\nabla}\times\boldsymbol{A} = \boldsymbol{\zeta}$. Thus, the equation and boundary conditions to be satisfied by $\boldsymbol{A}$ are

$$\boldsymbol{\nabla}\times\boldsymbol{\nabla}\times\boldsymbol{A} = \boldsymbol{\zeta}, \qquad \boldsymbol{n}\cdot\boldsymbol{\nabla}\times\boldsymbol{A}|_S = 0, \quad \boldsymbol{n}\times\boldsymbol{\nabla}\times\boldsymbol{A}|_S = \boldsymbol{n}\times(-\boldsymbol{\nabla}\varphi|_S + \boldsymbol{b}). \tag{3.11}$$

This set does not determine $\boldsymbol{A}$ uniquely, since $\boldsymbol{A}$ can be changed by the gradient of an arbitrary scalar function Ψ according to

$$\boldsymbol{A} \longrightarrow \boldsymbol{A} + \boldsymbol{\nabla}\Psi$$

without affecting the velocity $\boldsymbol{u}$. This transformation is called a *gauge transformation* and is possible because (3.5) contains only the curl of $\boldsymbol{A}$. The freedom associated with the gauge transformation allows to impose additional conditions on $\boldsymbol{A}$.

Part of the freedom is exploited so as to transform the (scalar) derivative boundary condition $\boldsymbol{n}\cdot\boldsymbol{\nabla}\times\boldsymbol{A}|_S = 0$ into a nonderivative boundary condition which fixes the tangential components of $\boldsymbol{A}$. To see how this is possible, we first observe that $\boldsymbol{n}\cdot\boldsymbol{\nabla}\times\boldsymbol{A}|_S = \boldsymbol{n}\cdot\boldsymbol{\nabla}_S\times\boldsymbol{A}_S$, where $\boldsymbol{\nabla}_S$ denotes the *surface gradient* operator over the boundary S and $\boldsymbol{A}_S = \boldsymbol{A}|_S$, see section 3.4.1 for details. In the following, the subscript index S is used to denote scalar and vector functions which are defined only on the surface S. (Note, however, that the data specifying the boundary values are never written with the subscript S, even though they are necessarily defined only on the boundary S.)

On the other hand, the evaluation of $\boldsymbol{n}\cdot\boldsymbol{\nabla}_S\times\boldsymbol{A}_S$ involves only the tangential components of $\boldsymbol{A}_S$. Let $\boldsymbol{A}_S^{\tan}$ denote the tangential component of the trace of $\boldsymbol{A}$ on the surface S defined by the relation

$$\boldsymbol{A}_S^{\tan} = \boldsymbol{A}|_S - \boldsymbol{n}(\boldsymbol{n}\cdot\boldsymbol{A})|_S.$$

Then we can write

$$\boldsymbol{n}\cdot\boldsymbol{\nabla}\times\boldsymbol{A}|_S = \boldsymbol{n}\cdot\boldsymbol{\nabla}_S\times\boldsymbol{A}_S = \boldsymbol{n}\cdot\boldsymbol{\nabla}_S\times\boldsymbol{A}_S^{\tan},$$

a relation which enables one to transform a boundary condition for a vector field defined in the volume V into an equation over the surface S for a vector field defined only on S and tangential to S. Thus, the boundary condition (3.8) gives the *surface* differential equation

$$\boldsymbol{n}\cdot\boldsymbol{\nabla}_S\times\boldsymbol{A}_S^{\tan} = 0.$$

Now, a vector field defined on a surface S, which is tangential to S together with its surface curl, is the surface gradient of some scalar function defined on S (see Weatherburn [38], p. 230). It follows that the solution to the previous equation is $\boldsymbol{A}_S^{\tan} = \boldsymbol{\nabla}_S p_S$, where p_S can be any scalar function defined on S.

Let p be any extension of p_S to V, *i.e.*, any function defined in V such that $p|_S = p_S$. Then the vector field $\boldsymbol{A} - \boldsymbol{\nabla}p$ is such that

$$\begin{aligned}\boldsymbol{n}\times(\boldsymbol{A} - \boldsymbol{\nabla}p)|_S &= \boldsymbol{n}\times\boldsymbol{A}|_S - \boldsymbol{n}\times\boldsymbol{\nabla}p|_S\\ &= \boldsymbol{n}\times\boldsymbol{A}_S^{\tan} - \boldsymbol{n}\times\boldsymbol{\nabla}_S p_S\\ &= \boldsymbol{n}\times(\boldsymbol{A}_S^{\tan} - \boldsymbol{\nabla}_S p_S) = 0.\end{aligned}$$

Since the vector field $\boldsymbol{A}-\boldsymbol{\nabla}p$ satisfies the same equation and boundary conditions in (3.11) satisfied by the vector potential $\boldsymbol{A}$, the arbitrariness of p_S allows to impose on $\boldsymbol{A}$ the additional boundary condition

$$\boldsymbol{n}\times\boldsymbol{A}|_S = 0. \tag{3.12}$$

After imposing this boundary condition, the potential $\boldsymbol{A}$ remains still indeterminate up to the gradient of any scalar function which vanishes on S so as to still satisfy boundary condition (3.12). Thus, in order to define $\boldsymbol{A}$ uniquely, another condition must be considered. A very simple choice consists in imposing $\boldsymbol{A}$ to be solenoidal, namely,

$$\nabla \cdot \boldsymbol{A} = 0. \tag{3.13}$$

This choice eliminates the indeterminacy by the gradient of a function vanishing on S since that function should now be also harmonic and therefore identically zero in V, by virtue of its zero value on the boundary.

Condition (3.13) is called a *gauge condition*. In the theory of the electromagnetism, condition (3.13) it is usually denoted as the *Coulomb gauge*. In the present fluid dynamic context, we prefer to call (3.13) the *Euclid gauge*, since the choice (3.13) is invariant with respect to the group of the transformations of the three-dimensional Euclidean space. A more general gauge condition for $\boldsymbol{A}$ of the type $\nabla \cdot \boldsymbol{A} = B$, where B is an arbitrary function, can also be considered, which could be referred to as a *nonhomogeneous* Euclid gauge.

Consider now the two equations $\nabla \times \nabla \times \boldsymbol{A} = \boldsymbol{\zeta}$ and $\nabla \cdot \boldsymbol{A} = 0$. Due to the identity $\nabla \times \nabla \times \boldsymbol{\alpha} = -\nabla^2 \boldsymbol{\alpha} + \nabla(\nabla \cdot \boldsymbol{\alpha})$, they imply the equation $-\nabla^2 \boldsymbol{A} = \boldsymbol{\zeta}$. On the other hand, the divergence of this equation gives $-\nabla^2 \nabla \cdot \boldsymbol{A} = 0$, since $\nabla \cdot \boldsymbol{\zeta} = 0$. It follows that, to satisfy the Euclid gauge, it is sufficient to impose the boundary condition

$$\nabla \cdot \boldsymbol{A}|_S = 0. \tag{3.14}$$

In conclusion, the set comprising the equation and all of the boundary conditions to be satisfied by the vector potential subject to the Euclid gauge is:

$$-\nabla^2 \boldsymbol{A} = \boldsymbol{\zeta}, \qquad \boldsymbol{n} \times \boldsymbol{A}|_S = 0, \quad \boldsymbol{n} \times \nabla \times \boldsymbol{A}|_S = \boldsymbol{n} \times (-\nabla \varphi|_S + \boldsymbol{b}), \quad \nabla \cdot \boldsymbol{A}|_S = 0. \tag{3.15}$$

Here and in the following, when several boundary conditions and possibly conditions of integral type are imposed on a vector unknown, they will be indicated in the following order. The integral conditions are written first, followed by the boundary conditions prescribing the value of the unknown, which are in turn followed by the derivative boundary conditions. Furthermore, the boundary condition for the tangential components usually precedes that for the normal component, and similarly the condition involving the curl of the unknown is written before that involving its divergence.

3.3.2 Governing equations

The entire set of equations and conditions for $\boldsymbol{\zeta}$ (3.2)–(3.4), for φ (3.9), and for $\boldsymbol{A}$ (3.15), obtained above, provides an equivalent formulation of the primitive variable three-dimensional Navier–Stokes equations, as indicated by the following

Theorem 3.1. *The primitive variable Navier–Stokes problem (1.1)–(1.4) is equivalent to the problem for the variables* φ, $\boldsymbol{\zeta}$ *and* $\boldsymbol{A}$, *defined by the set of equations and conditions*

$$-\nabla^2\varphi = 0, \qquad \boldsymbol{n}\cdot\boldsymbol{\nabla}\varphi|_S = \boldsymbol{n}\cdot\boldsymbol{b};$$

$$\frac{\partial\boldsymbol{\zeta}}{\partial t} - \nu\nabla^2\boldsymbol{\zeta} + \boldsymbol{\nabla}\times[\boldsymbol{\zeta}\times(\boldsymbol{\nabla}\varphi + \boldsymbol{\nabla}\times\boldsymbol{A})] = 0, \qquad \boldsymbol{\zeta}|_{t=0} = \boldsymbol{\nabla}\times\boldsymbol{u}_0,\ \boldsymbol{\nabla}\cdot\boldsymbol{\zeta}|_S = 0,$$

$$-\nabla^2\boldsymbol{A} = \boldsymbol{\zeta}, \qquad \boldsymbol{n}\times\boldsymbol{A}|_S = 0,\ \boldsymbol{n}\times\boldsymbol{\nabla}\times\boldsymbol{A}|_S = \boldsymbol{n}\times(-\boldsymbol{\nabla}\varphi|_S + \boldsymbol{b}),\ \boldsymbol{\nabla}\cdot\boldsymbol{A}|_S = 0, \tag{3.16}$$

provided that the data $\boldsymbol{n}\cdot\boldsymbol{b}$ *and* $\boldsymbol{u}_0$ *satisfy the following conditions*

$$\oint \boldsymbol{n}\cdot\boldsymbol{b}\,dS = 0, \qquad \boldsymbol{\nabla}\cdot\boldsymbol{u}_0 = 0, \qquad \boldsymbol{n}\cdot\boldsymbol{b}|_{t=0} = \boldsymbol{n}\cdot\boldsymbol{u}_0|_S. \tag{3.17}$$

Proof. The implication is evident. Conversely, let $(\varphi, \boldsymbol{\zeta}, \boldsymbol{A})$ be a solution to the set of equations and conditions (3.16) with the data $\boldsymbol{n}\cdot\boldsymbol{b}$ and $\boldsymbol{u}_0$ satisfying the conditions (3.17). First observe that φ is uniquely defined up to an additive function of time $\Phi(t)$ since the solvability condition of the Neumann problem $\oint \boldsymbol{n}\cdot\boldsymbol{b}\,dS = 0$ is satisfied, for any $t > 0$, by the first condition in (3.17).

Let us define the vector field $\boldsymbol{v} = \boldsymbol{\nabla}\varphi + \boldsymbol{\nabla}\times\boldsymbol{A}$. Then $\boldsymbol{v}$ is solenoidal since φ is harmonic. Furthermore, $\boldsymbol{\nabla}\times\boldsymbol{v} = \boldsymbol{\nabla}\times\boldsymbol{\nabla}\times\boldsymbol{A} = -\nabla^2\boldsymbol{A} + \boldsymbol{\nabla}(\boldsymbol{\nabla}\cdot\boldsymbol{A}) = -\nabla^2\boldsymbol{A} = \boldsymbol{\zeta}$, since $\boldsymbol{A}$ is solenoidal. (The solenoidal character of $\boldsymbol{A}$ follows by taking the divergence of the equation $-\nabla^2\boldsymbol{A} = \boldsymbol{\zeta}$ and observing that $\boldsymbol{\nabla}\cdot\boldsymbol{A}|_S = 0$ and $\boldsymbol{\nabla}\cdot\boldsymbol{\zeta} = 0$; the latter follows in turn by observing that the divergence of the vorticity equation gives the heat equation for the variable $\boldsymbol{\nabla}\cdot\boldsymbol{\zeta}$ which satisfies the homogeneous initial and boundary conditions $\boldsymbol{\nabla}\cdot\boldsymbol{\zeta}|_{t=0} = \boldsymbol{\nabla}\cdot\boldsymbol{\nabla}\times\boldsymbol{u}_0 = 0$ and $\boldsymbol{\nabla}\cdot\boldsymbol{\zeta}|_S = 0$, respectively.) Substituting $\boldsymbol{\zeta} = \boldsymbol{\nabla}\times\boldsymbol{v}$ in the vorticity equation gives

$$\frac{\partial\boldsymbol{\nabla}\times\boldsymbol{v}}{\partial t} - \nu\nabla^2\boldsymbol{\nabla}\times\boldsymbol{v} + \boldsymbol{\nabla}\times[(\boldsymbol{\nabla}\times\boldsymbol{v})\times\boldsymbol{v}] = 0,$$

namely,

$$\boldsymbol{\nabla}\times\left[\frac{\partial\boldsymbol{v}}{\partial t} - \nu\nabla^2\boldsymbol{v} + (\boldsymbol{\nabla}\times\boldsymbol{v})\times\boldsymbol{v}\right] = 0,$$

which means

$$\frac{\partial\boldsymbol{v}}{\partial t} - \nu\nabla^2\boldsymbol{v} + (\boldsymbol{\nabla}\times\boldsymbol{v})\times\boldsymbol{v} = \boldsymbol{\nabla}Q,$$

Q being an arbitrary function. To identify the field $\boldsymbol{v}$ with the velocity field $\boldsymbol{u}$ solution of the primitive variable Navier–Stokes equations (and Q with $-P - \frac{1}{2}u^2$), the boundary and initial values of $\boldsymbol{v}$ must be shown to coincide with those prescribed in the boundary and initial conditions for $\boldsymbol{u}$. For the normal component of the velocity on the boundary it is:

$$\boldsymbol{n}\cdot\boldsymbol{v}|_S = \boldsymbol{n}\cdot(\boldsymbol{\nabla}\varphi + \boldsymbol{\nabla}\times\boldsymbol{A})|_S = \boldsymbol{n}\cdot\boldsymbol{\nabla}\varphi|_S + \boldsymbol{n}\cdot\boldsymbol{\nabla}\times\boldsymbol{A}|_S = \boldsymbol{n}\cdot\boldsymbol{b},$$

due to the Neumann condition satisfied by φ and since the boundary condition $\boldsymbol{n}\times\boldsymbol{A}|_S = 0$ satisfied by $\boldsymbol{A}$ implies $\boldsymbol{n}\cdot\boldsymbol{\nabla}\times\boldsymbol{A}|_S = 0$. For the tangential components of the velocity boundary value it is

$$\begin{aligned}\boldsymbol{n}\times\boldsymbol{v}|_S &= \boldsymbol{n}\times(\boldsymbol{\nabla}\varphi + \boldsymbol{\nabla}\times\boldsymbol{A})|_S = \boldsymbol{n}\times\boldsymbol{\nabla}\varphi|_S + \boldsymbol{n}\times\boldsymbol{\nabla}\times\boldsymbol{A}|_S \\ &= \boldsymbol{n}\times\boldsymbol{\nabla}\varphi|_S + \boldsymbol{n}\times(-\boldsymbol{\nabla}\varphi|_S + \boldsymbol{b}) = \boldsymbol{n}\times\boldsymbol{b}.\end{aligned}$$

Thus $\boldsymbol{v}|_S = \boldsymbol{b} = \boldsymbol{u}|_S$. Concerning the initial condition, one has to determine the initial value implied on $\boldsymbol{v}$ by imposing the initial condition for $\boldsymbol{\zeta}$ in the φ-$\boldsymbol{\zeta}$-$\boldsymbol{A}$ system. Let φ_0 and $\boldsymbol{A}_0$ be the solutions of the following problems:

$$-\nabla^2\varphi_0 = 0, \qquad\qquad \boldsymbol{n}\cdot\boldsymbol{\nabla}\varphi_0|_S = \boldsymbol{n}\cdot\boldsymbol{b}|_{t=0}$$

and

$$-\nabla^2\boldsymbol{A}_0 = \boldsymbol{\zeta}_0 = \boldsymbol{\nabla}\times\boldsymbol{u}_0, \qquad \boldsymbol{n}\times\boldsymbol{A}_0|_S = 0, \quad \boldsymbol{\nabla}\cdot\boldsymbol{A}_0|_S = 0,$$

respectively. By the assumed continuity of $\boldsymbol{n}\cdot\boldsymbol{b}(\boldsymbol{x}_S,t)$ as $t \to 0^+$ and since $\boldsymbol{\zeta}_0 = \boldsymbol{\zeta}|_{t=0}$, the well-posedness of the two previous elliptic problems implies that

$$\boldsymbol{v}|_{t=0} = \lim_{t\to 0^+}(\boldsymbol{\nabla}\varphi + \boldsymbol{\nabla}\times\boldsymbol{A}) = \boldsymbol{\nabla}\varphi_0 + \boldsymbol{\nabla}\times\boldsymbol{A}_0.$$

Now, $\boldsymbol{A}_0$ being solenoidal, the equation for $\boldsymbol{A}_0$ can be written as $\boldsymbol{\nabla}\times\boldsymbol{\nabla}\times\boldsymbol{A}_0 = \boldsymbol{\nabla}\times\boldsymbol{u}_0$ that is $\boldsymbol{\nabla}\times(\boldsymbol{\nabla}\times\boldsymbol{A}_0 - \boldsymbol{u}_0) = 0$, which means

$$\boldsymbol{\nabla}\times\boldsymbol{A}_0 - \boldsymbol{u}_0 = \boldsymbol{\nabla}\alpha,$$

for some function α, which is harmonic since $\boldsymbol{\nabla}\cdot\boldsymbol{\nabla}\times\boldsymbol{A}_0 = 0$ and $\boldsymbol{\nabla}\cdot\boldsymbol{u}_0 = 0$ by the second of the compatibility conditions (3.17). Taking the normal component of this equation on S gives:

$$\boldsymbol{n}\cdot(\boldsymbol{\nabla}\times\boldsymbol{A}_0 - \boldsymbol{u}_0)|_S = \boldsymbol{n}\cdot\boldsymbol{\nabla}\times\boldsymbol{A}_0|_S - \boldsymbol{n}\cdot\boldsymbol{u}_0|_S = -\boldsymbol{n}\cdot\boldsymbol{u}_0|_S = \boldsymbol{n}\cdot\boldsymbol{\nabla}\alpha|_S,$$

since the boundary condition $\boldsymbol{n}\times\boldsymbol{A}_0|_S = 0$ implies $\boldsymbol{n}\cdot\boldsymbol{\nabla}\times\boldsymbol{A}_0|_S = 0$. By the third of the compatibility conditions in (3.17), the relation above becomes $\boldsymbol{n}\cdot\boldsymbol{\nabla}\alpha|_S = -\boldsymbol{n}\cdot\boldsymbol{b}|_{t=0}$. Thus, α is harmonic and satisfies the same Neumann boundary condition, with a negative sign, satisfied by the other harmonic function φ_0, so that $\alpha = -\varphi_0 + A(t)$, with $A(t)$ an arbitrary function. Therefore $\boldsymbol{\nabla}\times\boldsymbol{A}_0 - \boldsymbol{u}_0 = -\boldsymbol{\nabla}\varphi_0$ and the initial value of $\boldsymbol{v}$ is $\boldsymbol{v}|_{t=0} = \boldsymbol{\nabla}\varphi_0 + \boldsymbol{\nabla}\times\boldsymbol{A}_0 = \boldsymbol{u}_0 = \boldsymbol{u}|_{t=0}$. Therefore $\boldsymbol{v} = \boldsymbol{u}$ and hence $Q = -P - \frac{1}{2}u^2 + C(t)$, where $C(t)$ is an arbitrary function. □

The boundary conditions associated with the vorticity and vector potential equations pose a problem similar to that encountered in the vorticity–stream function formulation for 2D flows: there are too many boundary conditions (five) for one variable ($\boldsymbol{A}$) and not enough (only one) for the other ($\boldsymbol{\zeta}$). On the other hand, the total number of boundary conditions for the $\boldsymbol{\zeta}$-$\boldsymbol{A}$ system is $1+5=6$

and is correct for supplementing the set of two second-order equations of vector type.

As in the two-dimensional situation, the system of $\boldsymbol{\zeta}$-$\boldsymbol{A}$ equations in (3.16) can be interpreted as a fourth-order biharmonic equation for the vector unknown $\boldsymbol{A}$, with $\boldsymbol{\zeta}$ being nothing but a short-hand notation for $-\nabla^2\boldsymbol{A}$ (cf. Bernardi 1979, Bendali, Dominguez and Gallic 1985, and Girault and Raviart 1986; for a formulation reducing the vector biharmonic equation of the Stokes problem to uncoupled scalar biharmonic equations see Ruas and Quartapelle 1990). Another possibility is to consider the $\boldsymbol{\zeta}$-$\boldsymbol{A}$ equations as a coupled systems of two second-order vector equations, a viewpoint that leads to a mixed variational formulations, as done by Fortin and Thomasset (1979), cf. also Girault and Raviart (1986). The next subsection will pursue instead the goal of regarding the vorticity equation as independent, letting aside the coupling due to the nonlinar term. An uncoupled formulation of the $\boldsymbol{\zeta}$-$\boldsymbol{A}$ equations will be derived by generalizing the projection argument of section 2.8 to the three-dimensional and vectorial situation under consideration.

3.3.3 Integral conditions for vorticity vector

Let $\varDelta$ denote the Laplace operator acting on vector fields and supplemented by the homogeneous version of the boundary conditions prescribed on $\boldsymbol{A}$ in problem (3.16), namely,

$$\varDelta = \left\{-\nabla^2, \quad \boldsymbol{n}\times\ldots|_S = 0,\ \boldsymbol{n}\times\boldsymbol{\nabla}\times\ldots|_S = 0,\ \boldsymbol{\nabla}\cdot\ldots|_S = 0\right\}. \tag{3.18}$$

Considering the particular case of $\boldsymbol{A}$ satisfying homogeneous boundary conditions, the fact that $\boldsymbol{\zeta}$ is such that $\boldsymbol{\zeta} = -\nabla^2\boldsymbol{A}$ with $\boldsymbol{n}\times\boldsymbol{A}|_S = 0$, $\boldsymbol{n}\times\boldsymbol{\nabla}\times\boldsymbol{A}|_S = 0$ and $\boldsymbol{\nabla}\cdot\boldsymbol{A}|_S = 0$, can be also stated in the form $\boldsymbol{\zeta} \in \mathcal{R}(\varDelta)$. The orthogonality theorem mentioned in section 2.7 states $\overline{\mathcal{R}(A)} = \mathcal{N}(A^\dagger)^\perp$, where A denotes any linear operator the range of which is a subspace of $\boldsymbol{L}^2(V)$ and $A^\dagger$ denotes its adjoint. Therefore the vorticity vector field must be such that

$$\boldsymbol{\zeta} \perp \mathcal{N}(\varDelta^\dagger). \tag{3.19}$$

To identify the adjoint operator $\varDelta^\dagger$ and its boundary conditions, it is necessary to resort to the *vector analogue* of Green identity for the operator ∇^2, which can be obtained as follows.

Let us consider the divergence theorem

$$\int \boldsymbol{\nabla}\cdot\boldsymbol{\alpha}\, dV = \oint \boldsymbol{n}\cdot\boldsymbol{\alpha}\, dS,$$

where $\boldsymbol{\alpha}$ is any well-behaved vector field defined in the volume V bounded by the closed surface S. Consider now the vector identities

$$\boldsymbol{\nabla}\cdot(\boldsymbol{A}\times\boldsymbol{\nabla}\times\boldsymbol{B}) = -\boldsymbol{A}\cdot\boldsymbol{\nabla}\times\boldsymbol{\nabla}\times\boldsymbol{B} + \boldsymbol{\nabla}\times\boldsymbol{A}\cdot\boldsymbol{\nabla}\times\boldsymbol{B},$$

$$\boldsymbol{\nabla}\cdot(\boldsymbol{A}\,\boldsymbol{\nabla}\cdot\boldsymbol{B}) = \boldsymbol{A}\cdot\boldsymbol{\nabla}(\boldsymbol{\nabla}\cdot\boldsymbol{B}) + \boldsymbol{\nabla}\cdot\boldsymbol{A}\ \boldsymbol{\nabla}\cdot\boldsymbol{B},$$

and use them to write the divergence theorem with $\boldsymbol{\alpha} = \boldsymbol{A}\times\boldsymbol{\nabla}\times\boldsymbol{B}$ and $\boldsymbol{\alpha} = \boldsymbol{A}\,\boldsymbol{\nabla}\cdot\boldsymbol{B}$, to give

$$\int(-\boldsymbol{A}\cdot\boldsymbol{\nabla}\times\boldsymbol{\nabla}\times\boldsymbol{B}+\boldsymbol{\nabla}\times\boldsymbol{A}\cdot\boldsymbol{\nabla}\times\boldsymbol{B})\,dV = \oint \boldsymbol{n}\times\boldsymbol{A}\cdot\boldsymbol{\nabla}\times\boldsymbol{B}\,dS,$$

$$\int(\boldsymbol{A}\cdot\boldsymbol{\nabla}(\boldsymbol{\nabla}\cdot\boldsymbol{B})+\boldsymbol{\nabla}\cdot\boldsymbol{A}\,\boldsymbol{\nabla}\cdot\boldsymbol{B})\,dV = \oint \boldsymbol{n}\cdot\boldsymbol{A}\,\boldsymbol{\nabla}\cdot\boldsymbol{B}\,dS,$$

respectively. By adding the two relations together and using the vector identity $-\boldsymbol{\nabla}\times\boldsymbol{\nabla}\times\boldsymbol{B}+\boldsymbol{\nabla}(\boldsymbol{\nabla}\cdot\boldsymbol{B}) = \nabla^2\boldsymbol{B}$, we obtain

$$\int(\boldsymbol{A}\cdot\nabla^2\boldsymbol{B}+\boldsymbol{\nabla}\times\boldsymbol{A}\cdot\boldsymbol{\nabla}\times\boldsymbol{B}+\boldsymbol{\nabla}\cdot\boldsymbol{A}\,\boldsymbol{\nabla}\cdot\boldsymbol{B})\,dV$$
$$= \oint(\boldsymbol{n}\times\boldsymbol{A}\cdot\boldsymbol{\nabla}\times\boldsymbol{B}+\boldsymbol{n}\cdot\boldsymbol{A}\,\boldsymbol{\nabla}\cdot\boldsymbol{B})\,dS.$$

If we write this relation again with $\boldsymbol{A}$ and $\boldsymbol{B}$ interchanged, and then subtract the new relation from the original one, the symmetric terms $\boldsymbol{\nabla}\times\boldsymbol{A}\cdot\boldsymbol{\nabla}\times\boldsymbol{B}$ and $\boldsymbol{\nabla}\cdot\boldsymbol{A}\,\boldsymbol{\nabla}\cdot\boldsymbol{B}$ cancel out and we obtain the following *vector Green identity*

$$\int(\boldsymbol{A}\cdot\nabla^2\boldsymbol{B}-\boldsymbol{B}\cdot\nabla^2\boldsymbol{A})\,dV$$
$$= \oint(\boldsymbol{n}\times\boldsymbol{A}\cdot\boldsymbol{\nabla}\times\boldsymbol{B}-\boldsymbol{n}\times\boldsymbol{B}\cdot\boldsymbol{\nabla}\times\boldsymbol{A}+\boldsymbol{n}\cdot\boldsymbol{A}\,\boldsymbol{\nabla}\cdot\boldsymbol{B}-\boldsymbol{n}\cdot\boldsymbol{B}\,\boldsymbol{\nabla}\cdot\boldsymbol{A})\,dS. \quad (3.20)$$

The left-hand side of this identity shows immediately that the operator $\Delta^\dagger$ adjoint to Δ is still $-\nabla^2$. The boundary conditions to be associated with $\Delta^\dagger$ are determined by requiring that the boundary integral on the right-hand side of Green identity be zero. By taking into account the boundary conditions supplementing the original operator Δ (see (3.18)), it is readily seen that only the integral $\oint \boldsymbol{n}\cdot\boldsymbol{A}\,\boldsymbol{\nabla}\cdot\boldsymbol{B}\,dS$ remains out of the four boundary terms. It follows that the adjoint operator is completely characterized as:

$$\Delta^\dagger = \left\{-\nabla^2, \quad \boldsymbol{\nabla}\cdot\ldots|_S = 0\right\}. \quad (3.21)$$

Then the linear space or subspace $\mathcal{N}(\Delta^\dagger)$ consists in the vector fields $\boldsymbol{\eta}$ solution to the harmonic problem

$$-\nabla^2\boldsymbol{\eta} = 0, \qquad \boldsymbol{\nabla}\cdot\boldsymbol{\eta}|_S = 0. \quad (3.22)$$

The linear space $\mathcal{N}(\Delta^\dagger)$ can be given an explicit characterization by supplementing problem (3.22) with the additional (nonhomogeneous) boundary condition $\boldsymbol{n}\times\boldsymbol{\eta}|_S \neq 0$. In this work, the vector fields $\boldsymbol{h}$ satisfying the Laplace equation $-\nabla^2\boldsymbol{h} = 0$ are called *harmonic*, so that any vector field solution to the problem (3.22) will be referred to as a *solenoidal harmonic* field[1].

[1]This nomenclature is different from that found in some mathematical texts, where a vector field is said to be harmonic when it is both irrotational and solenoidal, see, *e.g.*, [8].

The orthogonality condition (3.19) for $\boldsymbol{\zeta}$ can then be expressed in the form

$$\int \boldsymbol{\zeta}\cdot\boldsymbol{\eta}\, dV = 0, \tag{3.23}$$

for any solenoidal harmonic vector field $\boldsymbol{\eta}$. Going back to nonhomogeneous boundary conditions, this orthogonality condition becomes

$$\int \boldsymbol{\zeta}\cdot\boldsymbol{\eta}\, dV = \oint \boldsymbol{n}\times(-\boldsymbol{\nabla}\varphi|_S + \boldsymbol{b})\cdot\boldsymbol{\eta}\, dS. \tag{3.24}$$

Now, the "number" of linearly independent vector fields $\boldsymbol{\eta}$ is twice the "number" of boundary points, as shown by the first, nonhomogeneous, boundary condition in (3.22). Therefore, the vorticity integral conditions (3.24) together with the boundary condition $\boldsymbol{\nabla}\cdot\boldsymbol{\zeta}|_S = 0$ provide the correct number of conditions needed to make the parabolic problem for the vector unknown $\boldsymbol{\zeta}$ complete. It follows that the φ-$\boldsymbol{\zeta}$-$\boldsymbol{A}$ formulation of the three-dimensional Navier–Stokes equations can be written in the following uncoupled form, apart from the nonlinearity,

$$\begin{aligned}
&-\nabla^2\varphi = 0, && \boldsymbol{n}\cdot\boldsymbol{\nabla}\varphi|_S = \boldsymbol{n}\cdot\boldsymbol{b};\\
&-\nu\nabla^2\boldsymbol{\zeta} + \frac{\partial\boldsymbol{\zeta}}{\partial t} + \boldsymbol{\nabla}\times[\boldsymbol{\zeta}\times(\boldsymbol{\nabla}\varphi + \boldsymbol{\nabla}\times\boldsymbol{A})] = 0, && \boldsymbol{\zeta}|_{t=0} = \boldsymbol{\nabla}\times\boldsymbol{u}_0,\\
&\int \boldsymbol{\zeta}\cdot\boldsymbol{\eta}\, dV = \oint \boldsymbol{n}\times(-\boldsymbol{\nabla}\varphi|_S + \boldsymbol{b})\cdot\boldsymbol{\eta}\, dS, && \boldsymbol{\nabla}\cdot\boldsymbol{\zeta}|_S = 0,\\
&\big[\, -\nabla^2\boldsymbol{\eta} = 0, && \boldsymbol{\nabla}\cdot\boldsymbol{\eta}|_S = 0; \,\big]\\
&-\nabla^2\boldsymbol{A} = \boldsymbol{\zeta}, && \boldsymbol{n}\times\boldsymbol{A}|_S = 0,\quad \boldsymbol{\nabla}\cdot\boldsymbol{A}|_S = 0.
\end{aligned} \tag{3.25}$$

In the event that a more general gauge condition of the type $\boldsymbol{\nabla}\cdot\boldsymbol{A} = B$ is considered, with B representing some given function, the equation for the vector potential and its scalar derivative boundary condition change as follows:

$$-\nabla^2\boldsymbol{A} = \boldsymbol{\zeta} - \boldsymbol{\nabla}B, \qquad \boldsymbol{\nabla}\cdot\boldsymbol{A}|_S = B|_S,$$

all the other boundary conditions remaining unaltered. Then, the same Green identity (3.20) together with the complete set of boundary conditions for $\boldsymbol{A}$ lead to the integral condition

$$\int (\boldsymbol{\zeta} - \boldsymbol{\nabla}B)\cdot\boldsymbol{\eta}\, dV = \oint [\boldsymbol{n}\times(-\boldsymbol{\nabla}\varphi|_S + \boldsymbol{b})\cdot\boldsymbol{\eta} - B|_S\, \boldsymbol{n}\cdot\boldsymbol{\eta}]\, dS.$$

Using the vector identity $(\boldsymbol{\nabla}B)\cdot\boldsymbol{\eta} = \boldsymbol{\nabla}\cdot(B\boldsymbol{\eta}) - B\,\boldsymbol{\nabla}\cdot\boldsymbol{\eta}$ and the divergence theorem, and since $\boldsymbol{\nabla}\cdot\boldsymbol{\eta} = 0$, the two terms involving B and appearing on both sides of this relation can be shown to coincide. Thus, the integral condition becomes

$$\int \boldsymbol{\zeta}\cdot\boldsymbol{\eta}\, dV = \oint \boldsymbol{n}\times(-\boldsymbol{\nabla}\varphi|_S + \boldsymbol{b})\cdot\boldsymbol{\eta}\, dS,$$

which is identical to (3.24). Therefore the vorticity integral condition is shown to be invariant with respect to the considered gauge transformation.

It is important to note that, in three dimensions, the vorticity equation is supplemented by conditions of both boundary and integral type, and that the integral conditions do couple the three vector components of the vorticity vector.

It may appear a paradox that, in order to eliminate the pressure and the incompressibility condition, a representation of the three-dimensional equations is constructed with the two vector fields $\boldsymbol{\zeta}$ and $\boldsymbol{A}$ as unknowns, which must be both solenoidal, not to speak of the auxiliary harmonic fields $\boldsymbol{\eta}$, which also have to be solenoidal. It is important to note, however, that the solenoidal character of these three different vector fields can be assured by imposing the condition of zero divergence only on the boundary. In this respect the uncoupled formulation (3.25) achieves the theoretically relevant result of segregating the solenoidality condition to a set possessing a lower dimensionality than that of the fluid domain. It should also be noticed that a boundary condition like $\nabla \cdot \boldsymbol{\eta}|_S = 0$ has not to be imposed on the auxiliary vector fields $\boldsymbol{w}$ which replace the harmonic fields $\boldsymbol{\eta}$ in the Glowinski–Pironneau method. This point will be explained in the next section, in connection with another representation of the incompressible equations, characterized by a slightly different treatment of the boundary and integral conditions.

3.4 q_S-ζ-ψ formulation

A drawback of the φ-$\boldsymbol{\zeta}$-$\boldsymbol{A}$ formulation is that the velocity field is described in terms of two potentials. For instance, this choice makes it complicate to establish the relation existing between the φ-$\boldsymbol{\zeta}$-$\boldsymbol{A}$ formulation and the ζ-ψ one for two-dimensional flows. However it is possible to simplify the representation of the solenoidal velocity field and to express it in terms of a vector potential only, by assuming a different set of boundary conditions for it. The price to be paid for such a simplified velocity representation is the introduction of an elliptic problem over the boundary governing a surface scalar unknown: the elliptic equation is a Laplace–Beltrami one and its solution allows to determine the tangential components of the vector potential over the boundary. This surface problem has been introduced by El Dabagi and Pironneau (1986) in the context of a method for calculating the velocity field in the case of compressible inviscid flows. It represents a substantial simplification with respect to an earlier method due to Hirasaki and Hellums (1968), which requires the solution of a vector elliptic equation over the boundary. Following El Dabagi and Pironneau, the new vector potential will be called here *stream vector*. It will be denoted by $\boldsymbol{\psi}$ to distinguish it explicitly from the vector potential $\boldsymbol{A}$ which has been used together with the scalar potential φ in the previous section.

In the present section, the complete set of boundary conditions to be imposed on the stream vector $\boldsymbol{\psi}$ describing the motion of an incompressible viscous fluid will be derived. Such a stream vector will be shown to represent the actual three-dimensional counterpart of the stream function variable used in two-dimensional problems. The elliptic equation over the boundary for the surface scalar potential q_S will be derived after introducing some basic notions concerning the differential operators over a smooth surface parametrized by a system of orthogonal coordinates. The equivalence of the q_S-$\boldsymbol{\zeta}$-$\boldsymbol{\psi}$ formulation with the original Navier–Stokes problem for the primitive variables will be demonstrated by essentially the same argument followed for the φ-$\boldsymbol{\zeta}$-$\boldsymbol{A}$ equations. An uncoupled formulation of the q_S-$\boldsymbol{\zeta}$-$\boldsymbol{\psi}$ equations will be obtained by resorting to the vector Green identity already considered in this chapter.

A time-discretized version of the uncoupled equations is then introduced to develop the geometrical interpretation of the integral conditions for problems in three dimensions and to formulate a three-dimensional counterpart of the Glowinski–Pironneau method for the vorticity and stream function. In this connection, it turns out that a basic component of the Glowinski–Pironneau method generalized to three dimensions is the determination of solenoidal vector fields by forcing the value of their divergence to vanish on the boundary. An influence matrix method for imposing the corresponding derivative boundary condition is described, which relies on the solution of only Dirichlet problems for scalar unknowns. At the end of the section, some remarks concerning the determination of the pressure field, after the solution of the nonprimitive variable Navier–Stokes equations has been calculated, are made.

3.4.1 Differential operators over a surface and surface scalar potential

Instead of "discharging" the boundary value of the normal component of velocity entirely onto a scalar potential, let it be imposed on a stream vector $\boldsymbol{\psi}$ which is introduced to represent the solenoidal velocity field, according to the definition:

$$\boldsymbol{u} = \boldsymbol{\nabla}\times\boldsymbol{\psi}. \tag{3.26}$$

The boundary condition for the normal velocity assumes the form of a derivative scalar condition for $\boldsymbol{\psi}$, namely,

$$\boldsymbol{n}\cdot\boldsymbol{\nabla}\times\boldsymbol{\psi}|_S = \boldsymbol{n}\cdot\boldsymbol{b}. \tag{3.27}$$

Since the evaluation of the component of $\boldsymbol{\nabla}\times\boldsymbol{\psi}$ normal to S only involves the tangential components of $\boldsymbol{\psi}$ on S, one has

$$\boldsymbol{n}\cdot\boldsymbol{\nabla}\times\boldsymbol{\psi}|_S = \boldsymbol{n}\cdot\boldsymbol{\nabla}_S\times\boldsymbol{\psi}_S^{\tan},$$

where

$$\boldsymbol{\psi}_S^{\tan} = \boldsymbol{\psi}|_S - \boldsymbol{n}(\boldsymbol{n}\cdot\boldsymbol{\psi})|_S.$$

Therefore, the considered boundary condition can be written also in the following form

$$\boldsymbol{n}\cdot\nabla\times\boldsymbol{\psi}_S^{\tan} = \boldsymbol{n}\cdot\boldsymbol{b}.$$

In this way, condition (3.27) has been transformed into a *surface differential equation* for the unknown vector field $\boldsymbol{\psi}_S^{\tan}$ defined only on S and tangential to it. In order to be able to solve this problem, we need to recall some elements of the theory of vector differential operators acting on scalar and vector functions defined on a general surface. The classical presentation of Weatherburn [38] will be followed, with only some minor notational changes.

Let S be a surface in the three-dimensional Euclidean space and let (u, v) constitute a system of *curvilinear coordinates* for the points on S. The position of the points $\boldsymbol{r}_S \in S$ is then determined by the values of u and v according to the relation

$$\boldsymbol{r}_S = \boldsymbol{s}(u, v).$$

The vectors tangential to the curves v = constant and u = constant are given by the partial derivative of $\boldsymbol{s}$ with respect to u and to v, namely,

$$\boldsymbol{s}_u = \frac{\partial \boldsymbol{s}}{\partial u} \qquad \text{and} \qquad \boldsymbol{s}_v = \frac{\partial \boldsymbol{s}}{\partial v}.$$

The *fundamental magnitudes of the first order* of the surface S are the scalar products of these vectors, that is

$$\frac{\partial \boldsymbol{s}}{\partial u}\cdot\frac{\partial \boldsymbol{s}}{\partial u}, \qquad \frac{\partial \boldsymbol{s}}{\partial u}\cdot\frac{\partial \boldsymbol{s}}{\partial v}, \qquad \frac{\partial \boldsymbol{s}}{\partial v}\cdot\frac{\partial \boldsymbol{s}}{\partial v}.$$

For *orthogonal* coordinates, $\frac{\partial \boldsymbol{s}}{\partial u}\cdot\frac{\partial \boldsymbol{s}}{\partial v} = 0$ and one introduces the *scale factors* of the orthogonal coordinate system

$$k_u = \left|\frac{\partial \boldsymbol{s}}{\partial u}\right| \qquad \text{and} \qquad k_v = \left|\frac{\partial \boldsymbol{s}}{\partial v}\right|,$$

which allow to define the *unit vectors* $\hat{\boldsymbol{u}}_S$ and $\hat{\boldsymbol{v}}_S$ tangential to S, by means of the relations:

$$\frac{\partial \boldsymbol{s}}{\partial u} = k_u\,\hat{\boldsymbol{u}}_S \qquad \text{and} \qquad \frac{\partial \boldsymbol{s}}{\partial v} = k_v\,\hat{\boldsymbol{v}}_S.$$

If the coordinate system (u, v) is orthogonal, the three unit vectors $\hat{\boldsymbol{u}}_S$, $\hat{\boldsymbol{v}}_S$ and $\boldsymbol{n}$ are mutually orthogonal over the entire surface S, which is assumed to be smooth everywhere.

It is also necessary to consider the second derivatives of $\boldsymbol{s}$ with respect to u and v, that is

$$\boldsymbol{s}_{uu} = \frac{\partial^2 \boldsymbol{s}}{\partial u^2}, \qquad \boldsymbol{s}_{uv} = \frac{\partial^2 \boldsymbol{s}}{\partial u \partial v}, \qquad \boldsymbol{s}_{vv} = \frac{\partial^2 \boldsymbol{s}}{\partial v^2}.$$

The components of these vectors in the direction of the normal to the surface are called *fundamental magnitudes of the second order* and allow to express the derivatives of $\boldsymbol{n}$ in terms of $\boldsymbol{s}_u$ and $\boldsymbol{s}_v$. In particular, for orthogonal coordinates of interest here, $\frac{\partial^2 \boldsymbol{s}}{\partial u \partial v} = 0$ and the two second-order fundamental magnitudes will be indicated by

$$U = \boldsymbol{n} \cdot \frac{\partial^2 \boldsymbol{s}}{\partial u^2} \qquad \text{and} \qquad V = \boldsymbol{n} \cdot \frac{\partial^2 \boldsymbol{s}}{\partial v^2}.$$

Any scalar function f_S defined on S can be described as a function of the orthogonal coordinates u and v, that is, $f_S = f_S(u, v)$. The subscript S is used to denote functions or vector fields defined only on the surface S. Similarly, any vector field $\boldsymbol{F}_S$ defined on S can be decomposed in terms of its surface components, according to the relation

$$\boldsymbol{F}_S(u, v) = F_{S,u}\,\hat{\boldsymbol{u}}_S + F_{S,v}\,\hat{\boldsymbol{v}}_S + F_{S,n}\,\boldsymbol{n},$$

where $F_{S,u} = F_{S,u}(u, v)$, $F_{S,v} = F_{S,v}(u, v)$ and $F_{S,n} = F_{S,n}(u, v)$.

The *surface gradient operator* is defined by

$$\boldsymbol{\nabla}_S = \hat{\boldsymbol{u}}_S \frac{1}{k_u}\frac{\partial}{\partial u} + \hat{\boldsymbol{v}}_S \frac{1}{k_v}\frac{\partial}{\partial v}.$$

Clearly, the application of this operator to any scalar function f_S of the surface coordinates gives a vector field tangential to the surface, namely,

$$\boldsymbol{\nabla}_S f_S = \hat{\boldsymbol{u}}_S \frac{1}{k_u}\frac{\partial f_S}{\partial u} + \hat{\boldsymbol{v}}_S \frac{1}{k_v}\frac{\partial f_S}{\partial v}.$$

On the other hand, the operator $\boldsymbol{\nabla}_S$ can be used to introduce the *surface divergence* and the *surface curl* of a vector field $\boldsymbol{F}_S$ defined on the surface (but not necessarily tangential to S). By applying the operator

$$\boldsymbol{\nabla}_S \cdot = \hat{\boldsymbol{u}}_S \cdot \frac{1}{k_u}\frac{\partial}{\partial u} + \hat{\boldsymbol{v}}_S \cdot \frac{1}{k_v}\frac{\partial}{\partial v}$$

to $\boldsymbol{F}_S = F_{S,u}\,\hat{\boldsymbol{u}}_S + F_{S,v}\,\hat{\boldsymbol{v}}_S + F_{S,n}\,\boldsymbol{n}$, one obtains:

$$\boldsymbol{\nabla}_S \cdot \boldsymbol{F}_S = \frac{1}{k_u k_v}\left[\frac{\partial}{\partial u}\left(k_v\, F_{S,u}\right) + \frac{\partial}{\partial v}\left(k_u\, F_{S,v}\right)\right] + F_{S,n}\, \boldsymbol{\nabla}_S \cdot \boldsymbol{n}.$$

Similarly, applying the operator

$$\boldsymbol{\nabla}_S \times = \hat{\boldsymbol{u}}_S \times \frac{1}{k_u}\frac{\partial}{\partial u} + \hat{\boldsymbol{v}}_S \times \frac{1}{k_v}\frac{\partial}{\partial v}$$

to the same field $\boldsymbol{F}_S$, one obtains:

$$\begin{aligned}\boldsymbol{\nabla}_S \times \boldsymbol{F}_S = {}& \frac{1}{k_u k_v}\left[\frac{\partial}{\partial u}\left(k_v\, F_{S,v}\right) - \frac{\partial}{\partial v}\left(k_u\, F_{S,u}\right)\right]\boldsymbol{n} \\ & + \frac{V}{k_v^2} F_{S,v}\,\hat{\boldsymbol{u}}_S - \frac{U}{k_u^2} F_{S,u}\,\hat{\boldsymbol{v}}_S + (\boldsymbol{\nabla}_S F_{S,n}) \times \boldsymbol{n}.\end{aligned}$$

An important consequence of this relation is that, if $\boldsymbol{F}_S = \boldsymbol{\nabla}_S f_S$, the normal component of $\boldsymbol{\nabla}_S \times \boldsymbol{F}_S$ vanishes identically: thus, the curl of a surface vector field which is the surface gradient of a scalar function is tangential to the surface. Also the converse holds: if a surface vector field is tangential to S together with its curl, then the vector field is the surface gradient of some scalar function.

We have also the important relationships:

$$\boldsymbol{n}\times\boldsymbol{\nabla} f|_S = \boldsymbol{n}\times\boldsymbol{\nabla}_S f_S, \qquad \text{where} \qquad f_S = f|_S,$$

and

$$\boldsymbol{n}\cdot\boldsymbol{\nabla}\times\boldsymbol{F}|_S = \boldsymbol{n}\cdot\boldsymbol{\nabla}_S\times\boldsymbol{F}_S, \qquad \text{where} \qquad \boldsymbol{F}_S = \boldsymbol{F}|_S.$$

Finally, it is possible to introduce the Laplace–Beltrami operator

$$\nabla_S^2 = \boldsymbol{\nabla}_S\cdot\boldsymbol{\nabla}_S = \frac{1}{k_u k_v}\left[\frac{\partial}{\partial u}\left(\frac{k_v}{k_u}\frac{\partial}{\partial u}\right) + \frac{\partial}{\partial v}\left(\frac{k_u}{k_v}\frac{\partial}{\partial v}\right)\right],$$

which is an elliptic operator acting on the scalar functions defined on S.

We now come back to the solution of the surface differential equation for $\boldsymbol{\psi}_S^{\tan}$ and, for notational convenience, indicate such an unknown by $\boldsymbol{a} = \boldsymbol{a}(\boldsymbol{x}_S) = \boldsymbol{\psi}_S^{\tan}(\boldsymbol{x}_S)$, so that the boundary condition for the normal velocity is written in terms of $\boldsymbol{a}$ as

$$\boldsymbol{n}\cdot\boldsymbol{\nabla}_S\times\boldsymbol{a} = \boldsymbol{n}\cdot\boldsymbol{b}. \tag{3.28}$$

Any vector field $\boldsymbol{a}$ defined only on S and tangential to S can be written in the form

$$\boldsymbol{a} = \boldsymbol{\nabla}_S p_S + \boldsymbol{\nabla}_S q_S\times\boldsymbol{n}, \tag{3.29}$$

where p_S and q_S are scalar functions which depend on $\boldsymbol{a}$.

(We notice incidentally that this representation of a vector field defined on a closed surface and tangential to it constitutes an orthogonal decomposition over the surface, in the sense that

$$\oint \boldsymbol{\nabla}_S p_S\cdot\boldsymbol{\nabla}_S q_S\times\boldsymbol{n}\,dS = 0.$$

In fact, by the divergence theorem, it results

$$\begin{aligned}\oint \boldsymbol{\nabla}_S p_S\cdot\boldsymbol{\nabla}_S q_S\times\boldsymbol{n}\,dS &= \oint \boldsymbol{n}\cdot\boldsymbol{\nabla}_S p_S\times\boldsymbol{\nabla}_S q_S\,dS\\ &= \int \boldsymbol{\nabla}\cdot(\boldsymbol{\nabla} p\times\boldsymbol{\nabla} q)\,dV = 0,\end{aligned}$$

where p and q are any extension of p_S and q_S to V.)

Equation (3.28) is solved exploiting the representation of $\boldsymbol{a}$ in terms of the surface functions p_S and q_S given by (3.29). First, we note that equation (3.28) leaves the function p_S completely indeterminate, since $\boldsymbol{n}\cdot\boldsymbol{\nabla}_S\times\boldsymbol{\nabla}_S p_S = 0$, by the observation made after introducing the operator $\nabla_S\times$. Equation (3.28) determines instead q_S. In fact, substituting the assumed expression of $\boldsymbol{a}$ given by (3.29) into $\nabla_S\times\boldsymbol{a}$ gives (cf. [38], p. 237)

$$\begin{aligned}\nabla_S\times\boldsymbol{a} &= \nabla_S\times(\nabla_S q_S)\\ &= (\boldsymbol{n}\cdot\nabla_S)\nabla_S q_S - \boldsymbol{n}(\nabla_S\cdot\nabla_S)q_S + (\nabla_S q_S)\,\nabla_S\cdot\boldsymbol{n} - (\nabla_S q_S\cdot\nabla_S)\boldsymbol{n}\\ &= -\boldsymbol{n}\,\nabla_S^2 q_S + (\nabla_S q_S)\,\nabla_S\cdot\boldsymbol{n} - (\nabla_S q_S\cdot\nabla_S)\boldsymbol{n},\end{aligned}$$

since $\boldsymbol{n}\cdot\nabla_S = 0$, ∇_S^2 being the Laplace–Beltrami operator defined before. Taking the normal component of this relation gives

$$\boldsymbol{n}\cdot\nabla_S\times\boldsymbol{a} = -\nabla_S^2 q_S,$$

since $\boldsymbol{n}\perp\nabla_S q_S$ and since both vectors $\partial\boldsymbol{n}/\partial u$ and $\partial\boldsymbol{n}/\partial v$ are perpendicular to $\boldsymbol{n}$ which is a vector of fixed length. As a consequence, equation (3.28) transforms into the following (surfacial) elliptic problem

$$-\nabla_S^2 q_S = \boldsymbol{n}\cdot\boldsymbol{b} \tag{3.30}$$

for the function q_S over the closed boundary S. If, in addition to the simple connectivity of V, one assumes that the boundary surface S consists of only one connected component to exclude the presence of "inclusion" in V, problem (3.30) can be solved uniquely, apart from an arbitrary additive constant Q (or function of time $Q(t)$, in the present fluid dynamical context), provided that the right-hand side of the equation satisfies the solvability condition $\oint \boldsymbol{n}\cdot\boldsymbol{b}\,dS = 0$. This condition is a consequence of the application of Green identity over a *closed* surface S, namely,

$$\oint(\psi_S\nabla_S^2\phi_S - \phi_S\nabla_S^2\psi_S)\,dS = 0, \tag{3.31}$$

when one takes $\psi_S(\boldsymbol{x}_S)\equiv 1$ and $\phi_S = q_S$ solution to the problem (3.30). Of course the solvability condition is satisfied here, being nothing else than the global condition (1.5).

3.4.2 Governing equations

Once the surface vector field $\boldsymbol{a}$ has been determined, it provides the proper boundary condition for the tangential components of the stream vector $\boldsymbol{\psi}$, in the form:

$$\boldsymbol{n}\times\boldsymbol{\psi}|_S = \boldsymbol{n}\times\boldsymbol{a}. \tag{3.32}$$

As far as the other boundary conditions for $\boldsymbol{\psi}$ are concerned, the imposition of the velocity tangential to the boundary requires $\boldsymbol{n}\times\nabla\times\boldsymbol{\psi}|_S = \boldsymbol{n}\times\boldsymbol{b}$, whereas the choice of the Euclid invariant gauge, *i.e.*, $\nabla\cdot\boldsymbol{\psi} = 0$, gives the scalar boundary condition $\boldsymbol{\nabla}\cdot\boldsymbol{\psi}|_S = 0$. Thus, the equation for the stream vector $\boldsymbol{\psi}$ and the full set of its boundary conditions are:

$$-\nabla^2\boldsymbol{\psi} = \boldsymbol{\zeta}, \qquad \boldsymbol{n}\times\boldsymbol{\psi}|_S = \boldsymbol{n}\times\boldsymbol{a}, \quad \boldsymbol{n}\times\boldsymbol{\nabla}\times\boldsymbol{\psi}|_S = \boldsymbol{n}\times\boldsymbol{b}, \quad \boldsymbol{\nabla}\cdot\boldsymbol{\psi}|_S = 0. \quad (3.33)$$

The number of conditions (five) and their kind is as in the φ-$\boldsymbol{\zeta}$-$\boldsymbol{A}$ representation, the only differences beeing the nonhomogeneous terms of the conditions for the tangential components of $\boldsymbol{\psi}$ and of $\boldsymbol{\nabla}\times\boldsymbol{\psi}$. The equivalence of the full set of equations and conditions for the variables q_S-$\boldsymbol{\zeta}$-$\boldsymbol{\psi}$ with the original primitive variable Navier–Stokes problem is stated in the following theorem, which is demonstrated very similarly to that valid for the φ-$\boldsymbol{\zeta}$-$\boldsymbol{A}$ representation.

Theorem 3.2. *The primitive variable Navier–Stokes problem (1.1)–(1.4) is equivalent to the problem for the variables q_S, $\boldsymbol{\zeta}$ and $\boldsymbol{\psi}$, defined by the set of equations and conditions*

$$-\nabla_S^2 q_S = \boldsymbol{n}\cdot\boldsymbol{b}, \qquad [\Longrightarrow \boldsymbol{a} = -\boldsymbol{n}\times\boldsymbol{\nabla}_S q_S\,]$$

$$\frac{\partial\boldsymbol{\zeta}}{\partial t} - \nu\nabla^2\boldsymbol{\zeta} + \boldsymbol{\nabla}\times(\boldsymbol{\zeta}\times\boldsymbol{\nabla}\times\boldsymbol{\psi}) = 0, \qquad \boldsymbol{\zeta}|_{t=0} = \boldsymbol{\nabla}\times\boldsymbol{u}_0, \quad \boldsymbol{\nabla}\cdot\boldsymbol{\zeta}|_S = 0,$$

$$-\nabla^2\boldsymbol{\psi} = \boldsymbol{\zeta}, \qquad \boldsymbol{n}\times\boldsymbol{\psi}|_S = \boldsymbol{n}\times\boldsymbol{a}, \quad \boldsymbol{n}\times\boldsymbol{\nabla}\times\boldsymbol{\psi}|_S = \boldsymbol{n}\times\boldsymbol{b}, \quad \boldsymbol{\nabla}\cdot\boldsymbol{\psi}|_S = 0, \quad (3.34)$$

provided that the data satisfy the three following conditions

$$\oint \boldsymbol{n}\cdot\boldsymbol{b}\,dS = 0, \qquad \boldsymbol{\nabla}\cdot\boldsymbol{u}_0 = 0,$$
$$\boldsymbol{n}\cdot\boldsymbol{b}|_{t=0} = \boldsymbol{n}\cdot\boldsymbol{u}_0|_S \quad \text{or} \quad \boldsymbol{n}\cdot\boldsymbol{\nabla}_S\times\boldsymbol{a}|_{t=0} = \boldsymbol{n}\cdot\boldsymbol{u}_0|_S. \quad (3.35)$$

Proof. We note first that the two compatibility conditions between the boundary and initial data in (3.35) are two different ways of writing the same condition, owing to the definition of $\boldsymbol{a}$ in terms of q_S and to the elliptic equation satisfied by q_S.

Coming to the proof of the Theorem, the direct implication is evident. Conversely, let $(\boldsymbol{\zeta}, \boldsymbol{\psi})$ be a solution to the set of equations and conditions (3.34) with the data $\boldsymbol{n}\cdot\boldsymbol{b}$ and $\boldsymbol{u}_0$ satisfying the conditions (3.35). Observe that q_S is uniquely defined up to an additive arbitrary function of time $Q(t)$, since the solvability condition $\oint \boldsymbol{n}\cdot\boldsymbol{b}\,dS = 0$ of the surface elliptic problem for q_S is satisfied by the first condition in (3.35). Then $\boldsymbol{a}$ is well defined in terms of q_S through relation $\boldsymbol{a} = -\boldsymbol{n}\times\boldsymbol{\nabla}_S q_S$.

Let us define the vector field $\boldsymbol{v} = \boldsymbol{\nabla}\times\boldsymbol{\psi}$. Then, $\boldsymbol{v}$ is solenoidal and $\boldsymbol{\nabla}\times\boldsymbol{v} = \boldsymbol{\nabla}\times\boldsymbol{\nabla}\times\boldsymbol{\psi} = -\nabla^2\boldsymbol{\psi} + \boldsymbol{\nabla}(\boldsymbol{\nabla}\cdot\boldsymbol{\psi}) = -\nabla^2\boldsymbol{\psi} = \boldsymbol{\zeta}$, since $\boldsymbol{\psi}$ is solenoidal, as it is $\boldsymbol{\zeta}$ for the same reasons seen in the proof of Theorem 3.1. Substituting $\boldsymbol{\zeta} = \boldsymbol{\nabla}\times\boldsymbol{v}$ in the vorticity equation gives

$$\frac{\partial\boldsymbol{\nabla}\times\boldsymbol{v}}{\partial t} - \nu\nabla^2\boldsymbol{\nabla}\times\boldsymbol{v} + \boldsymbol{\nabla}\times[(\boldsymbol{\nabla}\times\boldsymbol{v})\times\boldsymbol{v}] = 0,$$

namely,

$$\nabla\times\left[\frac{\partial \boldsymbol{v}}{\partial t}-\nu\nabla^2\boldsymbol{v}+(\nabla\times\boldsymbol{v})\times\boldsymbol{v}\right]=0,$$

which means

$$\frac{\partial \boldsymbol{v}}{\partial t}-\nu\nabla^2\boldsymbol{v}+(\nabla\times\boldsymbol{v})\times\boldsymbol{v}=\nabla Q,$$

for some function Q. To identify the field $\boldsymbol{v}$ with the velocity field $\boldsymbol{u}$ solution of the primitive variable Navier–Stokes equations (and Q with $-P-\frac{1}{2}u^2$), the boundary and initial values of $\boldsymbol{v}$ must be shown to coincide with those prescribed in the boundary and initial conditions for $\boldsymbol{u}$. For the normal component of the velocity on the boundary, it is:

$$\boldsymbol{n}\cdot\boldsymbol{v}|_S=\boldsymbol{n}\cdot\nabla\times\boldsymbol{\psi}|_S=\boldsymbol{n}\cdot\nabla_S\times\boldsymbol{a}=-\nabla_S^2 q_S=\boldsymbol{n}\cdot\boldsymbol{b}.$$

For the tangential components of the velocity on the boundary one has

$$\boldsymbol{n}\times\boldsymbol{v}|_S=\boldsymbol{n}\times\nabla\times\boldsymbol{\psi}|_S=\boldsymbol{n}\times\boldsymbol{b}.$$

Thus $\boldsymbol{v}|_S=\boldsymbol{b}=\boldsymbol{u}|_S$. Concerning the initial condition, one has to determine the initial value implied on $\boldsymbol{v}$ by the imposition of the initial condition for $\boldsymbol{\zeta}$ in the $\boldsymbol{\zeta}$-$\boldsymbol{\psi}$ system. Let $\boldsymbol{\psi}_0$ be the solution of the problem

$$-\nabla^2\boldsymbol{\psi}_0=\boldsymbol{\zeta}_0=\nabla\times\boldsymbol{u}_0,\qquad \boldsymbol{n}\times\boldsymbol{\psi}_0|_S=\boldsymbol{n}\times\boldsymbol{a}|_{t=0},\quad \nabla\cdot\boldsymbol{\psi}_0|_S=0.$$

By the assumed continuity of $\boldsymbol{n}\cdot\boldsymbol{b}=\boldsymbol{n}\cdot\nabla_S\times\boldsymbol{a}$ as $t\to 0^+$ and since $\boldsymbol{\zeta}_0=\boldsymbol{\zeta}|_{t=0}$, the well-posedness of the two previous elliptic problems implies that

$$\boldsymbol{v}|_{t=0}=\lim_{t\to 0^+}\nabla\times\boldsymbol{\psi}=\nabla\times\boldsymbol{\psi}_0.$$

Now, $\boldsymbol{\psi}_0$ being solenoidal, the equation for $\boldsymbol{\psi}_0$ can be written as $\nabla\times\nabla\times\boldsymbol{\psi}_0=\nabla\times\boldsymbol{u}_0$, that is, $\nabla\times(\nabla\times\boldsymbol{\psi}_0-\boldsymbol{u}_0)=0$. It follows that $\nabla\times\boldsymbol{\psi}_0-\boldsymbol{u}_0=\nabla\alpha$, for some function α, which is harmonic since $\nabla\cdot\nabla\times\boldsymbol{\psi}_0=0$ and $\nabla\cdot\boldsymbol{u}_0=0$ by the second of the compatibility conditions (3.35). Taking the normal component of this equation on S gives:

$$\begin{aligned}\boldsymbol{n}\cdot(\nabla\times\boldsymbol{\psi}_0-\boldsymbol{u}_0)|_S&=\boldsymbol{n}\cdot\nabla\times\boldsymbol{\psi}_0|_S-\boldsymbol{n}\cdot\boldsymbol{u}_0|_S\\&=\boldsymbol{n}\cdot\nabla_S\times\boldsymbol{a}|_{t=0}-\boldsymbol{n}\cdot\boldsymbol{u}_0|_S=\boldsymbol{n}\cdot\nabla\alpha|_S.\end{aligned}$$

By the third of the compatibility conditions (3.35), the relation just obtained becomes $\boldsymbol{n}\cdot\nabla\alpha|_S=0$, which, α being harmonic, gives $\alpha=A(t)$, with $A(t)$ an arbitrary function. Thus $\nabla\times\boldsymbol{\psi}_0=\boldsymbol{u}_0$, so that $\boldsymbol{v}|_{t=0}=\nabla\times\boldsymbol{\psi}_0=\boldsymbol{u}_0=\boldsymbol{u}|_{t=0}$. Therefore $\boldsymbol{v}=\boldsymbol{u}$ and hence $Q=-P-\frac{1}{2}u^2+C(t)$, where $C(t)$ is an arbitrary function. □

3.4.3 Split formulation

Starting from problem (3.34), the orthogonal projection argument of section 3.3.3 can be repeated, recalling the vector version of Green identity:

$$\int (\boldsymbol{\psi}\cdot\nabla^2\boldsymbol{\phi} - \boldsymbol{\phi}\cdot\nabla^2\boldsymbol{\psi})\,dV$$
$$= \oint (\boldsymbol{n}\times\boldsymbol{\psi}\cdot\boldsymbol{\nabla}\times\boldsymbol{\phi} - \boldsymbol{n}\times\boldsymbol{\phi}\cdot\boldsymbol{\nabla}\times\boldsymbol{\psi} + \boldsymbol{n}\cdot\boldsymbol{\psi}\,\boldsymbol{\nabla}\cdot\boldsymbol{\phi} - \boldsymbol{n}\cdot\boldsymbol{\phi}\,\boldsymbol{\nabla}\cdot\boldsymbol{\psi})\,dS. \quad (3.36)$$

Taking into account the boundary conditions satisfied by $\boldsymbol{\psi}$, the vorticity in the q_S-$\boldsymbol{\zeta}$-$\boldsymbol{\psi}$ representation can be shown to be subject to the integral conditions:

$$\int \boldsymbol{\zeta}\cdot\boldsymbol{\eta}\,dV = \oint (\boldsymbol{n}\times\boldsymbol{a}\cdot\boldsymbol{\nabla}\times\boldsymbol{\eta} + \boldsymbol{n}\times\boldsymbol{b}\cdot\boldsymbol{\eta})\,dS, \quad (3.37)$$

where $\boldsymbol{\eta}$ are the same solenoidal harmonic vector fields defined by problem (3.22).

It is noteworthy that these vorticity integral conditions are not affected by the possible presence of the term $\boldsymbol{\nabla}_S p_S$ (with p_S arbitrary) which appears in the expression (3.29) of $\boldsymbol{a}$. In fact, assume that this term is retained in the expression so that the tangential boundary condition for $\boldsymbol{\psi}$ becomes $\boldsymbol{n}\times\boldsymbol{\psi}|_S = \boldsymbol{n}\times\boldsymbol{\nabla}_S p_S + \boldsymbol{\nabla}_S q_S$ and a different stream vector is obtained depending on the arbitrary function p_S. The vorticity integral conditions do not change because the additional term present in the boundary integral of (3.37) vanishes, as shown by the following direct calculation employing the divergence theorem,

$$\begin{aligned}
\oint \boldsymbol{n}\times\boldsymbol{\nabla}_S p_S\cdot\boldsymbol{\nabla}\times\boldsymbol{\eta}\,dS &= \oint \boldsymbol{n}\times\boldsymbol{\nabla} p_S\cdot\boldsymbol{\nabla}\times\boldsymbol{\eta}\,dS \\
&= \oint \boldsymbol{n}\cdot\boldsymbol{\nabla} p_S\times\boldsymbol{\nabla}\times\boldsymbol{\eta}\,dS \\
&= \int \boldsymbol{\nabla}\cdot(\boldsymbol{\nabla} p\times\boldsymbol{\nabla}\times\boldsymbol{\eta})\,dV \\
&= \int [\boldsymbol{\nabla}\times(\boldsymbol{\nabla} p)\cdot\boldsymbol{\nabla}\times\boldsymbol{\eta} - \boldsymbol{\nabla} p\cdot\boldsymbol{\nabla}\times\boldsymbol{\nabla}\times\boldsymbol{\eta}]\,dV \\
&= 0,
\end{aligned}$$

where p is any extension of p_S to V, and where the harmonic and solenoidal character of $\boldsymbol{\eta}$ has been exploited in the last passage.

We note that, whereas the vorticity integral conditions occurring in the φ-$\boldsymbol{\zeta}$-$\boldsymbol{A}$ formulation have only an essential contribution, those associated with the q_S-$\boldsymbol{\zeta}$-$\boldsymbol{\psi}$ formulation have both an essential and a natural contribution. A difference seems to exist between the vorticity integral conditions occurring in the two formulations, because the respective boundary terms differ by the two terms

$$\oint \boldsymbol{n}\times\boldsymbol{a}\cdot\boldsymbol{\nabla}\times\boldsymbol{\eta}\,dS \qquad \text{and} \qquad \oint \boldsymbol{n}\times(-\boldsymbol{\nabla}\varphi|_S)\cdot\boldsymbol{\eta}\,dS.$$

However, these two integrals can be readily shown to be only two different ways of expressing the same quantity, due to the different representation of the velocity in the two formulations. In fact, using the boundary condition $\boldsymbol{n}\times\boldsymbol{\psi}|_S = \boldsymbol{n}\times\boldsymbol{a}$ and standard vector identities:

$$\begin{aligned}
\oint \boldsymbol{n}\times\boldsymbol{a}\cdot\nabla\times\boldsymbol{\eta}\, dS &= \oint \boldsymbol{n}\times\boldsymbol{\psi}|_S\cdot\nabla\times\boldsymbol{\eta}\, dS \\
&= \oint \boldsymbol{n}\cdot\boldsymbol{\psi}|_S\times\nabla\times\boldsymbol{\eta}\, dS \\
&= \int \nabla\cdot(\boldsymbol{\psi}\times\nabla\times\boldsymbol{\eta})\, dV \\
&= \int (\nabla\times\boldsymbol{\psi}\cdot\nabla\times\boldsymbol{\eta} - \boldsymbol{\psi}\cdot\nabla\times\nabla\times\boldsymbol{\eta})\, dV \\
&= \int \nabla\times\boldsymbol{\psi}\cdot\nabla\times\boldsymbol{\eta}\, dV,
\end{aligned}$$

since $\nabla\times\nabla\times\boldsymbol{\eta} = 0$. The stream vector $\boldsymbol{\psi}$ and the potentials φ-$\boldsymbol{A}$ represent the same velocity field, so that $\nabla\times\boldsymbol{\psi} = \nabla\varphi + \nabla\times\boldsymbol{A}$ and we can write

$$\oint \boldsymbol{n}\times\boldsymbol{a}\cdot\nabla\times\boldsymbol{\eta}\, dS = \int \nabla\times\boldsymbol{\psi}\cdot\nabla\times\boldsymbol{\eta}\, dV = \int (\nabla\varphi + \nabla\times\boldsymbol{A})\cdot\nabla\times\boldsymbol{\eta}\, dV.$$

Consider now the two terms in the last integral. For the first one, recalling that φ is harmonic and using the divergence theorem give:

$$\begin{aligned}
\int \nabla\varphi\cdot\nabla\times\boldsymbol{\eta}\, dV &= -\int \nabla\cdot(\nabla\varphi\times\boldsymbol{\eta})\, dV \\
&= -\oint \boldsymbol{n}\cdot\nabla\varphi|_S\times\boldsymbol{\eta}\, dS \\
&= -\oint \boldsymbol{n}\times\nabla\varphi|_S\cdot\boldsymbol{\eta}\, dS.
\end{aligned}$$

The second term is instead zero since, using again the equation $\nabla\times\nabla\times\boldsymbol{\eta} = 0$ and the divergence theorem,

$$\begin{aligned}
\int \nabla\times\boldsymbol{A}\cdot\nabla\times\boldsymbol{\eta}\, dV &= \int \nabla\cdot(\boldsymbol{A}\times\nabla\times\boldsymbol{\eta})\, dV \\
&= \oint \boldsymbol{n}\cdot\boldsymbol{A}|_S\times\nabla\times\boldsymbol{\eta}\, dS \\
&= \oint \boldsymbol{n}\times\boldsymbol{A}|_S\cdot\nabla\times\boldsymbol{\eta}\, dS \\
&= 0,
\end{aligned}$$

by virtue of the boundary condition $\boldsymbol{n}\times\boldsymbol{A}|_S = 0$. In conclusion,

$$\oint \boldsymbol{n}\times\boldsymbol{a}\cdot\nabla\times\boldsymbol{\eta}\, dS = \oint \boldsymbol{n}\times(-\nabla\varphi|_S)\cdot\boldsymbol{\eta}\, dS,$$

which demonstrates the identity of the vorticity integral conditions (3.37) and (3.24) in the q_S-$\boldsymbol{\zeta}$-$\boldsymbol{\psi}$ and φ-$\boldsymbol{\zeta}$-$\boldsymbol{A}$ formulations, respectively.

The vorticity conditions (3.37) allow to formulate the q_S-$\boldsymbol{\zeta}$-$\boldsymbol{\psi}$ equations in the following split form:

$$
\begin{aligned}
&-\nabla_S^2 q_S = \boldsymbol{n}\cdot\boldsymbol{b}, \qquad [\Longrightarrow \boldsymbol{a} = -\boldsymbol{n}\times\boldsymbol{\nabla}_S q_S\,] \\
&-\nu\nabla^2\boldsymbol{\zeta} + \frac{\partial\boldsymbol{\zeta}}{\partial t} + \boldsymbol{\nabla}\times(\boldsymbol{\zeta}\times\boldsymbol{\nabla}\times\boldsymbol{\psi}) = 0, \qquad \boldsymbol{\zeta}|_{t=0} = \boldsymbol{\nabla}\times\boldsymbol{u}_0, \\
&\int\boldsymbol{\zeta}\cdot\boldsymbol{\eta}\,dV = \oint(\boldsymbol{n}\times\boldsymbol{a}\cdot\boldsymbol{\nabla}\times\boldsymbol{\eta} + \boldsymbol{n}\times\boldsymbol{b}\cdot\boldsymbol{\eta})\,dS, \qquad \boldsymbol{\nabla}\cdot\boldsymbol{\zeta}|_S = 0, \\
&\Big[-\nabla^2\boldsymbol{\eta} = 0, \qquad \boldsymbol{n}\times\boldsymbol{\eta}|_S \neq 0, \qquad \boldsymbol{\nabla}\cdot\boldsymbol{\eta}|_S = 0;\,\Big] \\
&-\nabla^2\boldsymbol{\psi} = \boldsymbol{\zeta}, \qquad \boldsymbol{n}\times\boldsymbol{\psi}|_S = \boldsymbol{n}\times\boldsymbol{a}, \quad \boldsymbol{\nabla}\cdot\boldsymbol{\psi}|_S = 0.
\end{aligned}
\tag{3.38}
$$

This set of equations and conditions should be compared with the set in (3.25). The only difference existing between the problems for the stream vector and for the vector potential is the nonhomogeneous character of the respective boundary conditions for the tangential components.

Formulation (3.38) represents the genuine extension to three dimensions of the vorticity–stream function formulation for two-dimensional flows. When these 3D equations are specialized to the situation of a fluid with a motion parallel to a plane xy and not dependent on the z-coordinate, only the z-component of the vector fields $\boldsymbol{\zeta}$ and $\boldsymbol{\psi}$ survive, and the respective equations in (3.38) become the two-dimensional equations governing the scalar unknowns $\zeta_z = \zeta$ and $\psi_z = \psi$. Since both ζ and ψ do not depend on z any concern about the solenoidal chacter of the variables disappears in two dimensions. Concerning the boundary conditions, the specification of the two tangential components of $\boldsymbol{\psi}$, namely, $\boldsymbol{n}\times\boldsymbol{\psi}|_S = \boldsymbol{n}\times\boldsymbol{a}$, transforms into the single boundary condition $\psi|_S = a$, where $a = \int\boldsymbol{n}\cdot\boldsymbol{b}\,ds$. Therefore, while in three dimensions the determination of the (tangential) boundary values of the stream vector requires to solve a second-order (surface elliptic) problem, in two dimensions the value of the stream function on the boundary is obtained by a simple one-dimensional integration. Finally, the integral conditions for the vorticity vector field $\boldsymbol{\zeta}$ become the scalar vorticity integral conditions discussed at some length in chapter 2.

3.4.4 Time-discretization and orthogonal projection

Consider now a time-discretized version of the governing equations of problem (3.38), to be solved for the unknowns $\boldsymbol{\zeta} = \boldsymbol{\zeta}^{n+1}$ and $\boldsymbol{\psi} = \boldsymbol{\psi}^{n+1}$, namely,

$$\begin{aligned}
&(-\nabla^2+\gamma)\boldsymbol{\zeta} = \boldsymbol{f},\\
&\int \boldsymbol{\zeta}\cdot\boldsymbol{\eta}\, dV = \oint (\boldsymbol{n}\times\boldsymbol{a}\cdot\boldsymbol{\nabla}\times\boldsymbol{\eta} + \boldsymbol{n}\times\boldsymbol{b}\cdot\boldsymbol{\eta})\, dS, \qquad \boldsymbol{\nabla}\cdot\boldsymbol{\zeta}|_S = 0; \qquad (3.39)\\
&-\nabla^2\boldsymbol{\psi} = \boldsymbol{\zeta}, \qquad \boldsymbol{n}\times\boldsymbol{\psi}|_S = \boldsymbol{n}\times\boldsymbol{a}, \qquad \boldsymbol{\nabla}\cdot\boldsymbol{\psi}|_S = 0;
\end{aligned}$$

where $\gamma = 1/(\nu\Delta t)$, $\boldsymbol{f} = \gamma\boldsymbol{\zeta}^n - \nu^{-1}\boldsymbol{\nabla}\times(\boldsymbol{\zeta}^n\times\boldsymbol{\nabla}\times\boldsymbol{\psi}^n)$, $\boldsymbol{a} = \boldsymbol{a}(\boldsymbol{x}_S) = \boldsymbol{a}(\boldsymbol{x}_S, t^{n+1})$ and $\boldsymbol{b} = \boldsymbol{b}(\boldsymbol{x}_S) = \boldsymbol{b}(\boldsymbol{x}_S, t^{n+1})$.

As for the two-dimensional equations, it is possible to impose the integral conditions following a procedure that requires the solution of only boundary value problems for the operators $(-\nabla^2+\gamma)$ and $-\nabla^2$. Let $\boldsymbol{\mathcal{H}}^{\text{sol}}$ denote the space of the vector fields that are harmonic and solenoidal in V, that is,

$$\boldsymbol{\mathcal{H}}^{\text{sol}} = \left\{\boldsymbol{\eta}\,|\, -\nabla^2\boldsymbol{\eta} = 0,\ \boldsymbol{\nabla}\cdot\boldsymbol{\eta}|_S = 0\right\}. \qquad (3.40)$$

Furthermore, let $\boldsymbol{Z}_\gamma^{\text{sol}}$ be the space of the solenoidal vector fields solution to the equation $(-\nabla^2+\gamma)\boldsymbol{z} = \boldsymbol{g}$, $\gamma \geq 0$, for any solenoidal field $\boldsymbol{g}$, that is,

$$\boldsymbol{Z}_\gamma^{\text{sol}} = \left\{\boldsymbol{z}\,|\,(-\nabla^2+\gamma)\boldsymbol{z} = \boldsymbol{g},\ \boldsymbol{\nabla}\cdot\boldsymbol{z}|_S = 0,\ \text{any } \boldsymbol{g} \text{ solenoidal}\right\}. \qquad (3.41)$$

Let the operator of orthogonal projection onto $\boldsymbol{\mathcal{H}}^{\text{sol}}$ be introduced. It will be denoted by $I\!P_{\mathcal{N}(-\widetilde{\nabla}^2_{\nabla\cdot})}$, where $\mathcal{N}$ indicates the null space and $\widetilde{\nabla}^2_{\nabla\cdot}$ represents the Laplace operator acting on vector fields and supplemented only by the single boundary condition of zero divergence, *i.e.*, the tangential components are left arbitrary. Similarly to the analysis of the scalar equations in two dimensions, one introduces the linear space $\boldsymbol{\mathcal{M}}_\gamma^{\text{sol}}$ of the solenoidal *metaharmonic* vector fields $\boldsymbol{\zeta}'$ that are solution to the homogeneous equation $(-\nabla^2+\gamma)\boldsymbol{\zeta}' = 0$, namely,

$$\boldsymbol{\mathcal{M}}_\gamma^{\text{sol}} = \left\{\boldsymbol{\zeta}'\,|\,(-\nabla^2+\gamma)\boldsymbol{\zeta}' = 0,\ \boldsymbol{\nabla}\cdot\boldsymbol{\zeta}'|_S = 0\right\}. \qquad (3.42)$$

If the orthogonal projection operator $I\!P_{\mathcal{N}(-\widetilde{\nabla}^2_{\nabla\cdot})}$ is made to act only on the elements of $\boldsymbol{\mathcal{M}}_\gamma^{\text{sol}}$, a linear operator is obtained which maps $\boldsymbol{\mathcal{M}}_\gamma^{\text{sol}}$ onto $\boldsymbol{\mathcal{H}}^{\text{sol}}$. Such an operator is the *restriction* of $I\!P_{\mathcal{N}(-\widetilde{\nabla}^2_{\nabla\cdot})}$ to the linear space $\boldsymbol{\mathcal{M}}_\gamma^{\text{sol}}$ and will be denoted by P^γ, according to the following definition:

$$P^\gamma = I\!P_{\mathcal{N}(-\widetilde{\nabla}^2_{\nabla\cdot})}\Big|_{\mathcal{N}(-\widetilde{\nabla}^2_{\nabla\cdot}+\gamma)}. \qquad (3.43)$$

The restricted projection operator P^γ establishes a mapping between the two linear spaces $\boldsymbol{\mathcal{M}}_\gamma^{\text{sol}}$ and $\boldsymbol{\mathcal{H}}^{\text{sol}}$, which is proved to be one-to-one by the following theorem (see also Fig. 3.1).

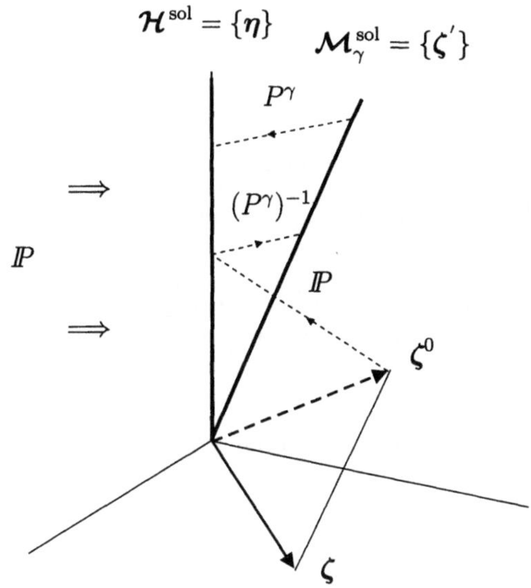

Figure 3.1: Construction of the vorticity vector field for homogeneous velocity boundary conditions.

Theorem 3.3. *Let $\widetilde{\nabla}^2_{\nabla\cdot}$ denote the Laplacian operator acting on vector fields and supplemented only by the single boundary condition of zero divergence and let $\mathcal{H}^{\text{sol}}$ be the space of the solenoidal vector fields harmonic in V. Furthermore, let $I\!P_{\mathcal{N}(-\widetilde{\nabla}^2_{\nabla\cdot})}$ denote the operator of orthogonal projection onto $\mathcal{H}^{\text{sol}}$. Then, the operator P^γ, representing the restriction of $I\!P_{\mathcal{N}(-\widetilde{\nabla}^2_{\nabla\cdot})}$ to the linear space $\mathcal{M}_\gamma^{\text{sol}}$ of the solenoidal vector fields metaharmonic in V, is one-to-one and therefore invertible.*

Proof. As in the scalar two-dimensional situation, the theorem is trivially true for $\gamma = 0$, so that only the case $\gamma > 0$ must be considered. To prove the invertibility of P^γ one has to show that $\mathcal{N}(P^\gamma) = 0$. This is equivalent to assert that there is no vector field $\boldsymbol{\zeta}' \neq 0$ in $\mathcal{M}_\gamma^{\text{sol}}$ orthogonal to the entire space $\mathcal{H}^{\text{sol}}$. We prove that for any given $\boldsymbol{\zeta}' \in \mathcal{M}_\gamma^{\text{sol}}$, with $\boldsymbol{\zeta}' \neq 0$ there is at least one field $\boldsymbol{\eta}' \in \mathcal{H}^{\text{sol}}$ such that

$$(\boldsymbol{\zeta}', \boldsymbol{\eta}') \neq 0.$$

For $\gamma > 0$, consider an arbitrary $\boldsymbol{\zeta}' \in \mathcal{M}_\gamma^{\text{sol}}$ such that $\boldsymbol{\zeta}' \neq 0$, and let $\boldsymbol{\eta}'$ be the particular vector field in $\mathcal{H}^{\text{sol}}$ with tangential components on the boundary coincident with those of $\boldsymbol{\zeta}'$, namely, such that $\boldsymbol{n}\times\boldsymbol{\eta}'|_S = \boldsymbol{n}\times\boldsymbol{\zeta}'|_S$. Since $\boldsymbol{\zeta}' = \gamma^{-1}\nabla^2\boldsymbol{\zeta}'$, using Green vector identity (3.36) and exploiting the harmonic character of η', as well as the solenoidal character of both η' and ζ', one obtains

$$
\begin{aligned}
(\zeta',\eta') &= \gamma^{-1}(\nabla^2\zeta',\eta') \\
&= \gamma^{-1}\Big[(\zeta',\nabla^2\eta') \\
&\quad + \oint(n\times\eta'\cdot\nabla\times\zeta' - n\times\zeta'\cdot\nabla\times\eta' + n\cdot\eta'\,\nabla\cdot\zeta' - n\cdot\zeta'\,\nabla\cdot\eta')\,dS\Big] \\
&= \gamma^{-1}\oint(n\times\eta'\cdot\nabla\times\zeta' - n\times\zeta'\cdot\nabla\times\eta')\,dS.
\end{aligned}
$$

Since the tangential components of ζ' and η' have been taken to be coincident on the boundary, $n\times\eta'$ can be interchanged with $n\times\zeta'$, so that, applying the divergence theorem,

$$
\begin{aligned}
(\zeta',\eta') &= \gamma^{-1}\oint(n\times\zeta'\cdot\nabla\times\zeta' - n\times\eta'\cdot\nabla\times\eta')\,dS \\
&= \gamma^{-1}\oint(n\cdot\zeta'\times\nabla\times\zeta' - n\cdot\eta'\times\nabla\times\eta')\,dS \\
&= \gamma^{-1}\int\big(\nabla\cdot(\zeta'\times\nabla\times\zeta') - \nabla\cdot(\eta'\times\nabla\times\eta')\big)dV \\
&= \gamma^{-1}\int\big(|\nabla\times\zeta'|^2 - \zeta'\cdot\nabla\times\nabla\times\zeta' - |\nabla\times\eta'|^2 + \eta'\cdot\nabla\times\nabla\times\eta'\big)dV.
\end{aligned}
$$

Exploiting again the solenoidal character of both ζ' and η' and taking into account the equations satisfied by ζ' and η', one obtains:

$$
\begin{aligned}
(\zeta',\eta') &= \gamma^{-1}\int\big(|\nabla\times\zeta'|^2 + \zeta'\cdot\nabla^2\zeta' - |\nabla\times\eta'|^2 - \eta'\cdot\nabla^2\eta'\big)dV \\
&= \gamma^{-1}\int\big(|\nabla\times\zeta'|^2 + \gamma|\zeta'|^2 - |\nabla\times\eta'|^2\big)\,dV.
\end{aligned}
$$

The Dirichlet principle generalized to the vector Laplace equation $-\nabla^2 F = 0$ gives

$$\int\big(|\nabla\times F|^2 + |\nabla\cdot F|^2\big)\,dV \le \int\big(|\nabla\times G|^2 + |\nabla\cdot G|^2\big)\,dV,$$

where G is any vector field satisfying the same boundary conditions imposed on the solution F of the Laplace equation. In the particular case of solenoidal vector fields, this principle, with F and G replaced respectively by η' and ζ', gives

$$\int|\nabla\times\eta'|^2\,dV \le \int|\nabla\times\zeta'|^2\,dV,$$

Using this inequality in the previous equation gives:

$$(\zeta',\eta') = \gamma^{-1}\int\big(|\nabla\times\zeta'|^2 + \gamma|\zeta'|^2 - |\nabla\times\eta'|^2\big)dV \ge \int|\zeta'|^2\,dV > 0,$$

since it has been assumed $\zeta' \neq 0$. Thus $(\zeta',\eta') > 0$ and there is no nontrivial subspace of $\mathcal{M}_\gamma^{\rm sol}$ orthogonal to $\mathcal{H}^{\rm sol}$. □

The invertibility of the restricted projection operator P^γ allows to express the stream vector ψ, solution to the system (3.39), according to the following *factorization*:

$$\psi = (-\nabla^2_{\nabla\cdot})^{-1}\big[I\!I - (P^\gamma)^{-1} I\!P_{\mathcal{N}(-\widetilde{\nabla}^2_{\nabla\cdot})}\big](-\nabla^2_{\nabla\cdot} + \gamma)^{-1} \boldsymbol{f}. \tag{3.44}$$

Here, $\nabla^2_{\nabla\cdot}$ denotes the Laplace operator supplemented by Dirichlet conditions for the two tangential components of the vector unknown and with the derivative boundary condition of zero divergence; instead $\widetilde{\nabla}^2_{\nabla\cdot}$ denotes the Laplace operator supplemented *only* with the boundary condition of zero divergence. Accordingly, $(-\nabla^2_{\nabla\cdot})^{-1}$ represents the formal solution of a mixed Dirichlet–Neumann problem for a vector unknown, and similarly for the operator $(-\nabla^2_{\nabla\cdot} + \gamma)^{-1}$. (For simplicity, homogeneous boundary conditions have been considered in the factorized expression above.)

3.4.5 Decomposition scheme and Glowinski–Pironneau method

A manner of performing the orthogonal projection required to satisfy the vector integral conditions consists in decomposing the solution $\zeta(\boldsymbol{x})$ to (3.39) in the following form:

$$\zeta(\boldsymbol{x}) = \zeta^0(\boldsymbol{x}) + \sum_{k'=1,2} \oint \zeta'_{k'}(\boldsymbol{x};\boldsymbol{\sigma}')\lambda_{k'}(\boldsymbol{\sigma}')\, dS(\boldsymbol{\sigma}'). \tag{3.45}$$

Here the vector fields $\zeta^0(\boldsymbol{x})$ and $\zeta'_{k'}(\boldsymbol{x};\boldsymbol{\sigma}')$, $k = 1,2$, for any $\boldsymbol{\sigma}' \in S$, give the same decomposition depicted in Figure 3.1 and which underlies the factorization (3.44). These fields are the solutions to the two following problems

$$(-\nabla^2 + \gamma)\zeta^0 = \boldsymbol{f}, \qquad \boldsymbol{n}\times\zeta^0|_S = 0, \quad \nabla\cdot\zeta^0|_S = 0, \tag{3.46}$$

and

$$(-\nabla^2 + \gamma)\zeta'_{k'} = 0, \qquad \boldsymbol{\tau}_h\cdot\zeta'_{k'}(\boldsymbol{x};\boldsymbol{\sigma}')|_S = \delta_{hk'}\delta^{(2)}(\boldsymbol{s}-\boldsymbol{\sigma}'), \quad \nabla\cdot\zeta'_{k'}|_S = 0, \tag{3.47}$$

where $\boldsymbol{\tau}_h$, $h = 1,2$, are the two unit vectors tangential to the surface S, δ_{hk} is the Kronecker delta and $\delta^{(2)}$ denotes the Dirac delta distribution over S. Thus, the decomposition involves a summation over the two tangential components of the vector field on the boundary. Correspondingly, the boundary unknown has two components for each point of S, namely, $\boldsymbol{\lambda}(\boldsymbol{\sigma}) = (\lambda_1(\boldsymbol{\sigma}), \lambda_2(\boldsymbol{\sigma}))$, with $\boldsymbol{\sigma} \in S$. Such an unknown is determined by imposing that ζ satisfy the integral conditions in (3.39), with respect to all the linearly independent harmonic vector fields $\boldsymbol{\eta}_k(\boldsymbol{x};\boldsymbol{\sigma})$, $k = 1,2$, for any $\boldsymbol{\sigma} \in S$, defined by the problem

$$-\nabla^2\boldsymbol{\eta}_k = 0, \qquad \boldsymbol{\tau}_h\cdot\boldsymbol{\eta}_k(\boldsymbol{x};\boldsymbol{\sigma})|_S = \delta_{hk}\,\delta^{(2)}(\boldsymbol{s}-\boldsymbol{\sigma}), \quad \nabla\cdot\boldsymbol{\eta}_k|_S = 0, \tag{3.48}$$

where $h, k = 1, 2$. This gives the linear problem

$$\sum_{k'=1,2} \oint A_{kk'}(\boldsymbol{\sigma}, \boldsymbol{\sigma}')\lambda_{k'}(\boldsymbol{\sigma}')\, dS(\boldsymbol{\sigma}') = \beta_k(\boldsymbol{\sigma}), \tag{3.49}$$

where

$$A_{kk'}(\boldsymbol{\sigma}, \boldsymbol{\sigma}') = \int \boldsymbol{\zeta}'_{k'}(\boldsymbol{x}; \boldsymbol{\sigma}') \cdot \boldsymbol{\eta}_k(\boldsymbol{x}; \boldsymbol{\sigma})\, dV \tag{3.50}$$

and

$$\begin{aligned}\beta_k(\boldsymbol{\sigma}) = &- \int \boldsymbol{\zeta}^0(\boldsymbol{x}) \cdot \boldsymbol{\eta}_k(\boldsymbol{x}; \boldsymbol{\sigma})\, dV \\ &+ \oint \Big[\boldsymbol{n}\times\boldsymbol{a}(\boldsymbol{x}_S) \cdot \boldsymbol{\nabla}\times\boldsymbol{\eta}_k(\boldsymbol{x}_S; \boldsymbol{\sigma}) + \boldsymbol{n}\times\boldsymbol{b}(\boldsymbol{x}_S) \cdot \boldsymbol{\eta}_k(\boldsymbol{x}_S; \boldsymbol{\sigma})\Big] dS.\end{aligned} \tag{3.51}$$

Such a computational scheme is therefore based on the solution of elliptic problems for the operators $-\nabla^2$ and $(-\nabla^2 + \gamma)$ with mixed Dirichlet–Neumann boundary conditions, together with the additional linear problem (3.49) to determine the unknown $\boldsymbol{\lambda}$. In the spatially discretized case, the matrix corresponding to the linear operator $A_{kk'}(\boldsymbol{\sigma}, \boldsymbol{\sigma}')$ is symmetric. As for plane problems, the operator $A_{kk'}(\boldsymbol{\sigma}, \boldsymbol{\sigma}')$ depends only on the geometry of the domain through the functions $\boldsymbol{\zeta}'$ and $\boldsymbol{\eta}$. Therefore, it can be determined once and for all at the beginning of the calculations and used subsequently to solve problems with different source terms and/or boundary data.

This procedure is not very convenient from the computational viewpoint since it requires to calculate and store the vector fields $\boldsymbol{\eta}_k$ in the computer memory. This inconvenience can be avoided, at the cost of doubling the total number of elliptic problem to be solved, by generalizing the method introduced by Glowinski and Pironneau for plane problems to the present context of three-dimensional vector equations. After the vorticity problems (3.46) and (3.47) have been solved, the following problems for the stream vector are considered and solved:

$$-\nabla^2\boldsymbol{\psi}^0 = \boldsymbol{\zeta}^0, \qquad \boldsymbol{n}\times\boldsymbol{\psi}^0|_S = \boldsymbol{n}\times\boldsymbol{a}, \quad \boldsymbol{\nabla}\cdot\boldsymbol{\psi}^0|_S = 0; \tag{3.52}$$

and

$$-\nabla^2\boldsymbol{\psi}'_{k'} = \boldsymbol{\zeta}'_{k'}, \qquad \boldsymbol{n}\times\boldsymbol{\psi}'_{k'}|_S = 0, \quad \boldsymbol{\nabla}\cdot\boldsymbol{\psi}'_{k'}|_S = 0. \tag{3.53}$$

Then, for each $\boldsymbol{\sigma} \in S$ the auxiliary vector field

$$\begin{aligned}&\boldsymbol{w}_k(\boldsymbol{x}; \boldsymbol{\sigma}) \text{ arbitrary in } V, \\ &\boldsymbol{\tau}_h \cdot \boldsymbol{w}_k(\boldsymbol{x}; \boldsymbol{\sigma})|_S = \delta_{hk}\,\delta^{(2)}(\boldsymbol{s} - \boldsymbol{\sigma}), \qquad \boldsymbol{n}\cdot\boldsymbol{w}_k(\boldsymbol{x}; \boldsymbol{\sigma})|_S = 0\end{aligned} \tag{3.54}$$

is introduced, where $h, k = 1, 2$. It is to be noted that the boundary conditions in the definition of $\boldsymbol{w}_k$ do not involve the derivative condition of zero divergence which is instead imposed on the harmonic vector fields $\boldsymbol{\eta}_k$.

The introduction of the auxiliary fields $\boldsymbol{w}_k$ allows to characterize the quantities $A_{kk'}$ and β_k in an equivalent form not involving the harmonic fields $\boldsymbol{\eta}_k$. In fact, using Eq. (3.53) and Green identity (3.36) with $\psi = \psi_{k'}'$ and $\phi = \eta_k$, since η_k is harmonic, one obtains

$$A_{kk'}(\sigma, \sigma') = \int (\zeta_{k'}' \cdot w_k - \nabla\times\psi_{k'}' \cdot \nabla\times w_k)\, dV. \tag{3.55}$$

A similar calculation can be performed to express β_k without using η_k, by introducing the solution ψ^0 of problem (3.52). The application of the same Green identity to expression (3.51) gives

$$\beta_k(\sigma) = -\int (\zeta^0 \cdot w_k - \nabla\times\psi^0 \cdot \nabla\times w_k)\, dV + \oint n\times b \cdot w_k\, dS. \tag{3.56}$$

The arbitrariness of the fields w_k at all the internal points of V is exploited by choosing $w_k = 0$ inside V so that the integration domain in the two previous relations becomes a narrow shell along the boundary. Once the linear problem $A\lambda = \beta$ has been solved, the sought solution (ζ, ψ) is determined by solving the two elliptic vector problems:

$$(-\nabla^2 + \gamma)\zeta = f, \qquad \tau_k \cdot \zeta|_S = \lambda_k,\ k = 1, 2, \quad \nabla\cdot\zeta|_S = 0, \tag{3.57}$$

and

$$-\nabla^2\psi = \zeta, \qquad n\times\psi|_S = n\times a, \quad \nabla\cdot\psi|_S = 0. \tag{3.58}$$

The entire procedure represents the three-dimensional counterpart of that described in section 2.9.3 for plane two-dimensional problems.

3.4.6 Variational solution of vector elliptic equations

To make the computational scheme above practicable in the framework of a finite-element-based spatial discretization, the various vector elliptic equations involved must be recast into a variational form. Consider, for example, the typical vorticity problem

$$(-\nabla^2 + \gamma)\zeta = g, \qquad n\times\zeta|_S = n\times c, \quad \nabla\cdot\zeta|_S = e, \tag{3.59}$$

where g is a given vector field defined in V whereas $n\times c$ and e are data prescribed on S. The variational form of problem (3.59) is obtained by multiplying the equation by suitable weighting vector functions φ and by integrating it over V. Using the vector identity $-\nabla^2\alpha = \nabla\times\nabla\times\alpha - \nabla(\nabla\cdot\alpha)$ and integrating by parts, one obtains

$$(\nabla\times\varphi, \nabla\times\zeta) + (\nabla\cdot\varphi, \nabla\cdot\zeta) + \gamma(\varphi, \zeta) = (\varphi, g) + \oint (n\times\varphi \cdot \nabla\times\zeta + n\cdot\varphi\, \nabla\cdot\zeta)\, dS. \tag{3.60}$$

In particular, for the boundary conditions of problem (3.59), the variational formulation reads:

Find $\boldsymbol{\zeta}$ with $\boldsymbol{n}\times\boldsymbol{\zeta}|_S = \boldsymbol{n}\times\boldsymbol{c}$ such that

$$(\nabla\times\boldsymbol{\varphi}, \nabla\times\boldsymbol{\zeta}) + (\nabla\cdot\boldsymbol{\varphi}, \nabla\cdot\boldsymbol{\zeta}) + \gamma(\boldsymbol{\varphi}, \boldsymbol{\zeta}) = (\boldsymbol{\varphi}, \boldsymbol{g}) + \oint \boldsymbol{n}\cdot\boldsymbol{\varphi}\, e\, dS, \qquad (3.61)$$

for any $\boldsymbol{\varphi}$ such that $\boldsymbol{n}\times\boldsymbol{\varphi}|_S = 0$. In this formulation, the tangential boundary conditions are *essential* whereas the derivative condition $\nabla\cdot\boldsymbol{\zeta}|_S = e$ plays the rôle of a *natural* boundary condition (see, *e.g.*, Strang and Fix [33]).

The important fact about this problem is that it constitutes a *coupled* system of equations, in the sense that the three Cartesian components of the unknown $\boldsymbol{\zeta}$ cannot be solved separately from each other. A similar coupled system is obtained when the vector Poisson equation for $\boldsymbol{\psi}$ is written in variational form. Therefore, while the introduction of the integral conditions for the vector vorticity allows to split, at least in the linear case, the vorticity equation from the stream vector equation, the problems for each of these two vector variables are characterized by a coupled system of equations for the three Cartesian components of the unknown. This is due to the presence of the derivative boundary conditions $\nabla\cdot\boldsymbol{\zeta}|_S = 0$ and $\nabla\cdot\boldsymbol{\psi}|_S = 0$, which are both required to assure the solenoidal character of the two fields. As pointed out previously, in the differential form of the problem, the coupling between the vector components occurs only on S, due to the boundary condition of zero divergence. The use of the variational formulation has the effect of reintroducing the coupling at the level of the (weak) governing equations. As a consequence, the numerical solution of the $\boldsymbol{\zeta}$-$\boldsymbol{\psi}$ (or $\boldsymbol{\zeta}$-$\boldsymbol{A}$) equations for three-dimensional problems in domains of arbitrary shape represents a by-far more formidable task as compared to the ζ-ψ equations in two dimensions, unless an uncoupling strategy can be adopted for solving vector elliptic equations with the divergence of the unknown prescribed on the boundary.

In the following, we describe a method for uncoupling the equations of the vector components, which is based on the *influence matrix* technique. The method reduces the solution of the vector elliptic problem (3.59) to a sequence of scalar Dirichlet problems plus an additional linear system for determining the unknown value of the normal component of $\boldsymbol{\zeta}$ on the boundary, $\omega = \boldsymbol{n}\cdot\boldsymbol{\zeta}|_S$, in terms of the available boundary datum $e = \nabla\cdot\boldsymbol{\zeta}|_S$ (Quartapelle and Muzzio, 1988). Using the superposition principle, let the solution $\boldsymbol{\zeta}$ of the problem (3.59) be expressed by the linear combination

$$\boldsymbol{\zeta}(\boldsymbol{x}) = \boldsymbol{\zeta}^0(\boldsymbol{x}) + \oint \boldsymbol{\zeta}'(\boldsymbol{x};\boldsymbol{\sigma}')\omega(\boldsymbol{\sigma}')\, dS(\boldsymbol{\sigma}'), \qquad (3.62)$$

where the fields $\boldsymbol{\zeta}^0$ and $\boldsymbol{\zeta}'$ are the solutions to the (purely) Dirichlet vector problems:

$$(-\nabla^2+\gamma)\boldsymbol{\zeta}^0 = \boldsymbol{g}, \qquad \boldsymbol{n}\times\boldsymbol{\zeta}^0|_S = \boldsymbol{n}\times\boldsymbol{c},\ \boldsymbol{n}\cdot\boldsymbol{\zeta}^0|_S = 0; \qquad (3.63a)$$

$$(-\nabla^2+\gamma)\boldsymbol{\zeta}' = 0, \qquad \boldsymbol{n}\times\boldsymbol{\zeta}'|_S = 0, \qquad \boldsymbol{n}\cdot\boldsymbol{\zeta}'|_S = \delta^{(2)}(\boldsymbol{s}-\boldsymbol{\sigma}'). \quad (3.63b)$$

(The symbols $\boldsymbol{\zeta}^0$ and $\boldsymbol{\zeta}'$ used here should not be confused with those employed in the decomposition introduced in the previous section to satisfy the vorticity integral conditions.) The scalar unknown ω defined on S is determined by imposing the linear combination (3.62) to satisfy the derivative boundary condition in (3.59), which gives

$$\nabla\cdot\boldsymbol{\zeta}^0|_S + \oint \nabla\cdot\boldsymbol{\zeta}'|_S\,\omega(\boldsymbol{\sigma}')\,dS(\boldsymbol{\sigma}') = e. \tag{3.64}$$

To employ such an equation in conjunction with a finite element approximation, the relationship must be expressed in a weak form. Thus, let us consider the auxiliary vector fields $\boldsymbol{y}(\boldsymbol{x};\boldsymbol{\sigma})$, for each $\boldsymbol{\sigma} \in S$, such that

$$\boldsymbol{y}(\boldsymbol{x};\boldsymbol{\sigma}) \text{ arbitrary in } V, \quad \boldsymbol{n}\times\boldsymbol{y}(\boldsymbol{x};\boldsymbol{\sigma})|_S = 0,\ \boldsymbol{n}\cdot\boldsymbol{y}(\boldsymbol{x};\boldsymbol{\sigma})|_S = \delta^{(2)}(\boldsymbol{s}-\boldsymbol{\sigma}), \tag{3.65}$$

and multiply the previous integral equation for ω by $\boldsymbol{n}\cdot\boldsymbol{y}|_S$. By integrating over S and interchanging the order of the two integrals, one obtains the linear problem

$$\oint B(\boldsymbol{\sigma},\boldsymbol{\sigma}')\omega(\boldsymbol{\sigma}')\,dS(\boldsymbol{\sigma}') = b(\boldsymbol{\sigma}), \tag{3.66}$$

where

$$B(\boldsymbol{\sigma},\boldsymbol{\sigma}') = \oint \nabla\cdot\boldsymbol{\zeta}'(\boldsymbol{x}_S;\boldsymbol{\sigma}')\,\boldsymbol{n}\cdot\boldsymbol{y}(\boldsymbol{x}_S;\boldsymbol{\sigma})\,dS, \tag{3.67a}$$

$$b(\boldsymbol{\sigma}) = \oint \Big[-\nabla\cdot\boldsymbol{\zeta}^0(\boldsymbol{x}_S) + e(\boldsymbol{x}_S)\Big]\boldsymbol{n}\cdot\boldsymbol{y}(\boldsymbol{x}_S;\boldsymbol{\sigma})\,dS. \tag{3.67b}$$

The divergence theorem can now be applied to characterize B and b in terms of volume integrals, instead of surface integrals. Using the vector identity $-\nabla^2\boldsymbol{\alpha} = \nabla\times\nabla\times\boldsymbol{\alpha} - \nabla(\nabla\cdot\boldsymbol{\alpha})$ and taking into account equations (3.63a) and (3.63b), one readily obtains

$$B(\boldsymbol{\sigma},\boldsymbol{\sigma}') = \int (\nabla\times\boldsymbol{\zeta}'\cdot\nabla\times\boldsymbol{y} + \nabla\cdot\boldsymbol{\zeta}'\,\nabla\cdot\boldsymbol{y})\,dV, \tag{3.68a}$$

$$b(\boldsymbol{\sigma}) = -\int (\nabla\times\boldsymbol{\zeta}^0\cdot\nabla\times\boldsymbol{y} + \nabla\cdot\boldsymbol{\zeta}^0\,\nabla\cdot\boldsymbol{y} - \boldsymbol{g}\cdot\boldsymbol{y})\,dV + \oint e\,\boldsymbol{n}\cdot\boldsymbol{y}\,dS \tag{3.68b}$$

Once the linear system $B\omega = b$ has been solved, the solution $\boldsymbol{\zeta}$ of the original elliptic vector problem can be obtained by solving a Dirichlet vector problem with ω as the normal boundary value and with the available datum $\boldsymbol{n}\times\boldsymbol{c}$ as tangential boundary condition.

The entire procedure and the analogous one leading to the linear system $C\pi = c$ for determining the boundary value of the normal component of the stream vector, $\pi = \boldsymbol{n}\cdot\boldsymbol{\psi}|_S$, so as to make $\boldsymbol{\psi}$ solenoidal, can be used in the decomposition scheme described in section 3.4.5. In this way, one can circumvent the coupling between the Cartesian components of $\boldsymbol{\zeta}$ and $\boldsymbol{\psi}$, brought about by the two derivative boundary conditions $\nabla\cdot\boldsymbol{\zeta}|_S = 0$ and $\nabla\cdot\boldsymbol{\psi}|_S = 0$. It is important to remark that, if such an uncoupling strategy is adopted for the solution of the

$\boldsymbol{\zeta}$-$\boldsymbol{\psi}$ equations subject to the vorticity integral conditions, the two linear systems $B\omega = b$ and $C\pi = c$ for determining the normal component of the vorticity and stream vector on the boundary have to be solved when calculating each pair $(\boldsymbol{\zeta}', \boldsymbol{\psi}')$ involved in the construction of the operator $A_{kk'}$ for the linear problem $A\boldsymbol{\lambda} = \boldsymbol{\beta}$. Therefore, the overall procedure for satisfying the vorticity integral conditions would require the preliminary determination and factorization of the matrices B and C in order to solve for the pair $(\boldsymbol{\zeta}', \boldsymbol{\psi}')$ in a subsequent phase. Once the matrix corresponding to the operator $A_{kk'}$ has been determined and factorized, the pair $(\boldsymbol{\zeta}^0, \boldsymbol{\psi}^0)$ is determined through the solution of the two linear systems $B\omega = b$ and $C\pi = c$, since $\boldsymbol{\zeta}^0$ and $\boldsymbol{\psi}^0$ must be solenoidal [cf. (3.46) and (3.52)]. At this point, one can determine the tangential components $\boldsymbol{\lambda} = (\lambda_1, \lambda_2)$ of the vorticity on the boundary by solving the linear problem $A\boldsymbol{\lambda} = \boldsymbol{\beta}$. The sought pair $(\boldsymbol{\zeta}, \boldsymbol{\psi})$ is finally obtained by imposing the now determined boundary values of the tangential vorticity, again through another solution of linear systems $B\omega = b$ and $C\pi = c$ to obtain $\boldsymbol{\zeta}$ and $\boldsymbol{\psi}$ solenoidal [cf. (3.57) and (3.58)].

The complete procedure is computationally expensive due to the "nesting" of the operators B and C, required to impose the solenoidality of the fields, within the construction of the operator A introduced for satisfying the vorticity integral conditions. On the other hand, the entire precedure is logically clear and, in principle, allows a direct, *i.e.*, noniterative solution of the three-dimensional equations for bounded domains of arbitrary shape.

3.4.7 Pressure determination

We conclude the section on the q_S-$\boldsymbol{\zeta}$-$\boldsymbol{\psi}$ formulation by describing how the pressure field can be calculated once the fields $\boldsymbol{\zeta}$ and $\boldsymbol{\psi}$ have been determined. The simplest method consists in the *line* integration of the momentum equation written in the following form, with $\boldsymbol{u} = \boldsymbol{\nabla}\times\boldsymbol{\psi}$,

$$-\boldsymbol{\nabla}P_{\text{tot}} = \frac{\partial\boldsymbol{\nabla}\times\boldsymbol{\psi}}{\partial t} + \nu\boldsymbol{\nabla}\times\boldsymbol{\zeta} + \boldsymbol{\zeta}\times\boldsymbol{\nabla}\times\boldsymbol{\psi},$$

where, as usual, $P_{\text{tot}} = P + \frac{1}{2}u^2 = P + \frac{1}{2}|\boldsymbol{\nabla}\times\boldsymbol{\psi}|^2$. Note that the viscous term does not appear in its standard form involving the Laplacian, but only as the first derivative of the vorticity. Of course, the line integral of the expression above determines the field P_{tot} only up to an arbitrary additive function of time. This line integration procedure is nothing but the extension to unsteady three-dimensional problems of the method introduced by Burggraf (1966) for steady flows in two dimensions, whicht has been mentioned in the previous chapter.

In spite of the conceptual simplicity of the aforementioned method, it is usually preferred to reconstruct the pressure field by solving a Poisson equation for P or P_{tot}. Such an equation is easily obtained from the momentum equation provided that the vorticity field has been already determined. In this case, the momentum equation can be written as

$$\frac{\partial \boldsymbol{u}}{\partial t} + \boldsymbol{\zeta} \times \boldsymbol{u} = -\boldsymbol{\nabla} P_{\text{tot}} - \nu \boldsymbol{\nabla} \times \boldsymbol{\zeta}.$$

Then, assuming also that the condition of incompressibility is satisfied, the divergence of the momentum equation gives

$$-\nabla^2 P_{\text{tot}} = \boldsymbol{\nabla} \cdot (\boldsymbol{\zeta} \times \boldsymbol{u}).$$

The boundary condition to be associated with this Poisson equation is derived by taking the normal component on the boundary of the momentum equation above, which, we insist, does not involve the Laplace operator. Using the velocity boundary condition and reordering the various term gives the following Neumann condition (cf., *e.g.*, Abdallah 1987):

$$\boldsymbol{n} \cdot \boldsymbol{\nabla} P_{\text{tot}}|_S = -\boldsymbol{n} \cdot \boldsymbol{\zeta} \times \boldsymbol{b}|_S - \nu \boldsymbol{n} \cdot \boldsymbol{\nabla} \times \boldsymbol{\zeta}|_S - \frac{\partial \boldsymbol{n} \cdot \boldsymbol{b}}{\partial t}.$$

It is important to remark that it is not possible to use such a Neumann condition as a means for developing a method for the solution of the Navier–Stokes equations in which the condition of incompressibility is enforced *via* an elliptic equation for pressure. In fact, even disregarding the nonlinear term, the Neumann condition requires a knowledge of the tangential components of the vorticity on S, to evaluate the viscous contribution $\nu \boldsymbol{n} \cdot \boldsymbol{\nabla} \times \boldsymbol{\zeta}|_S$. Therefore, the actual use of this condition would require that the vorticity field $\boldsymbol{\zeta}$ be known, which would require in turn that the Navier–Stokes problem had been already solved. In this connection, we anticipate that the appropriate conditions for the pressure which supplement the Poisson equation only in terms of the data of the problem will be discussed in chapters 5 and 6, dealing with the primitive variable incompressible equations with a discrete and continuous time dependence, respectively.

When the Neumann problem for the pressure is expressed in a variational form, the first term on the right-hand side of the boundary condition disappears by virtue of the integration by parts. The final variational form of the problem then reads: find $P_{\text{tot}} \in H^1(V)$ such that

$$(\boldsymbol{\nabla} Q, \boldsymbol{\nabla} P_{\text{tot}}) = -(\boldsymbol{\nabla} Q, \boldsymbol{\zeta} \times \boldsymbol{u}) - \nu \oint Q \, \boldsymbol{n} \cdot \boldsymbol{\nabla} \times \boldsymbol{\zeta} \, dS - \frac{d}{dt} \oint Q \, \boldsymbol{n} \cdot \boldsymbol{b} \, dS,$$

for any $Q \in H^1(V)$.

3.5 Irreducible vorticity integral conditions

The most important difference between the vorticity integral conditions for two-dimensional problems and the corresponding conditions established for the three-dimensional equations is that in three dimensions there are two integral conditions per each boundary point, whereas only one exists in two dimensions. This occurs because in three dimensions there are two tangential components of $\boldsymbol{\zeta}$ to be determined on no-slip boundaries, whereas in plane or axially symmetric problems the vorticity vector reduces to a scalar variable.

On the other hand, in chapter 5, the primitive variable Navier–Stokes equations will be shown to be equivalent, after a time discretization of nonfractional-step type, to a system of (second-order) elliptic equations for the variables velocity and pressure. Also, it will be shown that the same orthogonality argument used so far to derive the vorticity conditions leads to discover the conditions supplementing the Poisson equation for the pressure in nonfractional-step discretizations of the primitive variable equations. Needless to say, these pressure conditions are found to be of an integral character and to involve a linear space of suitable vector fields. The "number" of linearly independent fields, and hence of integral conditions, is found to correspond to the number of boundary points, so that these conditions serve the purpose of determining, jointly with the Laplace operator, the value of the pressure on no-slip boundaries. Of course, in the primitive variable representation of the Navier–Stokes equations no structural difference exists between two and three dimensions. Now the intriguing question arises: how can the "double" linear space involved in the vorticity integral conditions for three-dimensional flows match the half-sized space involved in the pressure integral conditions, when going from the nonprimitive variable formulation to the primitive variable one?

In this section this theoretical question is examined and an orthogonal decomposition of the space $\boldsymbol{\mathcal{H}}^{\text{sol}}$ is introduced which sheds some light on the transition between the two aforementioned representations of the Navier–Stokes equations. In fact, the proposed decomposition allows to halve the number of the vorticity integral conditions in three-dimensional problems and to isolate their *irreducible component* which is necessary to consider in incompressible viscous flows.

3.5.1 Orthogonal decomposition of the projection space

Let us recall the definition (3.40) of the space $\boldsymbol{\mathcal{H}}^{\text{sol}}$ involved in the vector vorticity integral conditions, namely,

$$\boldsymbol{\mathcal{H}}^{\text{sol}} = \left\{ \boldsymbol{\eta} \,|\, -\nabla^2 \boldsymbol{\eta} = 0, \quad \boldsymbol{\nabla} \cdot \boldsymbol{\eta}|_S = 0 \right\}, \tag{3.69}$$

whose elements are the vector fields $\boldsymbol{\eta}$ that are harmonic and solenoidal in V. Let us decompose each $\boldsymbol{\eta} \in \boldsymbol{\mathcal{H}}^{\text{sol}}$ as follows

$$\boldsymbol{\eta} = \boldsymbol{\nabla}\chi + \overline{\boldsymbol{\eta}}, \tag{3.70}$$

where the function χ is constructed directly as solution of the Neumann problem

$$-\nabla^2 \chi = 0, \qquad \boldsymbol{n} \cdot \boldsymbol{\nabla}\chi|_S = \boldsymbol{n} \cdot \boldsymbol{\eta}|_S.$$

The other component $\overline{\boldsymbol{\eta}}$ is thus given by $\overline{\boldsymbol{\eta}} = \boldsymbol{\eta} - \boldsymbol{\nabla}\chi$ and satisfies the boundary condition $\boldsymbol{n} \cdot \overline{\boldsymbol{\eta}}|_S = 0$ in addition to $\boldsymbol{\nabla} \cdot \overline{\boldsymbol{\eta}}|_S = 0$. The two components of $\boldsymbol{\eta}$ so introduced gives an *orthogonal* decomposition in the $\boldsymbol{L}^2$ sense of $\boldsymbol{\mathcal{H}}^{\text{sol}}$. In fact, using a well known identity together the condition $\boldsymbol{\nabla} \cdot \overline{\boldsymbol{\eta}} = 0$ ($\overline{\boldsymbol{\eta}} \in \boldsymbol{\mathcal{H}}^{\text{sol}}$) and the divergence theorem, one has

$$\int \boldsymbol{\nabla}\chi \cdot \overline{\boldsymbol{\eta}}\, dV = \int [\boldsymbol{\nabla} \cdot (\chi \overline{\boldsymbol{\eta}}) - \chi \boldsymbol{\nabla} \cdot \overline{\boldsymbol{\eta}}]\, dV = \oint \boldsymbol{n} \cdot \overline{\boldsymbol{\eta}}\, \chi\, dS = 0,$$

the last equality being a consequence of the boundary condition $\boldsymbol{n} \cdot \overline{\boldsymbol{\eta}}|_S = 0$.

Therefore, introducing the two linear spaces

$$\boldsymbol{\nabla}\mathcal{H} = \left\{ \boldsymbol{\nabla}\chi \,|\, -\nabla^2 \chi = 0 \right\} \tag{3.71}$$

and

$$\boldsymbol{\mathcal{H}}^{\text{sol}}_{\text{tan}} = \left\{ \overline{\boldsymbol{\eta}} \,|\, -\nabla^2 \overline{\boldsymbol{\eta}} = 0, \quad \boldsymbol{n} \cdot \overline{\boldsymbol{\eta}}|_S = 0, \quad \boldsymbol{\nabla} \cdot \overline{\boldsymbol{\eta}}|_S = 0 \right\}, \tag{3.72}$$

the decomposition (3.70) can be denoted in the form

$$\boldsymbol{\mathcal{H}}^{\text{sol}} = \boldsymbol{\nabla}\mathcal{H} \oplus \boldsymbol{\mathcal{H}}^{\text{sol}}_{\text{tan}}. \tag{3.73}$$

This orthogonal decomposition leads to a reduction of the integral conditions to be imposed on the vorticity, since the component of the conditions involving the harmonic fields in the subspace $\boldsymbol{\nabla}\mathcal{H}$ can be shown to be equivalent to a condition of boundary value type for the vorticity.

Consider $\boldsymbol{\eta} \in \boldsymbol{\nabla}\mathcal{H}$ and substitute $\boldsymbol{\eta} = \boldsymbol{\nabla}\chi$ in the integral condition (3.37), to give

$$\begin{aligned} \int \boldsymbol{\zeta} \cdot \boldsymbol{\nabla}\chi\, dV &= \oint [\boldsymbol{n} \times \boldsymbol{a} \cdot \boldsymbol{\nabla} \times (\boldsymbol{\nabla}\chi) + \boldsymbol{n} \times \boldsymbol{b} \cdot \boldsymbol{\nabla}\chi]\, dS \\ &= \oint \boldsymbol{n} \times \boldsymbol{b} \cdot \boldsymbol{\nabla}\chi\, dS = \oint \boldsymbol{n} \cdot \boldsymbol{b} \times \boldsymbol{\nabla}\chi\, dS. \end{aligned}$$

By the vector identity $\boldsymbol{\nabla} \cdot (\boldsymbol{\zeta}\chi) = (\boldsymbol{\nabla} \cdot \boldsymbol{\zeta})\chi + \boldsymbol{\zeta} \cdot \boldsymbol{\nabla}\chi$ with $\boldsymbol{\nabla} \cdot \boldsymbol{\zeta} = 0$, the left-hand side of the relation above can be written as $\int \boldsymbol{\nabla} \cdot (\boldsymbol{\zeta}\chi)\, dV$. Thus, using the identity $\boldsymbol{n} \cdot \boldsymbol{\nabla} \times (\boldsymbol{b}\chi) = \boldsymbol{n} \cdot (\boldsymbol{\nabla}_S \times \boldsymbol{b}\,\chi - \boldsymbol{b} \times \boldsymbol{\nabla}\chi)$ in the right-hand side gives:

$$\int \boldsymbol{\nabla} \cdot (\boldsymbol{\zeta}\chi)\, dV = \oint \boldsymbol{n} \cdot \boldsymbol{\nabla}_S \times \boldsymbol{b}\, \chi\, dS - \oint \boldsymbol{n} \cdot \boldsymbol{\nabla} \times (\boldsymbol{b}\chi)\, dS.$$

Now, the divergence theorem is used first to transform the left-hand side into the boundary integral $\oint \boldsymbol{n} \cdot \boldsymbol{\zeta}\chi\, dS$, and then to show that the second boundary integral on the right-hand side vanishes since $\int \boldsymbol{\nabla} \cdot \boldsymbol{\nabla} \times (\widehat{\boldsymbol{b}\chi})\, dV = 0$, where $\widehat{\boldsymbol{b}\chi}$ represents any extension of $\boldsymbol{b}\chi$ to V. It follows that

$$\oint \boldsymbol{n} \cdot \boldsymbol{\zeta}\chi\, dV = \oint \boldsymbol{n} \cdot \boldsymbol{\nabla}_S \times \boldsymbol{b}\, \chi\, dS.$$

By the arbitrariness of the boundary value of χ, this relation is equivalent to the following boundary condition for $\boldsymbol{\zeta}$

$$\boldsymbol{n}\cdot\boldsymbol{\zeta}|_S = \boldsymbol{n}\cdot\boldsymbol{\nabla}_S\times\boldsymbol{b}. \tag{3.74}$$

Therefore, the vorticity integral conditions with respect to the subspace $\boldsymbol{\nabla}\mathcal{H}$ are equivalent to the boundary condition (3.74) for the normal component of the vorticity. This derivation clearly shows that boundary condition (3.74) is not independent from the vorticity integral conditions. As a consequence, the integral condition (3.37) together with the boundary condition (3.74) do not provide a complete set of conditions for the vorticity (cf. Quartapelle and Valz-Gris 1981). It is the boundary condition (3.4) for divergence of $\boldsymbol{\zeta}$ that has to be combined with the integral condition (3.37) to obtain a complete set of conditions for $\boldsymbol{\zeta}$, see the formulation of Dennis and Quartapelle (1985) for problems flows in a spherical gap and also the discussion of Morino (1986).

If the boundary condition (3.74) for the normal component is used to supplement the vorticity equation, it is sufficient to satisfy the vorticity integral conditions only with respect to the space $\boldsymbol{\mathcal{H}}_{\text{tan}}^{\text{sol}}$. In this way one is led to the *irreducible component* of the integral conditions, called also *reduced* integral conditions, which assumes the form

$$\int \boldsymbol{\zeta}\cdot\boldsymbol{\overline{\eta}}\, dV = \oint (\boldsymbol{n}\times\boldsymbol{a}\cdot\boldsymbol{\nabla}\times\boldsymbol{\overline{\eta}} + \boldsymbol{n}\times\boldsymbol{b}\cdot\boldsymbol{\overline{\eta}})\, dS. \tag{3.75}$$

The result just obtained is important. It demonstrates that the concept of orthogonality is essential not only for deriving the proper conditions satisfied by the vorticity field but even in order to decompose the projection space and isolate the subspace associated with the component of the vorticity integral conditions which cannot be reduced to the form of standard conditions of boundary value type.

3.5.2 Uncoupled formulation with irreducible integral conditions

The complete set of equations and conditions governing the variables $\boldsymbol{\zeta}$ and $\boldsymbol{\psi}$, when the boundary condition for the normal component of vorticity is imposed, are

$$\begin{aligned}
&-\nu\nabla^2\boldsymbol{\zeta} + \frac{\partial\boldsymbol{\zeta}}{\partial t} + \boldsymbol{\nabla}\times(\boldsymbol{\zeta}\times\boldsymbol{\nabla}\times\boldsymbol{\psi}) = 0, && \boldsymbol{\zeta}|_{t=0} = \boldsymbol{\nabla}\times\boldsymbol{u}_0, \\
&\int \boldsymbol{\zeta}\cdot\boldsymbol{\overline{\eta}}\, dV = \oint (\boldsymbol{n}\times\boldsymbol{a}\cdot\boldsymbol{\nabla}\times\boldsymbol{\overline{\eta}} + \boldsymbol{n}\times\boldsymbol{b}\cdot\boldsymbol{\overline{\eta}})\, dS, && \boldsymbol{n}\cdot\boldsymbol{\zeta}|_S = \boldsymbol{n}\cdot\boldsymbol{\nabla}_S\times\boldsymbol{b}, \quad \boldsymbol{\nabla}\cdot\boldsymbol{\zeta}|_S = 0, \\
&\Big[-\nabla^2\boldsymbol{\overline{\eta}} = 0, && \boldsymbol{n}\cdot\boldsymbol{\overline{\eta}}|_S = 0, \quad \boldsymbol{\nabla}\cdot\boldsymbol{\overline{\eta}}|_S = 0; \Big] \\
&-\nabla^2\boldsymbol{\psi} = \boldsymbol{\zeta}, && \boldsymbol{n}\times\boldsymbol{\psi}|_S = \boldsymbol{n}\times\boldsymbol{a}, \quad \boldsymbol{\nabla}\cdot\boldsymbol{\psi}|_S = 0.
\end{aligned} \tag{3.76}$$

An interesting feature of such a formulation is that the same kind of boundary conditions are imposed on $\boldsymbol{\zeta}$ and $\overline{\boldsymbol{\eta}}$.

Considering now the case of a time discretized version of the problem (3.76), when the operation of orthogonal projection for satisfying the irreducible integral conditions comes into play, one has to introduce the a "reduced" space of metaharmonic solenoidal fields which are tangential to the boundary, namely,

$$\mathcal{M}^{\rm sol}_{\gamma,\rm tan} = \left\{ \boldsymbol{\zeta}' \,|\, (-\nabla^2 + \gamma)\boldsymbol{\zeta}' = 0, \quad \boldsymbol{n}\cdot\boldsymbol{\zeta}'|_S = 0, \quad \boldsymbol{\nabla}\cdot\boldsymbol{\zeta}'|_S = 0 \right\}.$$

It is natural to introduce the operator of orthogonal projection onto the linear space $\mathcal{H}^{\rm sol}_{\rm tan}$

$$\overline{I\!\!P} = I\!\!P_{\mathcal{H}^{\rm sol}_{\rm tan}},$$

and its restriction $\overline{P}^{\gamma}$ to $\mathcal{M}^{\rm sol}_{\gamma,\rm tan}$. Between the spaces $\mathcal{H}^{\rm sol}_{\rm tan}$ and $\mathcal{M}^{\rm sol}_{\gamma,\rm tan}$ there exists a geometrical relationship analogous to that existing between the spaces $\mathcal{H}^{\rm sol}$ and $\mathcal{M}^{\rm sol}_{\gamma}$ discussed in section 3.4.3 for the original integral conditions. The vorticity construction associated with the system (3.76) using the irreducible vorticity conditions is depicted in Figure 3.2. It is to be remarked that the dimensionality of the two spaces $\mathcal{H}^{\rm sol}_{\rm tan}$ and $\mathcal{M}^{\rm sol}_{\gamma,\rm tan}$ here indicated is half of that of the spaces $\mathcal{H}^{\rm sol}$ and $\mathcal{M}^{\rm sol}_{\gamma}$ represented in Figure 3.1.

Unfortunately, the analysis of the governing equations using the irreducible version of the vorticity integral conditions is made difficult in comparison with those based on the double integral conditions when a variational formulation of the problem is considered. In fact, the simultaneous imposition of two boundary conditions

$$\boldsymbol{n}\cdot\boldsymbol{\zeta}|_S = \boldsymbol{n}\cdot\boldsymbol{\nabla}_S\times\boldsymbol{b} \qquad \text{and} \qquad \boldsymbol{\nabla}\cdot\boldsymbol{\zeta}|_S = 0, \tag{3.77}$$

seems not possible within a variational framework for the operator $-\nabla^2$: the last term in the boundary integral of the variational equation (3.60) indicates that these two boundary conditions are mutually exclusive. As a consequence, giving a variational meaning to the (separate) vorticity problem contained in system (3.76) has proved so far to be an impossible task to the author.

This difficulty might be related to another subtle aspect, not completely understood, of the vorticity problem contained in (3.76). Let us consider a vector equation of the type

$$\frac{\partial \boldsymbol{\omega}}{\partial t} = \nu\nabla^2\boldsymbol{\omega} + \boldsymbol{s},$$

where $\boldsymbol{s}$ is a given solenoidal source term, supplemented by the following boundary and initial conditions

$$\boldsymbol{n}\cdot\boldsymbol{\omega}|_S = c, \qquad \boldsymbol{\nabla}\cdot\boldsymbol{\omega}|_S = e, \qquad \boldsymbol{\omega}|_{t=0} = \boldsymbol{\omega}_0,$$

where c, e and $\boldsymbol{\omega}_0$ are the boundary and initial data, respectively. If these data are chosen arbitrarily, the problem for $\boldsymbol{\omega}$ may admit no solution. Consider,

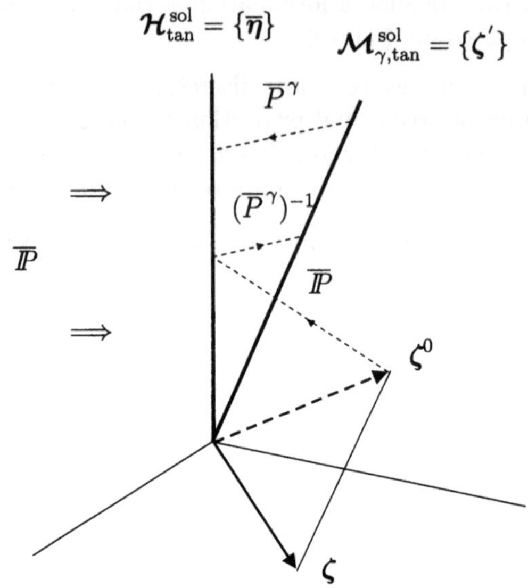

Figure 3.2: Construction of the vorticity vector field using the irreducible component of the vorticity integral conditions.

for instance, the particular case $e = 0$ and $\boldsymbol{\omega}_0 = \boldsymbol{\nabla}\times\boldsymbol{w}_0$ ($\boldsymbol{w}_0$ is an arbitrary vector field) so that $\boldsymbol{\nabla}\cdot\boldsymbol{\omega}_0 = 0$. Then, taking the divergence of the equation governing $\boldsymbol{\omega}$, one sees that $\boldsymbol{\nabla}\cdot\boldsymbol{\omega}$ satisfies the diffusion equation (since $\boldsymbol{\nabla}\cdot\boldsymbol{s} = 0$) supplemented by homogeneous boundary and initial conditions $\boldsymbol{\nabla}\cdot\boldsymbol{\omega}|_S = 0$ and $\boldsymbol{\nabla}\cdot\boldsymbol{\omega}|_{t=0} = 0$, respectively. It follows that $\boldsymbol{\nabla}\cdot\boldsymbol{\omega}$ is identically zero in V for all $t > 0$. On the other hand, the divergence theorem implies that

$$0 = \int \boldsymbol{\nabla}\cdot\boldsymbol{\omega}\, dV = \oint \boldsymbol{n}\cdot\boldsymbol{\omega}\, dS = \oint c\, dS,$$

by virtue of the first boundary condition imposed on $\boldsymbol{\omega}$. Therefore the boundary datum c cannot be prescribed independently from the other data of the problem for $\boldsymbol{\omega}$, for instance it cannot be $c > 0$ on the entire boundary S.

Such a counterexample does not imply necessarily that the problem for the vorticity $\boldsymbol{\zeta}$ in (3.76) has no solution, since in this case the form of the boundary datum for the normal component of $\boldsymbol{\zeta}$ is such that the condition on c obtained above is satisfied. In fact, one has

$$\oint \boldsymbol{n}\cdot\boldsymbol{\zeta}\, dS = \oint \boldsymbol{n}\cdot\boldsymbol{\nabla}_S\times\boldsymbol{b}\, dS = \int \boldsymbol{\nabla}\cdot\boldsymbol{\nabla}\times\widehat{\boldsymbol{b}}\, dV = 0,$$

where $\widehat{\boldsymbol{b}}$ denotes any extension of $\boldsymbol{b}$ to V. The vanishing of the integral is exactly what required by the divergence theorem and by the fact that $\boldsymbol{\nabla}\cdot\boldsymbol{\zeta} = 0$ in V

for all $t > 0$, as implied by the equation governing $\boldsymbol{\zeta}$ under the other boundary condition and the specified initial condition.

3.5.3 A representation of the irreducible projection space

We pursue the investigation of the nonprimitive variable formulation using the irreducible vorticity integral conditions by introducing a potentially interesting representation of the underlying linear space $\mathcal{H}_{\rm tan}^{\rm sol}$ which allows to express these conditions in a slightly different form.

Let us consider a field $\overline{\boldsymbol{\eta}} \in \mathcal{H}_{\rm tan}^{\rm sol}$ and assume that it can be expressed in terms of a vector potential, denoted by $\boldsymbol{\chi}$, by means of the relationship

$$\overline{\boldsymbol{\eta}} = \nabla\times\boldsymbol{\chi}.$$

The condition that $\overline{\boldsymbol{\eta}}$ be solenoidal is automatically satisfied while the boundary condition $\boldsymbol{n}\cdot\overline{\boldsymbol{\eta}}|_S = 0$ requires to impose the boundary condition $\boldsymbol{n}\cdot\nabla\times\boldsymbol{\chi}|_S = 0$.

To derive the equation governing $\boldsymbol{\chi}$, note that the equation of $\overline{\boldsymbol{\eta}}$, $-\nabla^2\overline{\boldsymbol{\eta}} = 0$, can also be written as $\nabla\times\nabla\times\overline{\boldsymbol{\eta}} = 0$, since $\overline{\boldsymbol{\eta}}$ is solenoidal. Thus $\nabla\times\overline{\boldsymbol{\eta}} = \nabla\eta$, where η is a harmonic function since the curl of a vector field is necessarily solenoidal. (Note that η has nothing to do with the magnitude of a vector field $\boldsymbol{\eta}$, *i.e.*, $\eta \neq |\boldsymbol{\eta}|$.)

Substituting in $\nabla\times\overline{\boldsymbol{\eta}} = \nabla\eta$ the representation of $\overline{\boldsymbol{\eta}}$ in terms of the vector potential $\boldsymbol{\chi}$ gives the equation $\nabla\times\nabla\times\boldsymbol{\chi} = \nabla\eta$. This equation together with the boundary condition $\boldsymbol{n}\cdot\nabla\times\boldsymbol{\chi}|_S = 0$ does not define the vector potential $\boldsymbol{\chi}$ uniquely, since $\boldsymbol{\chi}$ can be changed by the gauge transformation

$$\boldsymbol{\chi} \longrightarrow \boldsymbol{\chi} + \nabla\Upsilon,$$

where Υ is an arbitrary function. We rewrite the equation for $\boldsymbol{\chi}$ in the form $-\nabla^2\boldsymbol{\chi} + \nabla(\nabla\cdot\boldsymbol{\chi}) = \nabla\eta$ and assume to impose the gauge condition that $\nabla\cdot\boldsymbol{\chi}$ be equal to the harmonic function η—a choice denoted as *harmonic gauge*. Under such a supplementary condition the equation for $\boldsymbol{\chi}$ simplifies to $-\nabla^2\boldsymbol{\chi} = 0$. Nevertheless, the vector potential $\boldsymbol{\chi}$ still remains indeterminate by the gradient of an arbitrary harmonic function. To define $\boldsymbol{\chi}$ uniquely we assume to impose the boundary condition $\boldsymbol{n}\times\boldsymbol{\chi}|_S = 0$, which ensures the fulfillment of the original boundary condition $\boldsymbol{n}\cdot\nabla\times\boldsymbol{\chi}|_S = 0$ and which eliminates the residual indeterminacy of $\boldsymbol{\chi}$ (up to a constant). This last condition means that the vector field $\boldsymbol{\chi}$ must be *normal* to the boundary: thus it will be denoted by $\boldsymbol{\chi}$ so that $\boldsymbol{\chi}$ must satisfy the following equation and boundary conditions

$$-\nabla^2\boldsymbol{\chi} = 0, \qquad \boldsymbol{n}\times\boldsymbol{\chi}|_S = 0.$$

The representation which has bee introduced for $\overline{\boldsymbol{\eta}}$ gives the following decomposition of a field $\boldsymbol{\eta} \in \boldsymbol{\mathcal{H}}^{\text{sol}}$:

$$\boldsymbol{\eta} = \overline{\boldsymbol{\eta}} + \boldsymbol{\nabla}\chi = \boldsymbol{\nabla}\times\boldsymbol{\chi} + \boldsymbol{\nabla}\chi,$$

where the vector fields $\overline{\boldsymbol{\eta}}$ and $\boldsymbol{\chi}$ are both harmonic in V and the scalar function χ is harmonic. The field $\overline{\boldsymbol{\eta}}$ is tangential to the boundary S whereas $\boldsymbol{\chi}$ is normal. Note furthermore that $\overline{\boldsymbol{\eta}}$ is solenoidal whereas $\boldsymbol{\chi}$ needs not to be solenoidal, since $\boldsymbol{\nabla}\cdot\boldsymbol{\chi}$ is only requested to be harmonic.

The introduction of the fields $\boldsymbol{\chi}$ to represent the harmonic vector fields belonging to the subspace $\boldsymbol{\mathcal{H}}_{\text{tan}}^{\text{sol}}$ corresponds to the following orthogonal decomposition of $\boldsymbol{\mathcal{H}}^{\text{sol}}$

$$\boldsymbol{\mathcal{H}}^{\text{sol}} = \boldsymbol{\nabla}\times\boldsymbol{\mathcal{H}}_{\text{nor}} \oplus \boldsymbol{\nabla}\mathcal{H},$$

where $\boldsymbol{\mathcal{H}}_{\text{nor}}$ denotes the space of harmonic vector fields *normal* to the boundary. It is important to note that this result is a particular harmonic realization of the the Hodge orthogonal decomposition of vector fields defined over simply connected domains (Dautray and Lions [12, p. 314]). To see this connection, it is necessary to introduce the Hilbert spaces involved in the decomposition of the square-summable vector fields $\boldsymbol{L}^2(V)$ defined in a simply connected three-dimensional domain V with a suitably smooth boundary S. First, Sobolev spaces containing vector fields which are either solenoidal or irrotational must be defined according to:

$$\boldsymbol{J}^0(V) = \Big\{\boldsymbol{u} \in \boldsymbol{L}^2(V), \quad \boldsymbol{\nabla}\cdot\boldsymbol{u} = 0\Big\},$$

$$\boldsymbol{J}_0^0(V) = \Big\{\boldsymbol{u} \in \boldsymbol{L}^2(V), \quad \boldsymbol{\nabla}\cdot\boldsymbol{u} = 0, \quad \boldsymbol{n}\cdot\boldsymbol{u}|_S = 0\Big\},$$

$$\boldsymbol{K}^0(V) = \Big\{\boldsymbol{u} \in \boldsymbol{L}^2(V), \quad \boldsymbol{\nabla}\times\boldsymbol{u} = 0\Big\},$$

$$\boldsymbol{K}_0^0(V) = \Big\{\boldsymbol{u} \in \boldsymbol{L}^2(V), \quad \boldsymbol{\nabla}\times\boldsymbol{u} = 0, \quad \boldsymbol{n}\times\boldsymbol{u}|_S = 0\Big\}.$$

Then, one introduces the following spaces of harmonic functions and harmonic vector fields:

$$\mathcal{H}(V) = \Big\{u| -\nabla^2 u = 0\Big\},$$

$$\boldsymbol{\mathcal{H}}(V) = \Big\{\boldsymbol{u}| -\nabla^2 \boldsymbol{u} = 0\Big\},$$

$$\mathcal{H}^1(V) = \mathcal{H}(V) \cap H^1(V),$$

where $H^1(V)$ is the usual notation for Sobolev space of functions which are square summable with their first derivatives. Finally, it is necessary to introduce a Sobolev space of vector fields *normal* to the boundary defined by

$$\boldsymbol{H}_{\text{nor}}^1(V) = \Big\{\boldsymbol{u} \in \boldsymbol{H}^1(V), \quad \boldsymbol{n}\times\boldsymbol{u}|_S = 0\Big\}.$$

The fundamental theorem of orthogonal decomposition of $\boldsymbol{L}^2(V)$ is

$$\boldsymbol{L}^2(V) = \overbrace{\boldsymbol{J}_0^0(V) \oplus \nabla\mathcal{H}^1(V)}^{\boldsymbol{J}^0(V)} \oplus \boldsymbol{K}_0^0(V).$$

$$\boldsymbol{L}^2(V) = \boldsymbol{J}_0^0(V) \oplus \underbrace{\nabla\mathcal{H}^1(V) \oplus \boldsymbol{K}_0^0(V)}_{\boldsymbol{K}^0(V)}.$$

Furthermore, the spaces $\boldsymbol{J}^0(V)$, $\boldsymbol{J}_0^0(V)$, $\boldsymbol{K}^0(V)$ and $\boldsymbol{K}_0^0(V)$ can be characterized also as the curl and the gradient of standard Sobolev spaces in the following way

$$\boldsymbol{J}^0(V) = \nabla\times\boldsymbol{H}^1(V), \qquad \boldsymbol{K}^0(V) = \nabla H^1(V),$$

$$\boldsymbol{J}_0^0(V) = \nabla\times\boldsymbol{H}^1_{\rm nor}(V), \qquad \boldsymbol{K}_0^0(V) = \nabla H_0^1(V).$$

As a consequence, the previous decomposition can also be written in the following form

$$\boldsymbol{L}^2(V) = \overbrace{\nabla\times\boldsymbol{H}^1_{\rm nor}(V) \oplus \nabla\mathcal{H}^1(V)}^{\nabla\times\boldsymbol{H}^1(V)} \oplus \nabla H_0^1(V).$$

$$\boldsymbol{L}^2(V) = \nabla\times\boldsymbol{H}^1_{\rm nor}(V) \oplus \underbrace{\nabla\mathcal{H}^1(V) \oplus \nabla H_0^1(V)}_{\nabla H^1(V)}.$$

Therefore, the decomposition $\boldsymbol{\mathcal{H}}^{\rm sol} = \nabla\times\boldsymbol{\mathcal{H}}_{\rm nor} \oplus \nabla\mathcal{H}$ is the "image" of the decomposition of $\nabla\times\boldsymbol{H}^1(V)$ on the space of the harmonic fields.

Substituting $\overline{\boldsymbol{\eta}} = \nabla\times\boldsymbol{\chi}$ into the irreducible vorticity conditions (3.75), one obtains

$$\int \boldsymbol{\zeta}\cdot\nabla\times\boldsymbol{\chi}\,dV = \oint(\boldsymbol{n}\times\boldsymbol{a}\cdot\nabla\times\nabla\times\boldsymbol{\chi} + \boldsymbol{n}\times\boldsymbol{b}\cdot\nabla\times\boldsymbol{\chi})\,dS$$

Considering separately the first term of the boundary integral, one has, since $-\nabla^2\boldsymbol{\chi} = 0$,

$$\begin{aligned}\oint \boldsymbol{n}\times\boldsymbol{a}\cdot\nabla\times\nabla\times\boldsymbol{\chi}\,dS &= \oint \boldsymbol{n}\times\boldsymbol{a}\cdot\nabla(\nabla\cdot\boldsymbol{\chi})\,dS \\ &= \oint \boldsymbol{n}\cdot\boldsymbol{a}\times\nabla(\nabla\cdot\boldsymbol{\chi})\,dS \\ &= \oint \boldsymbol{n}\cdot[-\nabla\times(\boldsymbol{a}\,\nabla\cdot\boldsymbol{\chi}) + \nabla_S\times\boldsymbol{a}\,\nabla\cdot\boldsymbol{\chi}]\,dS \\ &= \oint \boldsymbol{n}\cdot\nabla_S\times\boldsymbol{a}\,\nabla\cdot\boldsymbol{\chi}\,dS,\end{aligned}$$

where the divergence theorem has been used in the last passage. Substituting this result into the complete expression of the integral conditions gives

$$\int \boldsymbol{\zeta}\cdot\nabla\times\boldsymbol{\chi}\,dV = \oint(\boldsymbol{n}\cdot\nabla_S\times\boldsymbol{a}\,\nabla\cdot\boldsymbol{\chi} + \boldsymbol{n}\times\boldsymbol{b}\cdot\nabla\times\boldsymbol{\chi})\,dS.$$

The order of the two terms on the right-hand side can be interchanged to give

$$\int \boldsymbol{\zeta} \cdot \nabla\times\boldsymbol{\chi}\, dV = \oint (\boldsymbol{n}\times\boldsymbol{b} \cdot \nabla\times\boldsymbol{\chi} + \boldsymbol{n} \cdot \nabla_S\times\boldsymbol{a}\, \nabla\cdot\boldsymbol{\chi})\, dS, \tag{3.78}$$

and the derivative boundary condition $\boldsymbol{n}\times\nabla\times\boldsymbol{\psi}|_S = \boldsymbol{n}\times\boldsymbol{b}$, can used to supplement the equation $-\nabla^2\boldsymbol{\psi} = \boldsymbol{\zeta}$ instead of the Dirichlet condition $\boldsymbol{n}\times\boldsymbol{\psi}|_S = \boldsymbol{n}\times\boldsymbol{a}$, in the final uncoupled formulation (there is only one integral condition for each boundary point and therfore these integral conditions cannot guarantee that two boundary conditions for velocity are imposed).

The complete set of equations and conditions based on the irreducible integral conditions and using the representation $\nabla\times\boldsymbol{\chi}$ of the projection space can be written in the form

$$\begin{aligned}
&-\nu\nabla^2\boldsymbol{\zeta} + \frac{\partial\boldsymbol{\zeta}}{\partial t} + \nabla\times(\boldsymbol{\zeta}\times\nabla\times\boldsymbol{\psi}) = 0, && \boldsymbol{\zeta}|_{t=0} = \nabla\times\boldsymbol{u}_0,\\
&\int \boldsymbol{\zeta} \cdot \nabla\times\boldsymbol{\chi}\, dV = \oint (\boldsymbol{n}\times\boldsymbol{b} \cdot \nabla\times\boldsymbol{\chi} + \boldsymbol{n} \cdot \nabla_S\times\boldsymbol{a}\, \nabla\cdot\boldsymbol{\chi})\, dS,\\
&\qquad \boldsymbol{n}\cdot\boldsymbol{\zeta}|_S = \boldsymbol{n}\cdot\nabla_S\times\boldsymbol{b}, && \nabla\cdot\boldsymbol{\zeta}|_S = 0,\\
&\big[\, -\nabla^2\boldsymbol{\chi} = 0, \qquad \boldsymbol{n}\times\boldsymbol{\chi}|_S = 0; \,\big]\\
&-\nabla^2\boldsymbol{\psi} = \boldsymbol{\zeta}, && \boldsymbol{n}\times\nabla\times\boldsymbol{\psi}|_S = \boldsymbol{n}\times\boldsymbol{b}, \ \ \nabla\cdot\boldsymbol{\psi}|_S = 0.
\end{aligned} \tag{3.79}$$

3.6 ζ-ϕ formulation

The stream vector $\boldsymbol{\psi}$ introduced in the previous section to represent the velocity field is characterizd by a particular choice of the boundary conditions which include the specification of the tangential components of $\boldsymbol{\psi}$. This choice makes it necessary to introduce the surface scalar potential q_S whose charge is to provide the tangential boundary values of the stream vector $\boldsymbol{\psi}$. The determination of the surface potential q_S requires to solve an elliptic equation on the boundary at each time $t > 0$ and this equation may be difficult in specific situations. Furthermore and more importantly, the boundary conditions chosen for $\boldsymbol{\psi}$ are not suitable for dealing with multiply connected domains. Without attempting to analyze the topological aspects involved by the multiply connected character of the domain, we will now investigate another characterization of the velocity field in terms of a different stream vector, denoted here by $\boldsymbol{\phi}$ for distinction, which is anticipated to be potentially attractive for three-dimensional domains of arbitrary connectedness.

3.6.1 Governing equations

Let $\boldsymbol{\phi}$ be a stream vector which represents the solenoidal velocity field $\boldsymbol{u}$, according to the definition

$$\boldsymbol{u} = \boldsymbol{\nabla}\times\boldsymbol{\phi}. \tag{3.80}$$

The velocity boundary condition $\boldsymbol{u}|_S = \boldsymbol{b}$ is imposed on $\boldsymbol{\phi}$ a as (derivative) boundary condition $\boldsymbol{\nabla}\times\boldsymbol{\phi}|_S = \boldsymbol{b}$, without any further differential elaboration. Of course, this condition can be rewritten separating its tangential and normal components, in the form:

$$\boldsymbol{n}\times\boldsymbol{\nabla}\times\boldsymbol{\phi}|_S = \boldsymbol{n}\times\boldsymbol{b} \qquad \text{and} \qquad \boldsymbol{n}\cdot\boldsymbol{\nabla}\times\boldsymbol{\phi}|_S = \boldsymbol{n}\cdot\boldsymbol{b}. \tag{3.81}$$

Taking the curl of (3.80) and recalling the vorticity definition, $\boldsymbol{\zeta} = \boldsymbol{\nabla}\times\boldsymbol{u}$, gives the equation for $\boldsymbol{\phi}$, namely,

$$\boldsymbol{\nabla}\times\boldsymbol{\nabla}\times\boldsymbol{\phi} = \boldsymbol{\zeta}. \tag{3.82}$$

Equation (3.82) and boundary conditions (3.81) do not define $\boldsymbol{\phi}$ uniquely, since $\boldsymbol{\phi}$ can be changed by adding the gradient of an arbitrary scalar function Θ according to

$$\boldsymbol{\phi} \longrightarrow \boldsymbol{\phi} + \boldsymbol{\nabla}\Theta,$$

which is a gauge transformation, as discussed in section 3.3.1. The freedom of the gauge transformation allows to impose additional conditions on $\boldsymbol{\phi}$. Instead of imposing the boundary condition for the tangential components of $\boldsymbol{\phi}$, as done for the stream vector $\boldsymbol{\psi}$ and the vector potential $\boldsymbol{A}$, we assume to impose on $\boldsymbol{\phi}$ the additional condition specifying its normal component on the boundary, in the form:

$$\boldsymbol{n}\cdot\boldsymbol{\phi}|_S = a, \tag{3.83}$$

where $a = a(\boldsymbol{x}_S, t)$ is a scalar function defined on S and completely arbitrary, for the moment being.

The equation (3.82) governing $\boldsymbol{\phi}$ and its boundary conditions (3.81) and (3.83) still leave $\boldsymbol{\phi}$ indeterminate by the gradient of a function, whose normal derivative must vanish on S so as to still satisfy the additional boundary condition (3.83). To define $\boldsymbol{\phi}$ uniquely, let us assume to impose the Euclid gauge $\boldsymbol{\nabla}\cdot\boldsymbol{\phi} = 0$. If $\boldsymbol{\phi}$ satisfies this gauge condition, equation (3.82) is equivalent to $-\nabla^2\boldsymbol{\phi} = \boldsymbol{\zeta}$; conversely, since $\boldsymbol{\nabla}\cdot\boldsymbol{\zeta} = 0$, the latter equation assures that the gauge condition can be satisfied simply by imposing the boundary condition $\boldsymbol{\nabla}\cdot\boldsymbol{\phi}|_S = 0$. However, the two boundary conditions $\boldsymbol{n}\cdot\boldsymbol{\phi}|_S = a$ and $\boldsymbol{\nabla}\cdot\boldsymbol{\phi}|_S = 0$ cannot be satisfied simultaneously unless the datum a satisfies a compatibility condition. In fact, by the divergence theorem it is

$$\oint a\, dS = \oint \boldsymbol{n}\cdot\boldsymbol{\phi}\, dS = \int \boldsymbol{\nabla}\cdot\boldsymbol{\phi}\, dV.$$

But, since $\boldsymbol{\phi}$ satisfies $-\nabla^2\boldsymbol{\phi} = \boldsymbol{\zeta}$, with $\boldsymbol{\nabla}\cdot\boldsymbol{\zeta} = 0$, and since $\boldsymbol{\nabla}\cdot\boldsymbol{\phi}|_S = 0$, it results $\boldsymbol{\nabla}\cdot\boldsymbol{\phi} = 0$, so that a must satisfy the single constraint

$$\oint a\, dS = 0. \tag{3.84}$$

We are thus led to the following set comprising the equation governing $\boldsymbol{\phi}$ and all of its boundary conditions:

$$-\nabla^2\boldsymbol{\phi} = \boldsymbol{\zeta}, \qquad \boldsymbol{n}\cdot\boldsymbol{\phi}|_S = a, \quad \boldsymbol{n}\times\boldsymbol{\nabla}\times\boldsymbol{\phi}|_S = \boldsymbol{n}\times\boldsymbol{b}, \quad \boldsymbol{\nabla}\cdot\boldsymbol{\phi}|_S = 0, \tag{3.85}$$
$$\boldsymbol{n}\cdot\boldsymbol{\nabla}\times\boldsymbol{\phi}|_S = \boldsymbol{n}\cdot\boldsymbol{b},$$

where the scalar datum a is subject to condition (3.84) and is for the rest arbitrary. There is a total of five boundary conditions, the same number found for both the $\boldsymbol{\psi}$ and $\boldsymbol{A}$ variables. However, the conditions are partially different since here two (scalar) conditions prescribe the normal components of the unknown and of its curl, in place of the single (vector) condition imposing the two tangential components of the unknown.

As a consequence of the different set of boundary conditions for $\boldsymbol{\phi}$, the argument enabling one to "discharge" two of the five boundary conditions for $\boldsymbol{\phi}$ on the vorticity field $\boldsymbol{\zeta}$ is different from that followed in the analysis of the other nonprimitive variable formulations. Essentially, the new argument proceeds in two steps. Firstly, the Euclid gauge $\boldsymbol{\nabla}\cdot\boldsymbol{\phi} = 0$ is shown to be automatically satisfied in problem (3.85) provided that the normal component of $\boldsymbol{\zeta}$ on the boundary is given by $\boldsymbol{n}\cdot\boldsymbol{\zeta}|_S = \boldsymbol{n}\cdot\boldsymbol{\nabla}_S\times\boldsymbol{b}$; this results makes such a vorticity condition an equivalent substitute for the derivative boundary condition $\boldsymbol{\nabla}\cdot\boldsymbol{\phi}|_S = 0$, obtained to impose the Euclid gauge. Secondly, the remaining four boundary conditions for $\boldsymbol{\phi}$ in problem (3.85) are shown provide (a single family of) integral conditions for the vorticity.

3.6.2 An equivalent ζ-ϕ formulation

We first note that the equation in (3.85) with $\boldsymbol{\nabla}\cdot\boldsymbol{\zeta} = 0$ together with the boundary condition $\boldsymbol{\nabla}\cdot\boldsymbol{\phi}|_S = 0$ imply $\boldsymbol{\nabla}\times\boldsymbol{\nabla}\times\boldsymbol{\phi} = \boldsymbol{\zeta}$. The normal component of the latter equation on the boundary is $\boldsymbol{n}\cdot\boldsymbol{\nabla}\times\boldsymbol{\nabla}\times\boldsymbol{\phi}|_S = \boldsymbol{n}\cdot\boldsymbol{\zeta}|_S$. But the evaluation of the normal component of the curl, *i.e.*, $\boldsymbol{n}\cdot\boldsymbol{\nabla}\times$, of whatever vector field, requires to know only its tangential components. The vector field to be considered here is $\boldsymbol{\nabla}\times\boldsymbol{\phi}|_S$. Its tangential components are $\boldsymbol{n}\times\boldsymbol{\nabla}\times\boldsymbol{\phi}|_S$ and they are prescribed by the second of the boundary conditions in (3.85), namely, $\boldsymbol{n}\times\boldsymbol{\nabla}\times\boldsymbol{\phi}|_S = \boldsymbol{n}\times\boldsymbol{b}$. It follows that the normal component of the vector equation $\boldsymbol{\nabla}\times\boldsymbol{\nabla}\times\boldsymbol{\phi} = \boldsymbol{\zeta}$ implies that the normal component of the vorticity on the boundary can be expressed in terms of $\boldsymbol{b}$ according to the relation

$$\boldsymbol{n}\cdot\boldsymbol{\zeta}|_S = \boldsymbol{n}\cdot\boldsymbol{\nabla}_S\times\boldsymbol{b}. \tag{3.86}$$

The important point is now that the fulfillment of such a condition in conjunction with the equation in (3.85) and the two boundary conditions $\boldsymbol{n}\cdot\boldsymbol{\phi}|_S = a$ and $\boldsymbol{n}\times\boldsymbol{\nabla}\times\boldsymbol{\phi}|_S = \boldsymbol{n}\times\boldsymbol{b}$ is sufficient to guarantee that $\boldsymbol{\phi}$ is solenoidal and thus also that $\boldsymbol{\nabla}\cdot\boldsymbol{\phi}|_S = 0$, as stated by the following theorem.

Theorem 3.4. *Let $\boldsymbol{\phi}$ be the vector field solution of the mixed Dirichlet–Neumann boundary value problem*

$$-\nabla^2\boldsymbol{\phi} = \boldsymbol{\zeta}, \qquad \boldsymbol{n}\cdot\boldsymbol{\phi}|_S = a, \quad \boldsymbol{n}\times\boldsymbol{\nabla}\times\boldsymbol{\phi}|_S = \boldsymbol{n}\times\boldsymbol{b},$$

with the data $\boldsymbol{\zeta}$ and a satisfying the conditions

$$\boldsymbol{\nabla}\cdot\boldsymbol{\zeta} = 0 \qquad \text{and} \qquad \oint a\, dS = 0,$$

and with $\boldsymbol{n}\cdot\boldsymbol{\zeta}|_S$ and $\boldsymbol{n}\times\boldsymbol{b}$ satisfying the following condition of compatibility

$$\boldsymbol{n}\cdot\boldsymbol{\zeta}|_S = \boldsymbol{n}\cdot\boldsymbol{\nabla}_S\times\boldsymbol{b}.$$

Then $\boldsymbol{\phi}$ is solenoidal.

Proof. Taking the divergence of the equation $-\nabla^2\boldsymbol{\phi} = \boldsymbol{\zeta}$ and using the first condition on the data, $\boldsymbol{\nabla}\cdot\boldsymbol{\zeta} = 0$, gives $-\nabla^2\boldsymbol{\nabla}\cdot\boldsymbol{\phi} = 0$, so that $\boldsymbol{\nabla}\cdot\boldsymbol{\phi}$ is a harmonic function. On the other hand, by vector identity $-\nabla^2\boldsymbol{\alpha} = \boldsymbol{\nabla}\times\boldsymbol{\nabla}\times\boldsymbol{\alpha} - \boldsymbol{\nabla}(\boldsymbol{\nabla}\cdot\boldsymbol{\alpha})$, the Poisson equation reads

$$\boldsymbol{\nabla}\times\boldsymbol{\nabla}\times\boldsymbol{\phi} - \boldsymbol{\nabla}(\boldsymbol{\nabla}\cdot\boldsymbol{\phi}) = \boldsymbol{\zeta}.$$

By taking the normal component of this equation on S and using the second boundary condition satisfied by $\boldsymbol{\phi}$, namely, $\boldsymbol{n}\times\boldsymbol{\nabla}\times\boldsymbol{\phi}|_S = \boldsymbol{n}\times\boldsymbol{b}$, gives

$$\boldsymbol{n}\cdot\boldsymbol{\nabla}\times\boldsymbol{\nabla}\times\boldsymbol{\phi}|_S - \boldsymbol{n}\cdot\boldsymbol{\nabla}(\boldsymbol{\nabla}\cdot\boldsymbol{\phi})|_S = \boldsymbol{n}\cdot\boldsymbol{\nabla}_S\times\boldsymbol{b} - \boldsymbol{n}\cdot\boldsymbol{\nabla}(\boldsymbol{\nabla}\cdot\boldsymbol{\phi})|_S = \boldsymbol{n}\cdot\boldsymbol{\zeta}|_S.$$

By the compatibility condition between the data $\boldsymbol{n}\cdot\boldsymbol{\zeta}|_S$ and $\boldsymbol{n}\times\boldsymbol{b}$, the last equality provides $\boldsymbol{n}\cdot\boldsymbol{\nabla}(\boldsymbol{\nabla}\cdot\boldsymbol{\phi})|_S = 0$, and, since $\boldsymbol{\nabla}\cdot\boldsymbol{\phi}$ is harmonic, $\boldsymbol{\nabla}\cdot\boldsymbol{\phi} = C$, where C is an arbitrary constant (or an arbitrary function of time if $\boldsymbol{\zeta}$ or $\boldsymbol{b}$ depend on time). Integrating such a relation over the domain V gives

$$C\int dV = \int \boldsymbol{\nabla}\cdot\boldsymbol{\phi}\, dV = \oint \boldsymbol{n}\cdot\boldsymbol{\phi}\, dS,$$

by the divergence theorem, so that, using the first boundary condition of $\boldsymbol{\phi}$,

$$C\int dV = \oint a\, dS.$$

The assumed condition $\oint a\, dS = 0$ for the datum a implies $C = 0$, since $\int dV > 0$. Therefore $\boldsymbol{\nabla}\cdot\boldsymbol{\phi} = 0$ and $\boldsymbol{\phi}$ is solenoidal. □

According to this theorem, the boundary condition $\boldsymbol{\nabla}\cdot\boldsymbol{\phi}|_S = 0$ stemming from the Euclid gauge condition can be dispensed with. It is automatically satisfied when the set of equations and boundary conditions for the variables $\boldsymbol{\zeta}$ and $\boldsymbol{\phi}$ are written in the form:

$$
\begin{aligned}
&\frac{\partial \boldsymbol{\zeta}}{\partial t} - \nu\nabla^2\boldsymbol{\zeta} + \boldsymbol{\nabla}\times(\boldsymbol{\zeta}\times\boldsymbol{\nabla}\times\boldsymbol{\phi}) = 0, && && \boldsymbol{\zeta}|_{t=0} = \boldsymbol{\nabla}\times\boldsymbol{u}_0, \\
& && \boldsymbol{n}\cdot\boldsymbol{\zeta}|_S = \boldsymbol{n}\cdot\boldsymbol{\nabla}_S\times\boldsymbol{b}, && \boldsymbol{\nabla}\cdot\boldsymbol{\zeta}|_S = 0, \\
&-\nabla^2\boldsymbol{\phi} = \boldsymbol{\zeta}, && \boldsymbol{n}\cdot\boldsymbol{\phi}|_S = a, && \boldsymbol{n}\times\boldsymbol{\nabla}\times\boldsymbol{\phi}|_S = \boldsymbol{n}\times\boldsymbol{b}, \\
& && && \boldsymbol{n}\cdot\boldsymbol{\nabla}\times\boldsymbol{\phi}|_S = \boldsymbol{n}\cdot\boldsymbol{b},
\end{aligned}
\tag{3.87}
$$

where the scalar datum a satisfies the condition $\oint a\, dS = 0$ and is otherwise arbitrary. In this formulation there are two boundary conditions for the vorticity and four for the stream vector. To have an uncoupled formulation it remains to show how to contrive the redundant conditions for $\boldsymbol{\phi}$ into a single, expectedly integral, condition for $\boldsymbol{\zeta}$.

3.6.3 Uncoupled formulation

Let us consider the integral

$$\int \boldsymbol{\zeta}\cdot\boldsymbol{\nabla}\times\boldsymbol{\chi}\, dV,$$

where $\boldsymbol{\zeta}$ is a vorticity field solution to problem (3.87) while $\boldsymbol{\chi}$ is a vector field left arbitrary, for the moment being. Since $\boldsymbol{\zeta} = -\nabla^2\boldsymbol{\phi}$, the vector identity $-\nabla^2\boldsymbol{\alpha} = \boldsymbol{\nabla}\times\boldsymbol{\nabla}\times\boldsymbol{\alpha} - \boldsymbol{\nabla}(\boldsymbol{\nabla}\cdot\boldsymbol{\alpha})$ allows to write the integral in the following form

$$
\begin{aligned}
\int \boldsymbol{\zeta}\cdot\boldsymbol{\nabla}\times\boldsymbol{\chi}\, dV &= \int (-\nabla^2\boldsymbol{\phi})\cdot\boldsymbol{\nabla}\times\boldsymbol{\chi}\, dV \\
&= \int \big\{(\boldsymbol{\nabla}\times\boldsymbol{\nabla}\times\boldsymbol{\phi})\cdot\boldsymbol{\nabla}\times\boldsymbol{\chi} - [\boldsymbol{\nabla}(\boldsymbol{\nabla}\cdot\boldsymbol{\phi})]\cdot\boldsymbol{\nabla}\times\boldsymbol{\chi}\big\}\, dV.
\end{aligned}
$$

The first term on the right-hand side can be expressed using the elementary vector identity $\boldsymbol{\nabla}\cdot(\boldsymbol{a}\times\boldsymbol{b}) = \boldsymbol{\nabla}\times\boldsymbol{a}\cdot\boldsymbol{b} - \boldsymbol{a}\cdot\boldsymbol{\nabla}\times\boldsymbol{b}$, to give

$$
\begin{aligned}
\int \boldsymbol{\zeta}\cdot\boldsymbol{\nabla}\times\boldsymbol{\chi}\, dV = \int \big\{&\boldsymbol{\nabla}\cdot[(\boldsymbol{\nabla}\times\boldsymbol{\phi})\cdot(\boldsymbol{\nabla}\times\boldsymbol{\chi})] \\
&+(\boldsymbol{\nabla}\times\boldsymbol{\phi})\cdot\boldsymbol{\nabla}\times\boldsymbol{\nabla}\times\boldsymbol{\chi} - [\boldsymbol{\nabla}(\boldsymbol{\nabla}\cdot\boldsymbol{\phi})]\cdot\boldsymbol{\nabla}\times\boldsymbol{\chi}\big\}\, dV.
\end{aligned}
$$

Let us assume now that the vector field $\boldsymbol{\chi}$ satisfies the equation $-\nabla^2\boldsymbol{\chi} = 0$. By the previous vector identity $\boldsymbol{\chi}$ satisfies also the equation $\boldsymbol{\nabla}\times\boldsymbol{\nabla}\times\boldsymbol{\chi} - \boldsymbol{\nabla}(\boldsymbol{\nabla}\cdot\boldsymbol{\chi}) = 0$ and the term $\boldsymbol{\nabla}\times\boldsymbol{\nabla}\times\boldsymbol{\chi}$ in the considered integral can be substituted by $\boldsymbol{\nabla}(\boldsymbol{\nabla}\cdot\boldsymbol{\chi})$, to give

$$
\begin{aligned}
\int \boldsymbol{\zeta}\cdot\boldsymbol{\nabla}\times\boldsymbol{\chi}\, dV = \int \big\{&\boldsymbol{\nabla}\cdot[(\boldsymbol{\nabla}\times\boldsymbol{\phi})\cdot(\boldsymbol{\nabla}\times\boldsymbol{\chi})] \\
&+(\boldsymbol{\nabla}\times\boldsymbol{\phi})\cdot\boldsymbol{\nabla}(\boldsymbol{\nabla}\cdot\boldsymbol{\chi}) - [\boldsymbol{\nabla}(\boldsymbol{\nabla}\cdot\boldsymbol{\phi})]\cdot\boldsymbol{\nabla}\times\boldsymbol{\chi}\big\}\, dV.
\end{aligned}
$$

It is important to remark that it would not be correct to take $\nabla\cdot\phi = 0$ in this expression, since we are deriving relations which are intended for conditioning the vorticity field *before* the solution of the stream vector equation. In other words, it is only after the correct vorticity conditions have been satisfied that the stream vector ϕ will come out exactly solenoidal.

The application of another elementary vector identity to the last two terms of the integral gives

$$\int \zeta\cdot\nabla\times\chi\, dV = \int \Big\{\nabla\cdot[(\nabla\times\phi)\cdot(\nabla\times\chi)] \\ +\nabla\cdot[(\nabla\times\phi)(\nabla\cdot\chi)] - \nabla\cdot[(\nabla\cdot\phi)(\nabla\times\chi)]\Big\}\, dV.$$

Finally, the divergence theorem applied to the three terms gives

$$\begin{aligned}\int \zeta\cdot\nabla\times\chi\, dV &= \oint (n\cdot\nabla\times\phi\times\nabla\times\chi + n\cdot\nabla\times\phi\;\nabla\cdot\chi - n\cdot\nabla\times\chi\;\nabla\cdot\phi)\, dS \\ &= \oint (n\times\nabla\times\phi\cdot\nabla\times\chi + n\cdot\nabla\times\phi\;\nabla\cdot\chi - n\cdot\nabla\times\chi\;\nabla\cdot\phi)\, dS.\end{aligned}$$

We observe now that the first two terms of the boundary integral involve two of the boundary conditions on ϕ in problem (3.87), whereas the third one contains the value $\nabla\cdot\phi|_S$ which is not prescribed by the statement of the problem. It is therefore necessary to impose on χ the boundary condition $n\cdot\nabla\times\chi|_S = 0$ in order to make the third term not to contribute to the boundary integral. Another boundary condition can be imposed on χ in order to have the correct number of linearly independent solutions to the equation $-\nabla^2\chi = 0$ needed to supplement the vorticity equation, together with its boundary conditions already available in (3.87). We assume to impose the vanishing of the tangential components of χ on the boundary, so that χ is *normal* to on the boundary and the boundary condition $n\cdot\nabla\times\chi|_S = 0$ is automatically satisfied. The vector field satisfying these two boundary condition will be denoted by χ and therefore it is solution to the problem

$$-\nabla^2\chi = 0, \qquad n\times\chi|_S = 0. \tag{3.88}$$

Therefore, one can conclude that the vorticity field ζ solution to problem (3.87) will satisfy the following integral condition

$$\int \zeta\cdot\nabla\times\chi\, dV = \oint (n\times b\cdot\nabla\times\chi + n\cdot b\,\nabla\cdot\chi)\, dS, \tag{3.89}$$

for any χ solution to problem (3.88).

It follows that the $\boldsymbol{\zeta}$-$\boldsymbol{\phi}$ equations can be formulated in the uncoupled form:

$$
\begin{aligned}
&-\nu\nabla^2\boldsymbol{\zeta}+\frac{\partial\boldsymbol{\zeta}}{\partial t}+\boldsymbol{\nabla}\times(\boldsymbol{\zeta}\times\boldsymbol{\nabla}\times\boldsymbol{\phi})=0, \qquad \boldsymbol{\zeta}|_{t=0}=\boldsymbol{\nabla}\times\boldsymbol{u}_0,\\
&\int\boldsymbol{\zeta}\cdot\boldsymbol{\nabla}\times\boldsymbol{\chi}\,dV=\oint(\boldsymbol{n}\times\boldsymbol{b}\cdot\boldsymbol{\nabla}\times\boldsymbol{\chi}+\boldsymbol{n}\cdot\boldsymbol{b}\,\boldsymbol{\nabla}\cdot\boldsymbol{\chi})\,dS,\\
&\qquad\qquad \boldsymbol{n}\cdot\boldsymbol{\zeta}|_S=\boldsymbol{n}\cdot\boldsymbol{\nabla}_S\times\boldsymbol{b}, \qquad \boldsymbol{\nabla}\cdot\boldsymbol{\zeta}|_S=0, \qquad\qquad (3.90)\\
&\Big[-\nabla^2\boldsymbol{\chi}=0, \qquad \boldsymbol{n}\times\boldsymbol{\chi}|_S=0;\Big]\\
&-\nabla^2\boldsymbol{\phi}=\boldsymbol{\zeta}, \qquad \boldsymbol{n}\cdot\boldsymbol{\phi}|_S=a, \qquad \boldsymbol{n}\times\boldsymbol{\nabla}\times\boldsymbol{\phi}|_S=\boldsymbol{n}\times\boldsymbol{b},
\end{aligned}
$$

always under the assumption that the condition $\oint a\,dS=0$ is satisfied. With respect to the corresponding uncoupled formulation (3.76) for the $\boldsymbol{\zeta}$-$\boldsymbol{\psi}$ variables based on the irreducible integral conditions there are several differences in the boundary conditions imposed on two stream vectors $\boldsymbol{\phi}$ and $\boldsymbol{\psi}$, and also on the vector fields $\boldsymbol{\chi}$ and $\overline{\boldsymbol{\eta}}$ for the vorticity integral conditions. The latter appear instead very similar and can be shown to be identical recalling how $\boldsymbol{a}$ is defined in terms of $\boldsymbol{n}\cdot\boldsymbol{b}$ through the surface potential q_S.

3.7 Conclusions

In this chapter, we have described a number of formulations of the incompressible Navier–Stokes equations expressed in terms of the vorticity and various potentials for the velocity. Complete statements of the problem, with each governing equation supplemented by its own independent conditions, have been presented. In all cases, it has been found that the transport equation governing the vorticity field is supplemented by conditions of an integral character which generalize those established for the scalar vorticity in two-dimensional flows and admit an analogous interpretation in terms of an orthogonal projection over a linear space of suitable vector fields.

In particular, two formulations of the incompressible equations, the first with the velocity represented in terms of both a scalar and a vector potential and the second with the velocity expressed through only a stream vector, have been shown to be characterized by integral conditions which determine the two tangential components of the vorticity on no-slip boundaries. A numerical scheme which implements these vorticity integral conditions for problems in arbitrarily shaped three-dimensional regions has also been described.

The integral conditions for the vorticity vector field have been further investigated to establish the connections with other conditions of boundary value type which can be considered to supplement the vorticity equation. It has been found that the integral conditions for the vector vorticity can be partly replaced by a boundary condition for the normal component of this variable, a component which is explicitly known in terms of the tangential boundary values of the velocity. A very interesting aspect of the reduction of the integral conditions is the fact that it is rooted in a Hodge orthogonal decomposition of the subspace of the harmonic vector fields involved by the vorticity integral conditions.

The reduced integral conditions so obtained have also found to be the one required in another vorticity–stream vector formulation with a stream vector satisfying boundary conditions suitable for dealing with multiply connected domains.

From such an investigation, a rather coherent picture emerges which indicates, beyond any possible doubt, the necessity of considering the projection integral conditions for a theoretical analysis and understanding of the interaction of vorticity with no-slip boundaries.

On the other hand, a distinct difficulty has been encountered in the formulations of the three-dimensional incompressible Navier–Stokes equations in which the vorticity is subject to the reduced set of integral conditions. In this case the vorticity transport equation is supplemented also by two (scalar) conditions of boundary value type which impose the normal component of the unknown and the vanishing of its divergence to ensure the solenoidal character of the vorticity field. Within a standard variational framework for the Laplace or diffusion operator, these two boundary conditions appear to be mutually exclusive and this fact has so far precluded the development of numerical schemes implementing the reduced form of the vorticity integral conditions for general three-dimensional problems.

Chapter 4

Vorticity–velocity representation

4.1 Introduction

There are several reasons for attempting to formulate the incompressible Navier–Stokes equations using the hybrid (primitive and nonprimitive) variables, velocity and vorticity.

- First, as pointed out originally by Lighthill, this pair of variables is the most suited from the viewpoint of a fluid dynamic description of incompressible viscous flows. In fact the vorticity is governed by an extensively studied and well understood dynamical equation while the velocity field, which by definition embodies the kinematical aspect of the problem, can be related to the vorticity field by a very simple elliptic equation.

- The variety of boundary conditions that can be chosen for the vector potential or the stream vectors by virtue of the nonuniqueness of the velocity representation is avoided since the velocity is supplemented by unique boundary conditions.

- The vorticity–velocity formulation can match the equations of the boundary layer theory more easily than the primitive variable formulation.

- The study of the inviscid limit of the Navier–Stokes equations to the Euler equations for incompressible flows can be made easier by the elimination of the pressure variable.

- A complete understanding of vortex methods in the presence of no-slip boundaries could be eventually achieved.

The use of the vorticity variable together with the velocity to compute the motion of an incompressible viscous fluid has been considered for the first time by Fasel

(1976) to investigate the stability of boundary layers in two dimensions and by Dennis, Ingham and Cook (1979) to derive a numerical method for computing steady-state three-dimensional flows.

Since then, several investigations have been conducted on hybrid variable formulations of the Navier–Stokes equations and many numerical schemes have been proposed for the solution of problems both in two and three dimensions. Surprisingly enough, in most of the published studies the issue of the conditions assuring the equivalence of the $\boldsymbol{\zeta}$-$\boldsymbol{u}$ equations with the original $\boldsymbol{u}$-P equations has not received the due attention. Disregarding the methods in which the velocity is calculated as solution of a system of first-order equations or involving the inversion of the operator $\nabla\times\nabla\times$, the boundary conditions adopted in the various numerical schemes for the $\boldsymbol{\zeta}$-$\boldsymbol{u}$ equations have been found to be very critical for the success of the method (see, *e.g.*, Napolitano and Catalano 1991).

Remarkable exceptions to this lack of analysis of the conditions assuring an equivalent formulation of the Navier–Stokes equations in terms of the variables vorticity and velocity are the work by Daube, Guermond and Sellier (1991) dealing with the formulations for multiply connected domains in two and three dimensions, and the work by Daube (1992) describing an influence matrix method for two-dimensional problems using finite differences on a *staggered* grid.

From the viewpoint adopted in this study of obtaining an uncoupled formulation of the equations, the $\boldsymbol{\zeta}$-$\boldsymbol{u}$ representation is expected to be faced with difficulties similar to those encountered by nonprimitive variable formulations, as far as the specification of the conditions for the vorticity is concerned. In this respect, it is somewhat disconcerting to observe that, in all the literature on the numerical solution of the $\boldsymbol{\zeta}$-$\boldsymbol{u}$ or ζ-$\boldsymbol{u}$ equations, the possibility for the vorticity to be subject to conditions of an integral character, analogous to that established for the vorticity–stream function equations since more than twenty years, has not even been mentioned.

In this chapter, we will investigate the vorticity–velocity representation of the incompressible Navier–Stokes equations considering the case of both two- and three-dimensional problems. We will demonstrate the basic theorems establishing the equivalence of these formulations with the primitive variable equations and describe also the uncoupled form of the equations governing the vorticity and velocity variables. Needless to say, it will be found that the vorticity field, both in two and three dimensions, must be supplemented by integral conditions in order for the vorticity equation to achieve an independent meaning with respect to the Poisson equation for the velocity, assuming of course to neglect the nonlinearity. A detailed description of the Glowinski–Pironneau method for the ζ-$\boldsymbol{u}$ equations for plane two-dimensional problems will be given. Finally, we will point out some difficulties that still remain to be solved and that have prevented so far, to our knowledge, a successfull implementation of such a formulation by means of finite elements or finite differences on *nonstaggered* grids.

4.2 Three-dimensional equations

4.2.1 Governing equations

Starting once more from the primitive variable Navier–Stokes equations (1.1)–(1.4), we introduce the vorticity vector

$$\boldsymbol{\zeta} = \boldsymbol{\nabla}\times\boldsymbol{u}, \tag{4.1}$$

whose evolution is found to be determined by the vorticity equation

$$\frac{\partial \boldsymbol{\zeta}}{\partial t} + \boldsymbol{\nabla}\times(\boldsymbol{\zeta}\times\boldsymbol{u}) = \nu\nabla^2\boldsymbol{\zeta}. \tag{4.2}$$

An alternative way of writing this equation is obtained by expanding the non-linear term by means of the vector identity

$$\boldsymbol{\nabla}\times(\boldsymbol{\zeta}\times\boldsymbol{u}) = (\boldsymbol{u}\cdot\boldsymbol{\nabla})\boldsymbol{\zeta} - (\boldsymbol{\zeta}\cdot\boldsymbol{\nabla})\boldsymbol{u} + \boldsymbol{\zeta}\,\boldsymbol{\nabla}\cdot\boldsymbol{u} - \boldsymbol{u}\,\boldsymbol{\nabla}\cdot\boldsymbol{\zeta}.$$

But $\boldsymbol{\nabla}\cdot\boldsymbol{\zeta} = 0$ and $\boldsymbol{\nabla}\cdot\boldsymbol{u} = 0$, hence the vorticity equation assumes the form

$$\frac{\partial \boldsymbol{\zeta}}{\partial t} + (\boldsymbol{u}\cdot\boldsymbol{\nabla})\boldsymbol{\zeta} = (\boldsymbol{\zeta}\cdot\boldsymbol{\nabla})\boldsymbol{u} + \nu\nabla^2\boldsymbol{\zeta},$$

which displays, in the order, the terms describing the advection, the stretching and the viscous diffusion of the vorticity field. From the initial condition of velocity, $\boldsymbol{u}|_{t=0} = \boldsymbol{u}_0$, it is immediate to derive the initial condition for $\boldsymbol{\zeta}$:

$$\boldsymbol{\zeta}|_{t=0} = \boldsymbol{\nabla}\times\boldsymbol{u}_0. \tag{4.3}$$

Adding to equations (4.1)–(4.2) and to the initial condition (4.3) the continuity equation (1.2), $\boldsymbol{\nabla}\cdot\boldsymbol{u} = 0$, as well as the velocity boundary conditions (1.3), $\boldsymbol{u}|_S = \boldsymbol{b}$, we obtain a complete set of equations and conditions for the unknowns $\boldsymbol{\zeta}$ and $\boldsymbol{u}$. This set is demonstrated to be equivalent to the original Navier–Stokes problem by the following theorem.

Theorem 4.1. *The primitive variable Navier–Stokes problem (1.1)–(1.4) is equivalent to the problem for the variables $\boldsymbol{\zeta}$ and $\boldsymbol{u}$, defined by the set of equations and conditions*

$$\begin{aligned}
&\frac{\partial \boldsymbol{\zeta}}{\partial t} - \nu\nabla^2\boldsymbol{\zeta} + \boldsymbol{\nabla}\times(\boldsymbol{\zeta}\times\boldsymbol{u}) = 0, && \boldsymbol{\zeta}|_{t=0} = \boldsymbol{\nabla}\times\boldsymbol{u}_0,\\
&\boldsymbol{\nabla}\times\boldsymbol{u} = \boldsymbol{\zeta}, \qquad \boldsymbol{\nabla}\cdot\boldsymbol{u} = 0, && \boldsymbol{u}|_S = \boldsymbol{b},
\end{aligned} \tag{4.4}$$

assuming the following conditions on the data

$$\oint \boldsymbol{n}\cdot\boldsymbol{b}\,dS = 0, \qquad \boldsymbol{\nabla}\cdot\boldsymbol{u}_0 = 0, \qquad \boldsymbol{n}\cdot\boldsymbol{b}|_{t=0} = \boldsymbol{n}\cdot\boldsymbol{u}_0|_S. \tag{4.5}$$

Proof. The implication is evident. Conversely, let $(\boldsymbol{\zeta}, \boldsymbol{v})$ be a solution to the set (4.4) with the data $\boldsymbol{u}_0$ and $\boldsymbol{n}\cdot\boldsymbol{b}$ satisfying the conditions (4.5). (The velocity field is indicated by $\boldsymbol{v}$ instead of $\boldsymbol{u}$ to make explicit the distinction between the solution of the hybrid variable system (4.4) and the solution of the primitive variable equations (1.1)–(1.4).)

First, let us note that $\boldsymbol{v}$ is uniquely defined for all $t > 0$ because the problem $\boldsymbol{\nabla}\times\boldsymbol{v} = \boldsymbol{\zeta}$, $\boldsymbol{\nabla}\cdot\boldsymbol{v} = 0$, $\boldsymbol{n}\cdot\boldsymbol{v}|_S = \boldsymbol{n}\cdot\boldsymbol{b}$, admits a unique solution provided that $\boldsymbol{\nabla}\cdot\boldsymbol{\zeta} = 0$ and $\oint \boldsymbol{n}\cdot\boldsymbol{b}\,dS = 0$, for all $t > 0$. Both conditions are here satisfied by the assumptions: the latter coincides with the first condition in (4.5), whereas the former, *i.e.*, $\boldsymbol{\nabla}\cdot\boldsymbol{\zeta} = 0$, is satisfied for all $t > 0$ by virtue of the first equation satisfied by $\boldsymbol{v}$, namely, $\boldsymbol{\nabla}\times\boldsymbol{v} = \boldsymbol{\zeta}$. Then, the vorticity equation means

$$\frac{\partial \boldsymbol{\nabla}\times\boldsymbol{v}}{\partial t} - \nu\nabla^2\boldsymbol{\nabla}\times\boldsymbol{v} + \boldsymbol{\nabla}\times[(\boldsymbol{\nabla}\times\boldsymbol{v})\times\boldsymbol{v}] = 0,$$

that is

$$\boldsymbol{\nabla}\times\left[\frac{\partial \boldsymbol{v}}{\partial t} - \nu\nabla^2\boldsymbol{v} + (\boldsymbol{\nabla}\times\boldsymbol{v})\times\boldsymbol{v}\right] = 0,$$

which means

$$\frac{\partial \boldsymbol{v}}{\partial t} - \nu\nabla^2\boldsymbol{v} + (\boldsymbol{\nabla}\times\boldsymbol{v})\times\boldsymbol{v} = \boldsymbol{\nabla}Q,$$

where Q is some function. To identify the field $\boldsymbol{v}$ with the velocity field $\boldsymbol{u}$ solution of the primitive variable Navier–Stokes equations (and Q with $-P - \frac{1}{2}u^2$), it is necessary to show that the initial value of $\boldsymbol{v}$ implied by the set (4.4) coincides with $\boldsymbol{u}_0$. The well-posedness of the problem $\boldsymbol{\nabla}\times\boldsymbol{v} = \boldsymbol{\zeta}$, $\boldsymbol{\nabla}\cdot\boldsymbol{v} = 0$, $\boldsymbol{n}\cdot\boldsymbol{v}|_S = \boldsymbol{n}\cdot\boldsymbol{b}$ (under the conditions $\boldsymbol{\nabla}\cdot\boldsymbol{\zeta} = 0$ and $\oint \boldsymbol{n}\cdot\boldsymbol{b}\,dS = 0$), together with the initial condition $\boldsymbol{\zeta}|_{t=0} = \boldsymbol{\nabla}\times\boldsymbol{u}_0$ and the assumed continuity of $\boldsymbol{n}\cdot\boldsymbol{b}$ as a function of time as $t \to 0^+$, implies that $\boldsymbol{v}|_{t=0} = \boldsymbol{v}_0$, where $\boldsymbol{v}_0$ is the solution of the problem

$$\boldsymbol{\nabla}\times\boldsymbol{v}_0 = \boldsymbol{\zeta}_0 = \boldsymbol{\nabla}\times\boldsymbol{u}_0, \qquad \boldsymbol{\nabla}\cdot\boldsymbol{v}_0 = 0, \qquad \boldsymbol{n}\cdot\boldsymbol{v}_0|_S = \boldsymbol{n}\cdot\boldsymbol{b}|_{t=0}.$$

The first equation means that $\boldsymbol{\nabla}\times(\boldsymbol{v}_0 - \boldsymbol{u}_0) = 0$, namely, $\boldsymbol{v}_0 - \boldsymbol{u}_0 = \boldsymbol{\nabla}\alpha$, for some function α, which is harmonic since $\boldsymbol{\nabla}\cdot\boldsymbol{v}_0 = 0$ and also $\boldsymbol{\nabla}\cdot\boldsymbol{u}_0 = 0$ by the second of the compatibility conditions (4.5). Taking the normal component of this equation on S gives:

$$\boldsymbol{n}\cdot(\boldsymbol{v}_0 - \boldsymbol{u}_0)|_S = \boldsymbol{n}\cdot\boldsymbol{v}_0|_S - \boldsymbol{n}\cdot\boldsymbol{u}_0|_S = \boldsymbol{n}\cdot\boldsymbol{b}|_{t=0} - \boldsymbol{n}\cdot\boldsymbol{u}_0|_S = \boldsymbol{n}\cdot\boldsymbol{\nabla}\alpha|_S.$$

Then, by the third of the compatibility conditions (4.5), $\boldsymbol{n}\cdot\boldsymbol{\nabla}\alpha|_S = 0$, so that α = constant. It follows that $\boldsymbol{v}_0 = \boldsymbol{u}_0$ and hence $\boldsymbol{v} = \boldsymbol{u}$, which completes the proof. □

The equation relating $\boldsymbol{u}$ to $\boldsymbol{\zeta}$ in the formulation above is of first order and its solution cannot be attempted by means of standard numerical techniques. It is however possible to make use of the second first-order equation for $\boldsymbol{u}$, which imposes that the velocity is solenoidal, to derive the second-order elliptic equation $-\nabla^2 \boldsymbol{u} = \boldsymbol{\nabla}\times\boldsymbol{\zeta}$, which can be solved by standard techniques. By means of such a transformation, one obtains a vorticity–velocity formulation relying only on second-order (vector) equations, as stated by the following theorem.

Theorem 4.2. *The set of equations (4.4) is equivalent to the following system of two second-order equations:*

$$\begin{aligned}
&\frac{\partial \boldsymbol{\zeta}}{\partial t} - \nu\nabla^2\boldsymbol{\zeta} + \boldsymbol{\nabla}\times(\boldsymbol{\zeta}\times\boldsymbol{u}) = 0, && \boldsymbol{\zeta}|_{t=0} = \boldsymbol{\nabla}\times\boldsymbol{u}_0, \\
& && \boldsymbol{n}\cdot\boldsymbol{\zeta}|_S = \boldsymbol{n}\cdot\boldsymbol{\nabla}_S\times\boldsymbol{b}, \quad \boldsymbol{\nabla}\cdot\boldsymbol{\zeta}|_S = 0, \\
&-\nabla^2\boldsymbol{u} = \boldsymbol{\nabla}\times\boldsymbol{\zeta}, && \boldsymbol{u}|_S = \boldsymbol{b}, \qquad \boldsymbol{\nabla}\cdot\boldsymbol{u}|_S = 0.
\end{aligned} \tag{4.6}$$

Proof. The implication is a consequence of the following four facts:

i) the elliptic equation $-\nabla^2\boldsymbol{u} = \boldsymbol{\zeta}$ follows from taking the curl of the equation $\boldsymbol{\zeta} = \boldsymbol{\nabla}\times\boldsymbol{u}$, using the vector identity $\boldsymbol{\nabla}\times\boldsymbol{\nabla}\times\boldsymbol{\alpha} = -\nabla^2\boldsymbol{\alpha} + \boldsymbol{\nabla}(\boldsymbol{\nabla}\cdot\boldsymbol{\alpha})$ and the incompressibility condition;

ii) the first boundary condition for the vorticity $\boldsymbol{n}\cdot\boldsymbol{\zeta}|_S = \boldsymbol{n}\cdot\boldsymbol{\nabla}_S\times\boldsymbol{b}$ is simply the component of the equation $\boldsymbol{\zeta} = \boldsymbol{\nabla}\times\boldsymbol{u}$ normal to S, in which the velocity boundary condition $\boldsymbol{n}\times\boldsymbol{u}|_S = \boldsymbol{n}\times\boldsymbol{b}$ has been taken ito account to express $\boldsymbol{n}\cdot\boldsymbol{\nabla}\times\boldsymbol{u}$ in terms of $\boldsymbol{n}\cdot\boldsymbol{\nabla}_S\times\boldsymbol{b}$;

iii) the second boundary condition for the vorticity $\boldsymbol{\nabla}\cdot\boldsymbol{\zeta}|_S = 0$ is a consequence of the equation $\boldsymbol{\nabla}\times\boldsymbol{u} = \boldsymbol{\zeta}$ which implies $\boldsymbol{\nabla}\cdot\boldsymbol{\zeta} = 0$ in the entire domain V, including the boundary S.

iv) the derivative boundary condition $\boldsymbol{\nabla}\cdot\boldsymbol{u}|_S = 0$ is the trace of the equation of continuity.

Conversely, let $(\boldsymbol{\omega}, \boldsymbol{v})$ be a solution of the set of equations and conditions (4.6). Then, $\boldsymbol{v}$ is solenoidal since $\boldsymbol{\nabla}\cdot(-\nabla^2\boldsymbol{v}) = -\nabla^2(\boldsymbol{\nabla}\cdot\boldsymbol{v}) = \boldsymbol{\nabla}\cdot\boldsymbol{\nabla}\times\boldsymbol{\omega} = 0$ and $\boldsymbol{\nabla}\cdot\boldsymbol{v}|_S = 0$. On the other hand, by the identity $-\nabla^2\boldsymbol{\alpha} = \boldsymbol{\nabla}\times\boldsymbol{\nabla}\times\boldsymbol{\alpha} - \boldsymbol{\nabla}(\boldsymbol{\nabla}\cdot\boldsymbol{\alpha})$, it is

$$-\nabla^2\boldsymbol{v} = \boldsymbol{\nabla}\times\boldsymbol{\nabla}\times\boldsymbol{v} - \boldsymbol{\nabla}(\boldsymbol{\nabla}\cdot\boldsymbol{v}) = \boldsymbol{\nabla}\times\boldsymbol{\nabla}\times\boldsymbol{v} = \boldsymbol{\nabla}\times\boldsymbol{\omega},$$

that is, $\boldsymbol{\nabla}\times(\boldsymbol{\nabla}\times\boldsymbol{v}-\boldsymbol{\omega}) = 0$ or equivalently $\boldsymbol{\nabla}\times\boldsymbol{v} = \boldsymbol{\omega}+\nabla\alpha$, for some function α. Now, $\boldsymbol{\omega}$ is solenoidal for any $t > 0$ because taking the divergence of the vorticity equation gives the heat equation for $\boldsymbol{\nabla}\cdot\boldsymbol{\omega}$ supplemented by the initial condition $\boldsymbol{\nabla}\cdot\boldsymbol{\omega}|_{t=0} = \boldsymbol{\nabla}\cdot\boldsymbol{\nabla}\times\boldsymbol{u}_0 = 0$ and the boundary condition $\boldsymbol{\nabla}\cdot\boldsymbol{\omega}|_S = 0$, which are both homogeneous. Thus, taking the divergence of the equation $\boldsymbol{\nabla}\times\boldsymbol{v} = \boldsymbol{\omega}+\boldsymbol{\nabla}\alpha$ gives $\nabla^2\alpha = 0$. Considering the normal component of the same equation on the boundary gives $\boldsymbol{n}\cdot\boldsymbol{\nabla}\times\boldsymbol{v}|_S = \boldsymbol{n}\cdot\boldsymbol{\omega}|_S + \boldsymbol{n}\cdot\nabla\alpha|_S$. By the boundary conditions for the tangential components of $\boldsymbol{v}$ and for the normal component of $\boldsymbol{\omega}$, it results

$\boldsymbol{n}\cdot\boldsymbol{\nabla}\alpha|_S = 0$. Therefore, $\alpha =$ constant so that $\boldsymbol{\nabla}\times\boldsymbol{v} = \boldsymbol{\omega}$. Thus $\boldsymbol{\omega}$ and $\boldsymbol{v}$ satisfy the same equations and conditions satisfied by $\boldsymbol{\zeta}$ and $\boldsymbol{u}$, so that $\boldsymbol{\omega} = \boldsymbol{\zeta}$ and $\boldsymbol{v} = \boldsymbol{u}$. □

The formulation of the $\boldsymbol{\zeta}$-$\boldsymbol{u}$ equations for the general case of multiply connected domains is discussed by Daube, Guermond and Sellier (1991).

4.2.2 Uncoupled formulation

The analysis of the $\boldsymbol{\zeta}$-$\boldsymbol{u}$ formulation is still incomplete unless one is capable of supplementing the vorticity equation with its own complete set of conditions, so as to obtain a problem for this variable independent from the determination of velocity, at least in the linear case. This goal can be achieved resorting to the same kind of argument involving an orthogonal projection which has been considered in the nonprimitive variable formulations. In fact, similarly to the previous analyses, there is one boundary condition in excess for the velocity whereas one condition is lacking for the vorticity. The argument must however be adapted to the present situation of the $\boldsymbol{\zeta}$-$\boldsymbol{u}$ equations, since the redundant boundary conditions are slightly different.

Let Δ be the Laplace operator acting on vector fields and supplemented by the boundary conditions which are the homogeneous version of the conditions prescribed on $\boldsymbol{u}$ in problem (4.6), namely,

$$\Delta = \left\{-\nabla^2, \quad \boldsymbol{n}\times\ldots|_S = 0,\ \boldsymbol{n}\cdot\ldots|_S = 0,\ \boldsymbol{\nabla}\cdot\ldots|_S = 0\right\}. \tag{4.7}$$

(Note that here the symbol Δ has a different meaning with respect to that in section 3.3.3.) To identify the operator $\Delta^\dagger$, adjoint to Δ, one introduces the vector Green identity for the operator $-\nabla^2$, namely,

$$\int(\boldsymbol{u}\cdot\nabla^2\boldsymbol{v} - \boldsymbol{v}\cdot\nabla^2\boldsymbol{u})\,dV$$
$$= \oint(\boldsymbol{n}\times\boldsymbol{u}\cdot\boldsymbol{\nabla}\times\boldsymbol{v} - \boldsymbol{n}\times\boldsymbol{v}\cdot\boldsymbol{\nabla}\times\boldsymbol{u} + \boldsymbol{n}\cdot\boldsymbol{u}\,\boldsymbol{\nabla}\cdot\boldsymbol{v} - \boldsymbol{n}\cdot\boldsymbol{v}\,\boldsymbol{\nabla}\cdot\boldsymbol{u})\,dS. \tag{4.8}$$

The examination of this identity shows immediately that the only boundary term remaining after taking into account the boundary conditions associated with Δ is $\boldsymbol{n}\times\boldsymbol{v}\cdot\boldsymbol{\nabla}\times\boldsymbol{u}$. It follows that the boundary condition to be associated with $\Delta^\dagger$ is $\boldsymbol{n}\times\ldots|_S = 0$, so that

$$\Delta^\dagger = \left\{-\nabla^2, \quad \boldsymbol{n}\times\ldots|_S = 0\right\}. \tag{4.9}$$

Considering now the particular case of $\boldsymbol{u}$ satisfying homogeneous boundary conditions, the fact that $\boldsymbol{\zeta}$ is such that $\boldsymbol{\nabla}\times\boldsymbol{\zeta} = -\nabla^2\boldsymbol{u}$, with $\boldsymbol{u}|_S = 0$ and $\boldsymbol{\nabla}\cdot\boldsymbol{u}|_S = 0$, can also be stated in the form $\boldsymbol{\nabla}\times\boldsymbol{\zeta} \in \mathcal{R}(\Delta)$. Then, the orthogonality theorem mentioned in section 2.8, namely,

$$\overline{\mathcal{R}(A)} = \mathcal{N}(A^\dagger)^\perp, \tag{4.10}$$

means that the vorticity vector field must be such that

$$\nabla\times\boldsymbol{\zeta}\perp\mathcal{N}(\Delta^\dagger). \tag{4.11}$$

The null space $\mathcal{N}(\Delta^\dagger)$ is characterized as follows

$$\mathcal{N}(\Delta^\dagger)=\left\{\boldsymbol{\chi}\,|\,-\nabla^2\boldsymbol{\chi}=0,\quad \boldsymbol{n}\times\boldsymbol{\chi}|_S=0\right\}. \tag{4.12}$$

The form of the boundary conditions on (4.12) means that the linear space involved by the vorticity conditions in the $\boldsymbol{\zeta}$-$\boldsymbol{u}$ representation has a "number" of linearly independent element $\boldsymbol{\chi}$ equal to that of the points of the boundary S. Moreover, the problem defining $\boldsymbol{\chi}$ in this formulation coincides with that which defines the space $\mathcal{H}_{\rm nor}$ for the orthogonal projection in the nonprimitive variable formulations using the reduced integral conditions.

The orthogonality condition in the $\boldsymbol{\zeta}$-$\boldsymbol{u}$ formulation can be expressed in the form

$$\int\nabla\times\boldsymbol{\zeta}\cdot\boldsymbol{\chi}\,dV=0, \tag{4.13}$$

for any harmonic vector field $\boldsymbol{\chi}$ defined by problem (4.12). Going back to the situation of nonhomogeneous boundary conditions, the vorticity condition above becomes the following projection condition:

$$\int\nabla\times\boldsymbol{\zeta}\cdot\boldsymbol{\chi}\,dV=\oint(\boldsymbol{n}\times\boldsymbol{b}\cdot\nabla\times\boldsymbol{\chi}+\boldsymbol{n}\cdot\boldsymbol{b}\,\nabla\cdot\boldsymbol{\chi})\,dS. \tag{4.14}$$

A major difference in the vorticity integral conditions for the $\boldsymbol{\zeta}$-$\boldsymbol{u}$ representation with respect to those appearing in the nonprimitive variable formulations is that *both* terms comparing in the boundary integral provide a natural contribution to the integral conditions. The final uncoupled form of the vorticity–velocity equations for three-dimensional flows reads

$$\begin{aligned}
&-\nu\nabla^2\boldsymbol{\zeta}+\frac{\partial\boldsymbol{\zeta}}{\partial t}+\nabla\times(\boldsymbol{\zeta}\times\boldsymbol{u})=0, && \boldsymbol{\zeta}|_{t=0}=\nabla\times\boldsymbol{u}_0,\\
&\int\nabla\times\boldsymbol{\zeta}\cdot\boldsymbol{\chi}\,dV=\oint(\boldsymbol{n}\times\boldsymbol{b}\cdot\nabla\times\boldsymbol{\chi}+\boldsymbol{n}\cdot\boldsymbol{b}\,\nabla\cdot\boldsymbol{\chi})\,dS, && \boldsymbol{n}\cdot\boldsymbol{\zeta}|_S=\boldsymbol{n}\cdot\nabla_S\times\boldsymbol{b},\quad \nabla\cdot\boldsymbol{\zeta}|_S=0,\\
&\left[\,-\nabla^2\boldsymbol{\chi}=0,\qquad \boldsymbol{n}\times\boldsymbol{\chi}|_S=0;\,\right]\\
&-\nabla^2\boldsymbol{u}=\nabla\times\boldsymbol{\zeta}, && \boldsymbol{u}|_S=\boldsymbol{b}.
\end{aligned} \tag{4.15}$$

It can be interesting to compare this formulation with any of those using the nonprimitive variables in three dimensions. The first observation concerns the character of the boundary conditions associated with the Poisson equation for the velocity: here we have a purely Dirichlet problem for the vector unknown, instead of the mixed Dirichlet–Neumann or purely Neumann conditions found for the vector potential or the stream vector in the nonprimitive variable representations.

Thus, the vector elliptic problem gives here three uncoupled Dirichlet problems for the vector components of velocity. The fact that all the boundary conditions for the velocity are required to solve the second vector equation of the system is in accordance with the natural character of both terms $\boldsymbol{n}\times\boldsymbol{b}$ and $\boldsymbol{n}\cdot\boldsymbol{b}$ appearing in the integral conditions. When the fulfillment of the integral conditions is achieved by means of the Glowinski–Pironneau method, the two aforementioned boundary data for the velocity will affect the vorticity field indirectly through the boundary conditions imposed on the field of the type $\boldsymbol{u}^0$. Correspondingly, also the projection space $\{\boldsymbol{\chi}\}$ can be characterized by boundary conditions of mere Dirichlet type $\boldsymbol{n}\times\boldsymbol{\chi}|_S = 0$ and $\boldsymbol{n}\cdot\boldsymbol{\chi}|_S \neq 0$, so that it could be determined in principle by solving uncoupled Dirichlet problems for the Cartesian components of $\boldsymbol{\chi}$.

In spite of all these favourable features, the $\boldsymbol{\zeta}$-$\boldsymbol{u}$ formulation presents the following disadvantage. The vorticity equation of this formulation occurs to be supplemented by the pair of boundary conditions

$$\boldsymbol{n}\cdot\boldsymbol{\zeta}|_S = \boldsymbol{n}\cdot\boldsymbol{\nabla}_S\times\boldsymbol{b} \qquad \text{and} \qquad \boldsymbol{\nabla}\cdot\boldsymbol{\zeta}|_S = 0.$$

Apparently, these conditions cannot be imposed simultaneously in the framework of a variational formulation for vector second-order elliptic or parabolic equations. Due to this difficulty, we are not able to obtain the Glowinski–Pironneau form of the $\boldsymbol{\zeta}$-$\boldsymbol{u}$ equations in three dimensions, although we firmly believe that a sort of it should be possible, provided that the aforementioned boundary conditions could be given some variational meaning.

4.3 Two-dimensional equations

4.3.1 Governing equations

The vorticity–velocity formulation can be used also to represent the motion of an incompressible fluid in two dimensions. For such a kind of flows, the vorticity vector reduces to its component ζ normal to plane of the fluid motion, namely,

$$\zeta = \boldsymbol{\nabla}\times\boldsymbol{u}\cdot\boldsymbol{k} \qquad \text{or} \qquad \zeta\,\boldsymbol{k} = \boldsymbol{\nabla}\times\boldsymbol{u},$$

where $\boldsymbol{k}$ denotes the unit vector normal to that plane and $\boldsymbol{\nabla} = (\partial/\partial x, \partial/\partial y)$. Then, the vorticity equation assumes the simpler form

$$\left(\frac{\partial\zeta}{\partial t} - \nu\nabla^2\zeta\right)\boldsymbol{k} + \boldsymbol{\nabla}\times(\zeta\,\boldsymbol{k}\times\boldsymbol{u}) = 0. \tag{4.16}$$

The velocity field $\boldsymbol{u}$ must satisfy the two first-order equations:

$$\boldsymbol{\nabla}\times\boldsymbol{u} = \zeta\,\boldsymbol{k} \qquad \text{and} \qquad \boldsymbol{\nabla}\cdot\boldsymbol{u} = 0. \tag{4.17}$$

The system (4.16)–(4.17) is supplemented by the initial and boundary conditions

$$\zeta|_{t=0} = \boldsymbol{\nabla}\times\boldsymbol{u}_0\cdot\boldsymbol{k}, \qquad \boldsymbol{u}|_S = \boldsymbol{b}, \tag{4.18}$$

with the data satisfying the usual compatibility conditions

$$\oint \boldsymbol{n}\cdot\boldsymbol{b}\,ds = 0, \qquad \boldsymbol{\nabla}\cdot\boldsymbol{u}_0 = 0, \qquad \boldsymbol{n}\cdot\boldsymbol{b}|_{t=0} = \boldsymbol{n}\cdot\boldsymbol{u}_0|_S, \tag{4.19}$$

where ds denotes the length of the element of the boundary S.

Theorem 4.3. *The set of equations and conditions (4.16)–(4.18) with the data satisfying the compatibility conditions (4.19) is equivalent to the primitive variable Navier–Stokes equations (1.1)–(1.4) in two dimensions.*

This theorem can be proved much in the same manner as its three-dimensional counterpart (Theorem 4.1), by noticing that the problem, in two dimensions,

$$\boldsymbol{\nabla}\times\boldsymbol{v} = \zeta\,\boldsymbol{k}, \qquad \boldsymbol{\nabla}\cdot\boldsymbol{v} = 0, \qquad \boldsymbol{n}\cdot\boldsymbol{v}|_S = \boldsymbol{n}\cdot\boldsymbol{b}$$

is well-posed provided that $\oint \boldsymbol{n}\cdot\boldsymbol{b}\,ds = 0$ and defines the vector field $\boldsymbol{v}$ uniquely.

The vorticity–velocity problem for two-dimensional flows can be reduced to a system of second-order equations according to a theorem, which is slightly different from its three-dimensional analogue (Theorem 4.2).

Theorem 4.4. *The set of equations and conditions (4.16)–(4.18) is equivalent to the following system of two second-order equations for the variables ζ and $\boldsymbol{u}$:*

$$\begin{aligned}
&\left(\frac{\partial\zeta}{\partial t} - \nu\nabla^2\zeta\right)\boldsymbol{k} + \boldsymbol{\nabla}\times(\zeta\,\boldsymbol{k}\times\boldsymbol{u}) = 0, && \zeta|_{t=0} = \boldsymbol{\nabla}\times\boldsymbol{u}_0\cdot\boldsymbol{k},\\
& && \int \zeta\,dV = \oint \boldsymbol{\tau}\cdot\boldsymbol{b}\,ds,\\
&-\nabla^2\boldsymbol{u} = \boldsymbol{\nabla}\zeta\times\boldsymbol{k}, && \boldsymbol{u}|_S = \boldsymbol{b}, \quad \boldsymbol{\nabla}\cdot\boldsymbol{u}|_S = 0,
\end{aligned} \tag{4.20}$$

where $\boldsymbol{\tau}$ denotes the unit vector tangential to the boundary.

Proof. The implication is a consequence of the three facts:

i) taking the curl of the equation $\boldsymbol{\nabla}\times\boldsymbol{u} = \zeta\,\boldsymbol{k}$ gives $\boldsymbol{\nabla}\times\boldsymbol{\nabla}\times\boldsymbol{u} = -\nabla^2\boldsymbol{u} + \boldsymbol{\nabla}(\boldsymbol{\nabla}\cdot\boldsymbol{u}) = \boldsymbol{\nabla}\times(\zeta\,\boldsymbol{k}) = \boldsymbol{\nabla}\zeta\times\boldsymbol{k}$, having used the equation of continuity;

ii) the derivative boundary condition $\boldsymbol{\nabla}\cdot\boldsymbol{u}|_S = 0$ is the trace of the equation of continuity;

iii) the (single) integral condition for the vorticity is obtained by integrating the equation $\boldsymbol{\nabla}\times\boldsymbol{u} = \zeta\,\boldsymbol{k}$ over the domain V. The application of Stokes theorem to the "surface" V with "contour" S gives

$$\int \zeta\,dV = \int \boldsymbol{\nabla}\times\boldsymbol{u}\cdot\boldsymbol{k}\,dV = \oint \boldsymbol{\tau}\cdot\boldsymbol{u}\,ds = \oint \boldsymbol{\tau}\cdot\boldsymbol{b}\,ds,$$

where the boundary condition for the tangential component of velocity has been used.

Conversely, let $(\omega, \boldsymbol{v})$ be a solution of the set of equations and conditions (4.20). We demonstrate that the pair $(\omega, \boldsymbol{v})$ coincides with the pair $(\zeta, \boldsymbol{u})$ solution to the set (4.16)–(4.18). First observe that $\boldsymbol{v}$ is solenoidal since $\boldsymbol{\nabla}\cdot(-\nabla^2\boldsymbol{v}) = -\nabla^2(\boldsymbol{\nabla}\cdot\boldsymbol{v}) = \boldsymbol{\nabla}\cdot(\boldsymbol{\nabla}\omega\times\boldsymbol{k}) = \boldsymbol{\nabla}\times\boldsymbol{\nabla}\omega\cdot\boldsymbol{k} - \boldsymbol{\nabla}\omega\cdot\boldsymbol{\nabla}\times\boldsymbol{k} = 0$ and $\boldsymbol{\nabla}\cdot\boldsymbol{v}|_S = 0$. On the other hand, the identity $-\nabla^2\boldsymbol{\alpha} = \boldsymbol{\nabla}\times\boldsymbol{\nabla}\times\boldsymbol{\alpha} - \boldsymbol{\nabla}(\boldsymbol{\nabla}\cdot\boldsymbol{\alpha})$ allows to write

$$-\nabla^2\boldsymbol{v} = \boldsymbol{\nabla}\times\boldsymbol{\nabla}\times\boldsymbol{v} - \boldsymbol{\nabla}(\boldsymbol{\nabla}\cdot\boldsymbol{v}) = \boldsymbol{\nabla}\times\boldsymbol{\nabla}\times\boldsymbol{v} = \boldsymbol{\nabla}\omega\times\boldsymbol{k} = \boldsymbol{\nabla}\times(\omega\,\boldsymbol{k}).$$

Reminding that all these equations refer to fields in two dimensions, the result above means

$$\boldsymbol{\nabla}\times\boldsymbol{v}\cdot\boldsymbol{k} - \omega = \text{constant}.$$

But, by the integral condition $\int \omega\, dV = \oint \boldsymbol{\tau}\cdot\boldsymbol{b}\, ds$, the constant can be shown to be zero, using Stokes theorem and the tangential boundary condition for $\boldsymbol{v}$. Therefore, $\boldsymbol{\nabla}\times\boldsymbol{v}\cdot\boldsymbol{k} = \omega$ and, all the equations and conditions in (4.16)–(4.18) being satisfied by $(\omega, \boldsymbol{v})$, it results $\omega = \zeta$ and $\boldsymbol{v} = \boldsymbol{u}$. □

A time-discretized version of the formulation (4.20) has been considered by Daube (1992) to obtain a finite difference methods using a staggered distribution of the dependent variables ζ and $\boldsymbol{u}$ over the spatial grid. In that work the additional boundary condition $\boldsymbol{\nabla}\cdot\boldsymbol{u}|_S = 0$ has been dealt with by means of the influence matrix technique. A second formulation still based on the influence matrix technique but using as additional boundary condition the relation $\zeta|_S = \boldsymbol{\nabla}\times\boldsymbol{u}\cdot\boldsymbol{k}|_S$ in place of $\boldsymbol{\nabla}\cdot\boldsymbol{u}|_S = 0$ has also been considered. The numerical comparisons show that the two methods provide identical result, within machine accuracy.

4.3.2 Uncoupled formulation

The boundary conditions for the velocity in the ζ-$\boldsymbol{u}$ formulation display the usual redundancy, in that three scalar conditions are obtained to supplement the velocity unknown solution of a Poisson equation in two dimensions. On the other hand, no condition is available for the scalar vorticity, apart from a *single* integral condition prescribing the average value of the vorticity field in terms of the tangential velocity along the boundary. To obtain vorticity conditions allowing an uncoupled formulation of the two equations, the orthogonal projection argument applied in section 4.2.2 can be repeated, considering this time the two-dimensional counterpart of vector Green identity (4.8). In this manner, one can obtain the vorticity integral conditions for two-dimensional flows in the form

$$\int \boldsymbol{\nabla}\zeta\times\boldsymbol{k}\cdot\boldsymbol{\chi}\, dV = \oint (\boldsymbol{n}\times\boldsymbol{b}\cdot\boldsymbol{\nabla}\times\boldsymbol{\chi} + \boldsymbol{n}\cdot\boldsymbol{b}\,\boldsymbol{\nabla}\cdot\boldsymbol{\chi})\, ds. \tag{4.21}$$

These conditions must be satisfied by ζ for any two-dimensional vector field $\boldsymbol{\chi}$ harmonic in V and normal to the boundary (cf. the definition (4.12)). Then, the complete set of uncoupled ζ-$\boldsymbol{u}$ equations for plane two-dimensional flows reads

$$\Big(-\nu\nabla^2\zeta + \frac{\partial\zeta}{\partial t}\Big)\boldsymbol{k} + \boldsymbol{\nabla}\times(\zeta\,\boldsymbol{k}\times\boldsymbol{u}) = 0, \qquad \zeta|_{t=0} = \boldsymbol{\nabla}\times\boldsymbol{u}_0\cdot\boldsymbol{k},$$
$$\int\boldsymbol{\nabla}\zeta\times\boldsymbol{k}\cdot\boldsymbol{\chi}\,dV = \oint(\boldsymbol{n}\times\boldsymbol{b}\cdot\boldsymbol{\nabla}\times\boldsymbol{\chi} + \boldsymbol{n}\cdot\boldsymbol{b}\,\boldsymbol{\nabla}\cdot\boldsymbol{\chi})\,ds, \qquad \int\zeta\,dV = \oint\boldsymbol{\tau}\cdot\boldsymbol{b}\,ds,$$
$$\Big[-\nabla^2\boldsymbol{\chi} = 0, \qquad \boldsymbol{\tau}\cdot\boldsymbol{\chi}|_S = 0;\Big]$$
$$-\nabla^2\boldsymbol{u} = \boldsymbol{\nabla}\zeta\times\boldsymbol{k}, \qquad \boldsymbol{u}|_S = \boldsymbol{b}. \tag{4.22}$$

We note that the vorticity equation and the projection integral conditions are not capable of determining the solution ζ uniquely because they involve this unknown only under derivative operators. It is the additional single integral condition for the vorticity, obtained by means of Stokes theorem, that allows the unique determination of ζ.

4.3.3 Glowinski–Pironneau method

We now consider a time-discretized version of the uncoupled equations for two-dimensional problems, namely,

$$(-\nabla^2+\gamma)\zeta = f, \qquad \int\zeta\,dV = \oint\boldsymbol{\tau}\cdot\boldsymbol{b}\,ds,$$
$$\int\boldsymbol{\nabla}\zeta\times\boldsymbol{k}\cdot\boldsymbol{\chi}\,dV = \oint(\boldsymbol{n}\times\boldsymbol{b}\cdot\boldsymbol{\nabla}\times\boldsymbol{\chi} + \boldsymbol{n}\cdot\boldsymbol{b}\,\boldsymbol{\nabla}\cdot\boldsymbol{\chi})\,ds; \tag{4.23}$$
$$-\nabla^2\boldsymbol{u} = \boldsymbol{\nabla}\zeta\times\boldsymbol{k}, \qquad \boldsymbol{u}|_S = \boldsymbol{b},$$

with an obvious "temporal" meaning of the various quantities indicated. It is not difficult to develop a method for imposing the integral conditions, belonging to the same class of the procedure introduced by Glowinski and Pironneau for the vorticity–stream function equations. Let the solution $\zeta(\boldsymbol{x})$ be decomposed in the usual form:

$$\zeta(\boldsymbol{x}) = \zeta^0(\boldsymbol{x}) + \oint\zeta'(\boldsymbol{x};\sigma')\lambda(\sigma')\,d\sigma'. \tag{4.24}$$

Here, the fields $\zeta^0(\boldsymbol{x})$ and $\zeta'(\boldsymbol{x};\sigma')$, for any $\sigma'\in S$, are the solutions to the following problems:

$$(-\nabla^2+\gamma)\zeta^0 = f, \qquad \zeta^0|_S = 0; \tag{4.25a}$$
$$(-\nabla^2+\gamma)\zeta' = 0, \qquad \zeta'(\boldsymbol{x};\sigma')|_S = \delta(s-\sigma'); \tag{4.25b}$$

where δ denotes the Dirac delta "function." The surface unknown $\lambda(\sigma)$, with $\sigma \in S$, is determined by imposing to ζ to satisfy the integral conditions in (4.23), which gives the linear problem

$$\oint B(\sigma,\sigma')\lambda(\sigma')\,d\sigma' = \beta(\sigma), \tag{4.26}$$

where

$$B(\sigma,\sigma') = \int \nabla\zeta'(\boldsymbol{x};\sigma')\times\boldsymbol{k}\cdot\boldsymbol{\chi}(\boldsymbol{x};\sigma)\,dV,$$

and

$$\begin{aligned}\beta(\sigma) = &-\int \nabla\zeta^0(\boldsymbol{x})\times\boldsymbol{k}\cdot\boldsymbol{\chi}(\boldsymbol{x};\sigma)\,dV \\ &+\oint \Big[\boldsymbol{n}\times\boldsymbol{b}(s)\cdot\boldsymbol{\chi}(\boldsymbol{x}_s;\sigma) + \boldsymbol{n}\cdot\boldsymbol{b}(s)\,\nabla\cdot\boldsymbol{\chi}(\boldsymbol{x}_s;\sigma)\Big]ds.\end{aligned}$$

Here, the vector field $\boldsymbol{\chi}(\boldsymbol{x};\sigma)$ is any solution to the following problem

$$-\nabla^2\boldsymbol{\chi} = 0, \qquad \boldsymbol{n}\cdot\boldsymbol{\chi}(\boldsymbol{x};\sigma)|_S = \delta(s-\sigma), \quad \boldsymbol{\tau}\cdot\boldsymbol{\chi}(\boldsymbol{x};\sigma)|_S = 0. \tag{4.27}$$

To eliminate the need of calculating and storing the harmonic vector fields $\boldsymbol{\chi}$, one can introduce the vector fields

$$\boldsymbol{v}(\boldsymbol{x};\sigma) \text{ arbitrary in } V, \qquad \boldsymbol{n}\cdot\boldsymbol{v}(\boldsymbol{x};\sigma)|_S = \delta(s-\sigma), \quad \boldsymbol{\tau}\cdot\boldsymbol{v}(\boldsymbol{x};\sigma)|_S = 0. \tag{4.28}$$

By means of vector Green identity for two-dimensional vector fields, it is possible to show that the quantities B and β can be characterized equivalently through the following expressions:

$$\begin{aligned}B(\sigma,\sigma') = &\int (\nabla\zeta'\times\boldsymbol{k}\cdot\boldsymbol{v} - \nabla\times\boldsymbol{u}'\cdot\nabla\times\boldsymbol{v} - \nabla\cdot\boldsymbol{u}'\,\nabla\cdot\boldsymbol{v})\,dV, \\ \beta(\sigma) = -&\int (\nabla\zeta^0\times\boldsymbol{k}\cdot\boldsymbol{v} - \nabla\times\boldsymbol{u}^0\cdot\nabla\times\boldsymbol{v} - \nabla\cdot\boldsymbol{u}^0\,\nabla\cdot\boldsymbol{v})\,dV,\end{aligned}$$

where we have introduced the solutions $\boldsymbol{u}^0$ and $\boldsymbol{u}'$ to the two problems

$$-\nabla^2\boldsymbol{u}^0 = \nabla\zeta^0\times\boldsymbol{k}, \qquad \boldsymbol{u}^0|_S = \boldsymbol{b}; \tag{4.29a}$$

$$-\nabla^2\boldsymbol{u}' = \nabla\zeta'\times\boldsymbol{k}, \qquad \boldsymbol{u}'|_S = 0. \tag{4.29b}$$

The arbitrariness of the fields $\boldsymbol{v}$ at all internal point of V can be exploited by choosing $\boldsymbol{v} = 0$ inside V so that the integration domain in the previous expressions becomes a narrow strip along the boundary. Of course, the linear problem $A\lambda = \beta$ is singular and the remaining integral condition

$$\int \zeta\,dV = \oint \boldsymbol{\tau}\cdot\boldsymbol{b}\,ds$$

should be used to eliminate the singularity. By imposing this condition on the vorticity expressed according to the decomposition (4.24), we obtain the single equation

$$\oint B_1(\sigma')\lambda(\sigma')\, d\sigma' = \beta_1, \tag{4.30}$$

where

$$\begin{aligned} B_1(\sigma') &= \int \zeta'\, dV, \\ \beta_1 &= -\int \zeta^0\, dV + \oint \boldsymbol{\tau}\cdot\boldsymbol{b}\, ds. \end{aligned}$$

Equation (4.30) is combined with the linear problem $B\lambda = \beta$ to give a nonsingular problem.

4.3.4 Discussion

The method defined by the equations (4.24)–(4.30) has been implemented by the author using four-noded quadrilateral elements with an equal-order bilinear interpolation of the variables ζ and $\boldsymbol{u}$. The numerical scheme so obtained has been applied to the solution of the well known driven cavity problem (Burggraf 1966). Unfortunately, in this implementation, the matrix B_h representing the discrete counterpart of the operator B turns out to be *doubly* singular. No explanation of such a strange occurrence has been found so that the feasibility of a finite element method for solving the uncoupled vorticity–velocity equations using equal-order interpolation of all variables remains to be demonstrated. Difficulties, maybe of a similar kind, have been encountered also by finite difference methods approximating the equations (4.20) on a nonstaggered grid (cf. Daube 1992). On the contrary, using a staggered distribution of the vorticity and velocity variables proves to be an effective means of avoiding the difficulty, both in two and three dimensions (see, *e.g.*, Napolitano and Pascazio 1991).

Chapter 5

Primitive variable formulation

5.1 Introduction

The methods for solving the time-dependent Navier–Stokes equations written in terms of the primitive variables velocity and pressure can be grouped in two main classes:

- methods which are based on a discretization in time of fractional-step type and characterized by the ortogonal projection onto the space of the solenoidal vector fields, in order to satisfy the incompressiblity condition, called *fractional-step projection method* or *fractional-step method* (Chorin 1968a-b and 1969, and Temam 1969a-b);
- methods relying upon standard, *i.e.*, nonfractional-step, time discretizations.

The fractional-step projection method occupies a very unique position within the domain of numerical methods for calculating incompressible viscous flows and, for this reason, it will be discussed in a separate chapter of this study. In particular, it will be shown that in a number of versions of the projection method the incompressibility condition is replaced by a Poisson equation for pressure. The precise form of the *boundary conditions* which are appropriate in each specific situation will be discussed.

The aim of the present chapter is to describe the second class of methods for the primitive variable unsteady Navier–Stokes equations. In nonfractional-step methods, the momentum and continuity equations are discretized in time according to the standard finite difference method without separating or splitting the treatment of viscosity from that of the incompressibility. This implies that the velocity and pressure fields are evaluated at the *same* time level so that no-slip velocity boundary condition is imposed simultaneously with the condition of incompressibility. A Poisson equation for pressure can be introduced in order to eliminate the continuity equation, and we will discuss only methods based on this assumption. It is important to keep in mind that the pressure Poisson equation considered in the fractional-step method is basically different from the Poisson equation derived in a nonfractional-step method: they should not be confused.

Unfortunately, the pressure Poisson equation introduced in a nonfractional-step method is *not completely* equivalent with the incompressibility condition, except in the absence of boundaries or with spatially periodic boundary conditions. Therefore, in order to ensure mass conservation, it is necessary to establish the equivalence of the primitive variable system containing the Poisson equation for pressure with the original Navier–Stokes problem. This goal has been achieved by Kleiser and Schumann (1980) who have shown that, to guarantee an equivalent formulation in the presence of solid walls, the derivative boundary condition $\nabla \cdot \boldsymbol{u}|_S = 0$ must be included in the system of equations and conditions.

This chapter and the next one intend to investigate the application of such a boundary condition to the numerical solution of the Navier–Stokes equations, and to provide its interpretation as far as the conditions effectively satisfied by the pressure are concerned. The present chapter addresses the issue within the framework of a *time-discretized* version of the equations and describes an uncoupled formulation for the numerical solution of the primitive variable equations. The next chapter will instead examine some mathematical consequences of the boundary condition of Kleiser and Schumann in the more general context of the unsteady equations with a *continuous* time dependence.

To complete our analysis of numerical schemes for the incompressible equations based on the solution of elliptic problems, the influence matrix method of Kleiser and Schumann for obtaining Chebyshev solutions to plane channel flows is described. A comparison of the numerical results calculated by this influence matrix method with those provided by the corresponding spectral method imposing integral conditions on pressure will be shown.

The present chapter contains also a brief description of the direct method introduced by Bristeau, Glowinski, Périaux and Pironneau (1987) for the solution of the primitive variable formulation of the Stokes problem. This method will be shown to be structurally very similar to the uncoupled solution method using the pressure integral conditions.

In the last section, following the analysis of Bristeau *et al.* (1987), the direct Stokes solver will be generalized to the compressible equations and a related uncoupled solution method still based on the concept of integral conditions will be sketched.

5.2 Time-discretized pressure–velocity equations

Consider a time-discretized version of the momentum equation (1.1) together with the continuity equation (1.2). Assume for simplicity that the nonlinear term is evaluated in an explicit manner whereas the linear term is taken into account in a fully implicit manner by a two-level integration scheme. If the unknown fields at the new time level are denoted by $\boldsymbol{u} = \boldsymbol{u}^{n+1}$ and $p = p^{n+1}$ ($p = \nu^{-1}P$), we have

$$\begin{aligned} (-\nabla^2 + \gamma)\boldsymbol{u} + \boldsymbol{\nabla} p &= \boldsymbol{g}, \\ \boldsymbol{\nabla}\cdot\boldsymbol{u} &= 0, \end{aligned} \tag{5.1}$$

where, as usual, $\gamma = 1/(\nu\Delta t)$ and $\boldsymbol{g} = \gamma\boldsymbol{u}^n - \nu^{-1}(\boldsymbol{u}^n\cdot\boldsymbol{\nabla})\boldsymbol{u}^n$. By taking the divergence of the (time-discretized) momentum equation and using the continuity equation, we obtain the following Poisson equation for p

$$\nabla^2 p = \boldsymbol{\nabla}\cdot\boldsymbol{g}. \tag{5.2}$$

The fulfillment of this equation and of the momentum equation does not assure that the incompressibility condition is satisfied. In fact, let $\boldsymbol{v}$ be a velocity field, not necessarily solenoidal, solution of the momentum equation with p solution of $\nabla^2 p = \boldsymbol{\nabla}\cdot\boldsymbol{g}$. By taking the divergence of the momentum equation one obtains $(-\nabla^2 + \gamma)\boldsymbol{\nabla}\cdot\boldsymbol{v} + \nabla^2 p = \boldsymbol{\nabla}\cdot\boldsymbol{g}$, that is, $(-\nabla^2 + \gamma)\boldsymbol{\nabla}\cdot\boldsymbol{v} = 0$. Thus, $\boldsymbol{\nabla}\cdot\boldsymbol{v}$ is only harmonic, but not necessarily equal to zero.

However, if one impose the boundary condition $\boldsymbol{\nabla}\cdot\boldsymbol{v}|_S = 0$, then the harmonicity of $\boldsymbol{\nabla}\cdot\boldsymbol{v}$ implies that $\boldsymbol{\nabla}\cdot\boldsymbol{v} = 0$ everywhere in V (Kleiser and Schumann 1980). This simple argument implies that the time-discretized Navier–Stokes problem can be written as a boundary value problem consisting of the following two elliptic equations:

$$\begin{aligned} (-\nabla^2 + \gamma)\boldsymbol{u} + \boldsymbol{\nabla} p &= \boldsymbol{g}, \\ -\nabla^2 p &= -\boldsymbol{\nabla}\cdot\boldsymbol{g}, \\ \boldsymbol{u}|_S = \boldsymbol{b}, \qquad \boldsymbol{\nabla}\cdot\boldsymbol{u}|_S &= 0, \end{aligned} \tag{5.3}$$

where, as usual, the global condition

$$\oint \boldsymbol{n}\cdot\boldsymbol{b}\, dS = 0 \tag{5.4}$$

is assumed to be satisfied. Note that here $\boldsymbol{b} = \boldsymbol{b}(\boldsymbol{x}_S) = \boldsymbol{b}(\boldsymbol{x}_S, t^{n+1})$.

The system (5.3) is faced with the same difficulty that has been encountered with the nonprimitive and hybrid variable representations of the Navier–Stokes equations: there are too many boundary condition for one variable, the velocity, and not enough for the other, the pressure. Thus, strictly speaking, problem (5.3) consisting of the two elliptic equations subject to the indicated boundary conditions has to be considered as a *coupled* system of equations, in the sense that the pressure and velocity should be solved simultaneously in consequence of a coupling brougth about by the redundant boundary conditions for $\boldsymbol{u}$.

5.3 Pressure integral conditions

To give the Poisson equation for pressure an independent individuality with respect to the velocity equation and thereby establish an uncoupled formulation of the set (5.3), it is necessary to derive conditions for the pressure field which do not depend on the values of $\boldsymbol{u}$ at points internal to the domain and which are only a function of the data $\boldsymbol{g}$ and $\boldsymbol{b}$ of the problem. This task can be accomplished by resorting once more to the fundamental orthogonality theorem

$$\overline{\mathcal{R}(A)} = \mathcal{N}(A^{\dagger})^{\perp}, \tag{5.5}$$

where A denotes the operator linking the variable subject to too many boundary conditions with the variable wanting conditions. The range of A is assumed to be a subspace of $\boldsymbol{L}^2(V)$ and $\perp$ indicates the orthogonal complement in $\boldsymbol{L}^2(V)$.

We first note that, by the momentum equation, the gradient of the pressure field must be such that $-\boldsymbol{\nabla} p + \boldsymbol{g} = (-\nabla^2 + \gamma)\boldsymbol{u}$, with $\boldsymbol{u}|_S = \boldsymbol{b}$ and $\boldsymbol{\nabla}\cdot\boldsymbol{u}|_S = 0$. Let $-\nabla^2_\gamma = (-\nabla^2 + \gamma)$ be the *metaharmonic* operator acting on the vector fields defined in V and let Δ_γ denote such an operator supplemented by the homogeneous version of the boundary conditions prescribed on $\boldsymbol{u}$ in (5.3), namely,

$$\Delta_\gamma = \Big\{(-\nabla^2 + \gamma), \quad \boldsymbol{n}\times \ldots|_S = 0,\ \boldsymbol{n}\cdot \ldots|_S = 0,\ \boldsymbol{\nabla}\cdot \ldots|_S = 0\Big\}. \tag{5.6}$$

(It can be noted that the boundary conditions associated with Δ_γ are coincident with those associated with the corresponding harmonic operator Δ occurring in the hybrid variable representation.)

Considering the particular case of homogeneous boundary conditions for $\boldsymbol{u}$, the previous characterization of $\boldsymbol{\nabla} p$ can also be expressed in the form

$$-\boldsymbol{\nabla} p + \boldsymbol{g} \in \mathcal{R}(\Delta_\gamma). \tag{5.7}$$

Then, the orthogonality theorem just recalled gives

$$-\boldsymbol{\nabla} p + \boldsymbol{g} \perp \mathcal{N}(\Delta_\gamma^{\dagger}). \tag{5.8}$$

To identify the operator $\Delta_\gamma^{\dagger}$, adjoint to Δ_γ, one considers vector Green identity for the operator ∇^2_γ, namely,

$$\int (\boldsymbol{u}\cdot\nabla^2_\gamma \boldsymbol{v} - \boldsymbol{v}\cdot\nabla^2_\gamma \boldsymbol{u})\, dV$$
$$= \oint (\boldsymbol{n}\times\boldsymbol{u}\cdot\boldsymbol{\nabla}\times\boldsymbol{v} - \boldsymbol{n}\times\boldsymbol{v}\cdot\boldsymbol{\nabla}\times\boldsymbol{u} + \boldsymbol{n}\cdot\boldsymbol{u}\,\boldsymbol{\nabla}\cdot\boldsymbol{v} - \boldsymbol{n}\cdot\boldsymbol{v}\,\boldsymbol{\nabla}\cdot\boldsymbol{u})\, dS. \tag{5.9}$$

The examination of this identity shows immediately that, after taking into account the boundary conditions associated with Δ_γ, only the term $\boldsymbol{n}\times\boldsymbol{v}\cdot\boldsymbol{\nabla}\times\boldsymbol{u}$ remains on the right hand side of Green identity (5.9). It follows that the boundary condition to be associated with $\Delta_\gamma^\dagger$ is $\boldsymbol{n}\times\ldots|_S = 0$, so that

$$\Delta_\gamma^\dagger = \Big\{(-\nabla^2+\gamma), \quad \boldsymbol{n}\times\ldots|_S = 0\Big\}. \tag{5.10}$$

Therefore, the null space $\mathcal{N}(\Delta_\gamma^\dagger)$ consists of the vector fields solution to the metaharmonic problem

$$(-\nabla^2+\gamma)\boldsymbol{\chi}_\gamma = 0, \qquad \boldsymbol{n}\times\boldsymbol{\chi}_\gamma|_S = 0, \quad \boldsymbol{n}\cdot\boldsymbol{\chi}_\gamma|_S \neq 0. \tag{5.11}$$

The boundary conditions imposed on $\boldsymbol{\chi}_\gamma$ mean that $\boldsymbol{\chi}_\gamma$ is normal to S, as the field $\boldsymbol{\chi}$ considered in connection with the vorticity conditions (cf. section 4.2.2). Thus the space of the metaharmonic fields $\{\boldsymbol{\chi}_\gamma\}$ is denoted by $\boldsymbol{\mathcal{M}}_{\gamma,\mathrm{nor}}$, and the condition (5.8) for the pressure gradient can also be written as

$$-\boldsymbol{\nabla}p+\boldsymbol{g} \perp \{\boldsymbol{\chi}_\gamma\} = \boldsymbol{\mathcal{M}}_{\gamma,\mathrm{nor}}. \tag{5.12}$$

The orthogonality condition in the p-$\boldsymbol{u}$ formulation is therefore expressed explicitly as

$$\int(-\boldsymbol{\nabla}p+\boldsymbol{g})\cdot\boldsymbol{\chi}_\gamma\, dV = 0, \tag{5.13}$$

for any vector field $\boldsymbol{\chi}_\gamma$ satisfying problem (5.11). Going back to the situation of nonhomogeneous boundary conditions, the orthogonality condition for pressure becomes the projection condition

$$\int(-\boldsymbol{\nabla}p+\boldsymbol{g})\cdot\boldsymbol{\chi}_\gamma\, dV = \oint(\boldsymbol{n}\times\boldsymbol{b}\cdot\boldsymbol{\nabla}\times\boldsymbol{\chi}_\gamma + \boldsymbol{n}\cdot\boldsymbol{b}\,\boldsymbol{\nabla}\cdot\boldsymbol{\chi}_\gamma)\, dS. \tag{5.14}$$

The important point about these integral conditions is that they depend only on the source term $\boldsymbol{g}$ of the momentum equation and on the value $\boldsymbol{b}$ of the velocity prescribed on the boundary. Furthermore, since the "number" of linearly independent metaharmonic fields $\boldsymbol{\chi}_\gamma$ is equal to that of boundary points, the projection integral conditions (5.14) provide the correct number of conditions to be combined with the Poisson equation to obtain a well determined problem for the pressure field, with p being defined, of course, up to an arbitrary additive constant. It follows that problem (5.3) can be reformulated equivalently in the uncoupled or split form (Quartapelle and Napolitano 1986)

$$\begin{aligned}
&-\nabla^2 p = -\boldsymbol{\nabla}\cdot\boldsymbol{g},\\
&-\int\boldsymbol{\nabla}p\cdot\boldsymbol{\chi}_\gamma\, dV = -\int\boldsymbol{g}\cdot\boldsymbol{\chi}_\gamma\, dV + \oint(\boldsymbol{n}\times\boldsymbol{b}\cdot\boldsymbol{\nabla}\times\boldsymbol{\chi}_\gamma + \boldsymbol{n}\cdot\boldsymbol{b}\,\boldsymbol{\nabla}\cdot\boldsymbol{\chi}_\gamma)\, dS;\\
&\Big[\ (-\nabla^2+\gamma)\boldsymbol{\chi}_\gamma = 0, \qquad\qquad \boldsymbol{n}\times\boldsymbol{\chi}_\gamma|_S = 0, \quad \boldsymbol{n}\cdot\boldsymbol{\chi}_\gamma|_S \neq 0;\ \Big]\\[2ex]
&(-\nabla^2+\gamma)\boldsymbol{u} = -\boldsymbol{\nabla}p+\boldsymbol{g}, \qquad\qquad \boldsymbol{u}|_S = \boldsymbol{b}.
\end{aligned} \tag{5.15}$$

The meaning of this uncoupled formulation is that, provided that the pressure field is determined subject to the integral conditions above, one can determine the velocity in a subsequent step, using $-\nabla p + \boldsymbol{g}$ as the source term for the elliptic equation with operator $(-\nabla^2 + \gamma)$ and Dirichlet conditions, obtaining a field which is automatically solenoidal.

A remarkable difference between the vorticity integral conditions established in the previous chapters and the pressure integral conditions just introduced is that the projection space associated with the former is defined prior to, and independently from, any time discretization of the governing equations whereas the projection space required by the latter is introduced after the time discretization of the primitive variable equations, as the dependence on the constant $\gamma = 1/(\nu\Delta t)$ in the notation $\boldsymbol{\chi}_\gamma$ indicates explicitly. A consequence of this fact is that the steady-state solutions calculated by uncoupled methods do not depend on the value of the time step Δt when ζ-ψ equations are considered, whereas are dependent on Δt when the primitive variables p and $\boldsymbol{u}$ are considered. This dependence is however very weak since it occurs through the pressure integral conditions and through the dependence on Δt of the fields $\boldsymbol{\chi}_\gamma$. Moreover, as it will be shown in the chapter on the fractional-step projection method, a similar dependence of the steady-state solutions on the value of Δt is encountered also in all methods of that kind whenever no-slip boundary conditions are specified.

5.4 Decomposition scheme

To obtain the pressure field satisfying the integral conditions in (5.15), the solution $p(\boldsymbol{x})$ is decomposed into its harmonic and nonharmonic components, as follows,

$$p(\boldsymbol{x}) = p^0(\boldsymbol{x}) + \oint p'(\boldsymbol{x};\boldsymbol{\sigma}')\lambda(\boldsymbol{\sigma}')\, dS(\boldsymbol{\sigma}'), \tag{5.16}$$

where $\lambda(\boldsymbol{\sigma}')$ is to be determined by imposing the integral conditions (5.15). The fields $p^0(\boldsymbol{x})$ and $p'(\boldsymbol{x};\boldsymbol{\sigma}')$, for any $\boldsymbol{\sigma}' \in S$, are the solutions to the problems:

$$-\nabla^2 p^0 = -\nabla\cdot\boldsymbol{g}, \qquad p^0|_S = 0; \tag{5.17a}$$

$$-\nabla^2 p' = 0, \qquad p'(\boldsymbol{x};\boldsymbol{\sigma}')|_S = \delta^{(2)}(\boldsymbol{s}-\boldsymbol{\sigma}'). \tag{5.17b}$$

Here $\delta^{(2)}(\boldsymbol{s})$ denotes the Dirac delta "function" over the boundary S, supposing that the domain V is three-dimensional. The geometrical meaning of such a decomposition and of the associated operation of orthogonal projection is very similar to that used to impose the vorticity integral conditions. Figure 5.1 shows the essential elements involved in the fulfillment of the pressure integral conditions, for the case of homogeneous data of the problem. The symbol $\mathbb{P}$ denotes the operator of orthogonal projection onto the linear space $\boldsymbol{\mathcal{M}}_{\gamma,\text{nor}} = \mathcal{N}(-\widetilde{\nabla}^2_{\times} + \gamma)$,

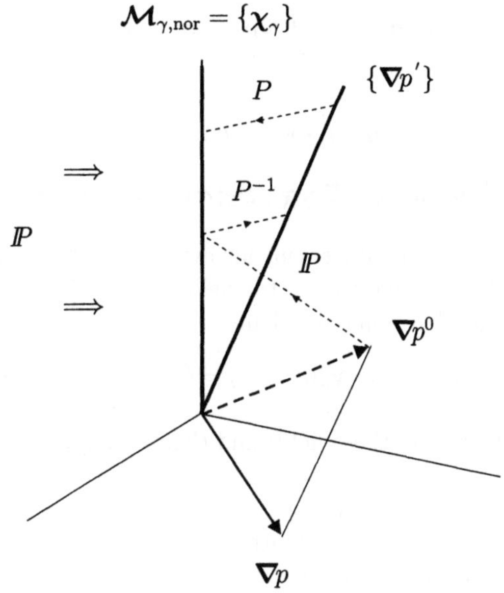

Figure 5.1: Schematic of the construction of the pressure field to impose the integral conditions, for homogeneous data of the problem.

that is, $\mathbb{P} = \mathbb{P}_{\mathcal{N}(-\widetilde{\nabla}^2_\times + \gamma)}$, where $\widetilde{\nabla}^2_\times$ is the Laplace operator acting on vector fields, supplemented with homogeneous boundary conditions for the tangential components ($\times$) but with unspecified normal component on S ($\tilde{\ }$). In addition, P denotes the restriction of $\mathbb{P}_{\mathcal{N}(-\widetilde{\nabla}^2_\times + \gamma)}$ to the linear space of the vector fields which are the gradient of the harmonic functions, namely

$$P = \mathbb{P}_{\mathcal{N}(-\widetilde{\nabla}^2_\times + \gamma)}\Big|_{\boldsymbol{\nabla}\mathcal{N}(-\widetilde{\nabla}^2)},$$

where $\widetilde{\nabla}^2$ represents the Laplace operator for scalar variables without boundary conditions. It can be interesting to note that in the transition from the nonprimitive variable representations to the primitive variable one, the harmonic and metaharmonic linear spaces interchange their respective rôles in the projective construction for satisfying the integral conditions.

The surface unknown $\lambda(\boldsymbol{\sigma})$, with $\boldsymbol{\sigma} \in S$, is determined by imposing the integral conditions in (5.15) on the function p given by the decomposition (5.16). This gives the linear problem

$$\oint B(\boldsymbol{\sigma}, \boldsymbol{\sigma}')\lambda(\boldsymbol{\sigma}')\, dS(\boldsymbol{\sigma}') = \beta(\boldsymbol{\sigma}), \tag{5.18}$$

where

$$B(\boldsymbol{\sigma},\boldsymbol{\sigma}') = -\int \boldsymbol{\nabla} p'(\boldsymbol{x};\boldsymbol{\sigma}')\cdot\boldsymbol{\chi}_\gamma(\boldsymbol{x};\boldsymbol{\sigma})\,dV, \tag{5.19}$$

and

$$\beta(\boldsymbol{\sigma}) = \int \Big[\boldsymbol{\nabla} p^0(\boldsymbol{x}) - \boldsymbol{g}(\boldsymbol{x})\Big]\cdot\boldsymbol{\chi}_\gamma(\boldsymbol{x};\boldsymbol{\sigma})\,dV \\ + \oint \Big[\boldsymbol{n}\times\boldsymbol{b}(\boldsymbol{x}_S)\cdot\boldsymbol{\nabla}\times\boldsymbol{\chi}_\gamma(\boldsymbol{x}_S;\boldsymbol{\sigma}) + \boldsymbol{n}\cdot\boldsymbol{b}(\boldsymbol{x}_S)\,\boldsymbol{\nabla}\cdot\boldsymbol{\chi}_\gamma(\boldsymbol{x}_S;\boldsymbol{\sigma})\Big]\,dS. \tag{5.20}$$

These expressions are not convenient for evaluating the quantities B and β numerically since they involve the vector fields $\boldsymbol{\chi}_\gamma(\boldsymbol{x};\boldsymbol{\sigma})$, for any $\boldsymbol{\sigma}\in S$, solution of the following metaharmonic problem

$$(-\nabla^2+\gamma)\boldsymbol{\chi}_\gamma = 0, \qquad \boldsymbol{n}\cdot\boldsymbol{\chi}_\gamma(\boldsymbol{x};\boldsymbol{\sigma})|_S = \delta^{(2)}(\boldsymbol{s}-\boldsymbol{\sigma}), \quad \boldsymbol{n}\times\boldsymbol{\chi}_\gamma(\boldsymbol{x};\boldsymbol{\sigma})|_S = 0. \tag{5.21}$$

The explicit evaluation of these fields and their storage in the computer memory can be avoided by means of a suitable generalization of the Glowinski–Pironneau method at the expense of increasing the number of elliptic equations to be solved. In place of the fields $\boldsymbol{\chi}_\gamma(\boldsymbol{x};\boldsymbol{\sigma})$, one introduces simpler vector fields $\boldsymbol{v}(\boldsymbol{x};\boldsymbol{\sigma})$ which are required to satisfy only boundary conditions identical to those in (5.21), namely,

$$\boldsymbol{v}(\boldsymbol{x};\boldsymbol{\sigma}) \text{ arbitrary in } V, \quad \boldsymbol{n}\cdot\boldsymbol{v}(\boldsymbol{x};\boldsymbol{\sigma})|_S = \delta^{(2)}(\boldsymbol{s}-\boldsymbol{\sigma}), \ \boldsymbol{n}\times\boldsymbol{v}(\boldsymbol{x};\boldsymbol{\sigma})|_S = 0. \tag{5.22}$$

After the Poisson equations (5.17), one has to solve the vector elliptic equations

$$(-\nabla^2+\gamma)\boldsymbol{u}^0 = -\boldsymbol{\nabla} p^0 + \boldsymbol{g}, \qquad \boldsymbol{u}^0|_S = \boldsymbol{b}; \tag{5.23a}$$

$$(-\nabla^2+\gamma)\boldsymbol{u}' = -\boldsymbol{\nabla} p', \qquad \boldsymbol{u}'|_S = 0; \tag{5.23b}$$

which are both supplemented by purely Dirichlet conditions. The quantities B and β can be obtained by equivalent expressions which involve the fields $\boldsymbol{v}$ instead of $\boldsymbol{\chi}_\gamma$. Let us consider first the expression of B. Substituting in (5.19) $\boldsymbol{\nabla} p'$ given by the equation (5.23b) for $\boldsymbol{u}'$ and using Green identity (5.9), we have

$$\begin{aligned} B(\boldsymbol{\sigma},\boldsymbol{\sigma}') &= -\int \boldsymbol{\nabla} p'\cdot\boldsymbol{\chi}_\gamma\,dV \\ &= \int (-\nabla^2+\gamma)\boldsymbol{u}'\cdot\boldsymbol{\chi}_\gamma\,dV \\ &= \int \boldsymbol{u}'\cdot(-\nabla^2+\gamma)\boldsymbol{\chi}_\gamma\,dV \\ &+ \oint (\boldsymbol{n}\times\boldsymbol{u}'\cdot\boldsymbol{\nabla}\times\boldsymbol{\chi}_\gamma - \boldsymbol{n}\times\boldsymbol{\chi}_\gamma\cdot\boldsymbol{\nabla}\times\boldsymbol{u}' + \boldsymbol{n}\cdot\boldsymbol{u}'\,\boldsymbol{\nabla}\cdot\boldsymbol{\chi}_\gamma - \boldsymbol{n}\cdot\boldsymbol{\chi}_\gamma\,\boldsymbol{\nabla}\cdot\boldsymbol{u}')\,dS. \end{aligned}$$

Taking into account that $\boldsymbol{\chi}_\gamma$ is solution of problem (5.21) and the (homogeneous) boundary conditions imposed on $\boldsymbol{u}'$, we obtain

$$B(\boldsymbol{\sigma},\boldsymbol{\sigma}') = \oint (-\boldsymbol{n}\cdot\boldsymbol{\chi}_\gamma\,\boldsymbol{\nabla}\cdot\boldsymbol{u}')\,dS = \oint (-\boldsymbol{n}\cdot\boldsymbol{v}\,\boldsymbol{\nabla}\cdot\boldsymbol{u}')\,dS,$$

where the identity of the boundary conditions imposed on the auxiliary fields $\boldsymbol{v}$ with those satisfied by $\boldsymbol{\chi}_\gamma$ has been exploited. The divergence theorem is now used to give

$$\begin{aligned} B(\boldsymbol{\sigma},\boldsymbol{\sigma}') &= \oint(-\boldsymbol{n}\cdot\boldsymbol{v}\,\boldsymbol{\nabla}\cdot\boldsymbol{u}')\,dS = \int\left[-\boldsymbol{\nabla}\cdot(\boldsymbol{v}\,\boldsymbol{\nabla}\cdot\boldsymbol{u}')\right]dV \\ &= \int\left[-\boldsymbol{\nabla}\cdot\boldsymbol{v}\,\boldsymbol{\nabla}\cdot\boldsymbol{u}' - \boldsymbol{v}\cdot\boldsymbol{\nabla}(\boldsymbol{\nabla}\cdot\boldsymbol{u}')\right]dV \\ &= \int\left[-\boldsymbol{\nabla}\cdot\boldsymbol{v}\,\boldsymbol{\nabla}\cdot\boldsymbol{u}' - \boldsymbol{v}\cdot(\boldsymbol{\nabla}\times\boldsymbol{\nabla}\times\boldsymbol{u}' + \nabla^2\boldsymbol{u}')\right]dV, \end{aligned}$$

by virtue of the identity $\boldsymbol{\nabla}(\boldsymbol{\nabla}\cdot\boldsymbol{\alpha}) = \boldsymbol{\nabla}\times\boldsymbol{\nabla}\times\boldsymbol{\alpha} + \nabla^2\boldsymbol{\alpha}$. The term containing $-\boldsymbol{v}\cdot\boldsymbol{\nabla}\times\boldsymbol{\nabla}\times\boldsymbol{u}'$ is integrated by parts, and this gives only the volume integral of $-\boldsymbol{\nabla}\times\boldsymbol{v}\cdot\boldsymbol{\nabla}\times\boldsymbol{u}'$ since the boundary term vanishes due to the boundary condition $\boldsymbol{n}\times\boldsymbol{u}'|_S = 0$. On the other hand, the last term $-\boldsymbol{v}\cdot\nabla^2\boldsymbol{u}'$ can be rewritten in the form $\boldsymbol{v}\cdot(-\nabla^2+\gamma)\boldsymbol{u}' - \gamma\,\boldsymbol{v}\cdot\boldsymbol{u}' = -\boldsymbol{v}\cdot\boldsymbol{\nabla}p' - \gamma\,\boldsymbol{v}\cdot\boldsymbol{u}'$. Thus, after some reordering of the terms, we finally obtain

$$B(\boldsymbol{\sigma},\boldsymbol{\sigma}') = -\int(\boldsymbol{\nabla}\times\boldsymbol{u}'\cdot\boldsymbol{\nabla}\times\boldsymbol{v} + \boldsymbol{\nabla}\cdot\boldsymbol{u}'\,\boldsymbol{\nabla}\cdot\boldsymbol{v} + \boldsymbol{\nabla}p'\cdot\boldsymbol{v} + \gamma\,\boldsymbol{u}'\cdot\boldsymbol{v})\,dV.$$

A similar calculation can be done for the quantity β in (5.20), using this time the equation and the boundary conditions for $\boldsymbol{u}^0$ stated in (5.23a). This readily gives the following characterization

$$\beta(\boldsymbol{\sigma}) = +\int(\boldsymbol{\nabla}\times\boldsymbol{u}^0\cdot\boldsymbol{\nabla}\times\boldsymbol{v} + \boldsymbol{\nabla}\cdot\boldsymbol{u}^0\,\boldsymbol{\nabla}\cdot\boldsymbol{v} + \boldsymbol{\nabla}p^0\cdot\boldsymbol{v} + \gamma\,\boldsymbol{u}^0\cdot\boldsymbol{v} - \boldsymbol{g}\cdot\boldsymbol{v})\,dV.$$

The arbitrariness of the fields $\boldsymbol{v}$ at all internal point of V can be exploited by choosing $\boldsymbol{v} = 0$ inside V, so that the integration domain in these expressions reduces to a narrow strip along the boundary, with obvious advantages from the viewpoint of the computational efficiency.

In contrast with the situation of the nonprimitive variables, the operator obtained from $B(\boldsymbol{\sigma},\boldsymbol{\sigma}')$ is not symmetric. Of course, the operator is not even definite because the pressure is determined by the set (5.15) only up to an arbitrary constant. This singularity is easily eliminated by imposing that pressure field has a prescribed average value over the domain V. Similarly, after the spatial disretization of the problem, the matrix corresponding to B is singular and the singularity can be eliminated by suppressing one of the equation of the linear system and imposing an arbitrary value of the pressure at the associated boundary point.

After the unknown λ has been determined from the linear problem $B\lambda = \beta$, the solution $(p,\boldsymbol{u})$ of problem (5.15) is finally obtained by solving the two elliptic equations:

$$-\nabla^2 p = -\boldsymbol{\nabla}\cdot\boldsymbol{g}, \qquad p|_S = \lambda; \tag{5.24a}$$

$$(-\nabla^2+\gamma)\boldsymbol{u} = -\boldsymbol{\nabla}p + \boldsymbol{g}, \qquad \boldsymbol{u}|_S = \boldsymbol{b}. \tag{5.24b}$$

We note that all the vector elliptic equations to be solved in this method give purely Dirichlet problems, so that uncoupled scalar Dirichlet problems for the Cartesian components of the vector unknowns have to be solved. A finite element implementation of this method using bilinear interpolation for both the pressure and velocity has been employed by Quartapelle and Napolitano (1986).

By summarizing, the overall procedure can be regarded as a method for obtaining the velocity field solution of the incompressible problem (5.15) through the following factorization scheme (cf. Quartapelle 1988)

$$\boldsymbol{u} = (-\nabla^2 + \gamma)^{-1}\Big[\mathbb{I} - P^{-1}\mathbb{P}_{\mathcal{N}(-\widetilde{\nabla}_{\times}^2 + \gamma)}\Big]\Big[1 + \boldsymbol{\nabla}(-\nabla^2)^{-1}\boldsymbol{\nabla}\cdot\Big]\boldsymbol{g}. \tag{5.25}$$

Here ∇^2 denotes the Laplace operator supplemented by Dirichlet condition, so that $(-\nabla^2)^{-1}$ represents formally the solution of a scalar Dirichlet problem, whereas $(-\nabla^2 + \gamma)^{-1}$ represents the solution of a vector Dirichlet problem. On the other hand, $\widetilde{\nabla}_{\times}^2$ denotes the Laplace operator acting on vector fields, supplemented with homogeneous boundary conditions for the tangential components, but with unspecified normal component on S.

5.5 Equations for plane channel flows

5.5.1 Uncoupled formulation

It is interesting to apply the formulation (5.15) to the equations governing the flow in a plane channel. Let us consider a straight channel between two parallel walls and let a Cartesian coordinate system be introduced with the x-axis normal to the walls, which are assumed to be located at $x = -a$ and $x = a$. Assuming that the flow field is spatially periodic in the two directions y and z parallel to the walls and with period L_y and L_z, respectively, the velocity field $\boldsymbol{u}(x, y, z)$ in the region $|x| \le a$ can be expanded in a double Fourier series

$$\boldsymbol{u}(x, y, z) = \sum_{k_y, k_z} \begin{bmatrix} u(x; k_y, k_z) \\ \mathrm{i}u_y(x; k_y, k_z) \\ \mathrm{i}u_z(x; k_y, k_z) \end{bmatrix} e^{\mathrm{i}(\kappa_y y + \kappa_z z)}, \tag{5.26}$$

where

$$\kappa_y = \frac{2\pi k_y}{L_y}, \qquad \kappa_z = \frac{2\pi k_z}{L_z},$$

and the summation extends over all the integers. The Laplace operator expressed in terms of this spectral representation becomes

$$-\nabla^2 \to -D^2 + \kappa_y^2 + \kappa_z^2 = -D^2 + \kappa^2,$$

where

$$D = \frac{d}{dx} \qquad \text{and} \qquad \kappa^2 = \kappa_y^2 + \kappa_z^2.$$

Let the metaharmonic field $\boldsymbol{\chi}_\gamma$ considered in problem (5.15) be expanded in a Fourier series similar (5.26). The tangential components of $\boldsymbol{\chi}_\gamma(x; k_y, k_z)$ vanish for $x = \pm a$ and satisfy the equations

$$(-D^2 + \kappa_\gamma^2)\chi_{\gamma,y} = 0 \qquad \text{and} \qquad (-D^2 + \kappa_\gamma^2)\chi_{\gamma,z} = 0,$$

where we have introduced the notation

$$\kappa_\gamma^2 = \kappa^2 + \gamma = \kappa_y^2 + \kappa_z^2 + \gamma.$$

Therefore, the tangential components vanish identically for $-a \le x \le a$ and the Fourier representation of $\boldsymbol{\chi}_\gamma$ reduces to

$$\boldsymbol{\chi}_\gamma(x, y, z) = \sum_{k_y, k_z} \chi_\gamma(x; k_y, k_z)\, e^{i(\kappa_y y + \kappa_z z)}\, \hat{\boldsymbol{x}}, \tag{5.27}$$

$\hat{\boldsymbol{x}}$ denoting the unit vector normal to the solid walls. By expanding the pressure field in double Fourier series, the Poisson equation in (5.15) gives the following ordinary differential equation for each Fourier coefficient $p(x; k_y, k_z)$

$$(-D^2 + \kappa^2)p = -Dg + \kappa_y g_y + \kappa_z g_z, \tag{5.28}$$

where g, g_y and g_z are the Fourier coefficients of the expansion for $\boldsymbol{g}$ similar to (5.26). The conditions supplementing the equation for each pressure mode $p(x; k_y, k_z)$ are obtained specializing the integral conditions in (5.15) to the present geometry and to the adopted Fourier representation. By virtue of (5.27) the pressure integral conditions assume the following one-dimensional form

$$\int_{-a}^{a} (Dp)\, \chi_\gamma\, dx = \int_{-a}^{a} g\, \chi_\gamma\, dx, \tag{5.29}$$

since the velocity is zero on the two walls. In this expression the function $\chi_\gamma(x; k_y, k_z)$ are any two linearly independent solution to the equation

$$(-D^2 + \kappa_\gamma^2)\chi_\gamma = 0, \tag{5.30}$$

for example, the two (normalized) solutions

$$\chi_\gamma^{(1)}(x; k_y, k_z) = \frac{\cosh(\kappa_\gamma x)}{\cosh(\kappa_\gamma a)}, \qquad \chi_\gamma^{(2)}(x; k_y, k_z) = \frac{\sinh(\kappa_\gamma x)}{\sinh(\kappa_\gamma a)}. \tag{5.31}$$

Then, for each pressure mode, excluding only the first mode with $k_y = k_z = 0$, the pressure coefficient $p(x; k_y, k_z)$ is determined as solution to the following "integral value problem"

$$\begin{gathered} (-D^2 + \kappa^2)p = -Dg + \kappa_y g_y + \kappa_z g_z, \\ \int_{-a}^{a} (Dp)\, \chi_\gamma^{(1)}\, dx = \int_{-a}^{a} g\, \chi_\gamma^{(1)}\, dx, \qquad \int_{-a}^{a} (Dp)\, \chi_\gamma^{(2)}\, dx = \int_{-a}^{a} g\, \chi_\gamma^{(2)}\, dx. \end{gathered} \tag{5.32}$$

When this pressure problem is solved approximately by means of a spectral method with Chebyshev polynomials, the imposition of the two integral conditions does not modify the profile of the matrix with respect to the imposition of usual conditions of boundary value type, since in any case these conditions give linear equations with all coefficients different from zero. Furthermore, using the two functions in (5.31) would also permit to decouple the equation system into its even and odd independent components, thus giving a maximal reduction of the required calculations.

The equation for the first pressure mode $p(x,0,0)$ is instead $-D^2p = -Dg$ and can be written $D(Dp - g) = 0$, which implies $Dp = g + K$, K being an integration constant. Imposing the integral condition with respect to the solution to the first-mode equation $(-D^2 + \gamma)\chi_\gamma = 0$ given by

$$\chi_\gamma(x;0,0) = \frac{\cosh(\sqrt{\gamma}\,x)}{\cosh(\sqrt{\gamma}\,a)}$$

shows that the constant K is zero. The integral condition with respect to the other linearly independent solution $\chi_\gamma(x;0,0)$ involving the hyperbolic sine does not say anything different from what already stated by the equation governing the first pressure mode. In other words, for the first mode the two integral conditions are not both linearly independent form the pressure equation itself. It follows that $p(x;0,0)$ is determined only up to an arbitrary constant, as required.

Once the pressure has been determined, one has to solve the modal equations for the three velocity components, namely,

$$\begin{aligned} (-D^2 + \kappa_\gamma^2)u &= -Dp + g, & u(\pm a) &= 0, \\ (-D^2 + \kappa_\gamma^2)u_y &= -\kappa_y p + g_y, & u_y(\pm a) &= 0, \\ (-D^2 + \kappa_\gamma^2)u_z &= -\kappa_z p + g_z, & u_z(\pm a) &= 0. \end{aligned} \tag{5.33}$$

This procedure reduces therefore the solution of the incompressible equations for three-dimensional flow in a plane channel to a sequence of second-order ordinary differential equations for scalar unknowns. The equation for each pressure coefficient is subject to two integral conditions, whereas those for the velocity coefficients are supplemented by conditions of standard boundary value type. This represents the separation of variables for the Stokes problem (5.15) in Cartesian domains with doubly periodic boundary conditions. Such a reduction of the governing equations is made possible by the fact that the metaharmonic fields $\boldsymbol{\chi}_\gamma$ can be determined explicitly, thus eliminating the need for the decomposition scheme considered in the previous section.

The same approach can also be used to analyze the three-dimensional equations of the time-discretized Stokes problem in spherical and cylindrical coordinates (Quartapelle and Verri 1992). In both cases a complete separation of variables is possible, leading to a set consisting only of uncoupled equations for scalar unknowns.

For spherical regions, the variables p and $\boldsymbol{u}$ can be expanded into a series of scalar and vector spherical harmonics, respectively. In terms of this representation, the general integral conditions for pressure can be expressed as one-dimensional integrals over the radial variable for the expansion coefficients of pressure. These integrals are found to involve the modified Bessel functions of half-odd order. In this way, the system of three-dimensional equations of the time-discretized Stokes problem is reduced to a sequence of uncoupled scalar second-order ordinary differential equations in the radial variable.

A similar set of uncoupled one-dimensional equations for scalar unknowns can be obtained also for cylindrical regions after the angular and axial dependencies are represented by means of a double Fourier series, assuming periodic boundary conditions along the axis of the cylinder. Here the one-dimensional integral conditions for the expansion coefficients of pressure are found to involve the modified Bessel function of integer order.

In this kind of methods, the integral conditions can be imposed directly on the expansion coefficients of the pressure since the expansion coefficients of χ_γ can be explicitly determined. As a result, the solenoidal velocity field can be calculated in a single step, without building up the auxiliary problems of the type $B\lambda = \beta$ described in section 5.4 or the analogous problem required by the influence matrix method to be explained in the next section.

The situation becomes more complicated for the case of the equations governing three-dimensional Stokes flows within cylindrical region of finite axial extent. When the Fourier series is used to represent the angular dependence, one still obtains a set of uncoupled and scalar equations for the expansion coefficients of pressure and velocity but now these equations are elliptic partial differential in two dimensions. As a consequence, it can become difficult to determine explicitly the Fourier coefficients of the metaharmonic field χ_γ to be used in the (Fourier transformed version of the) pressure integral conditions.

5.5.2 Influence matrix method

The influence matrix method has been introduced by Kleiser and Schumann (1980) to solve the primitive variable equations in the same plane channel considered in the previous section. The method exploits the superposition principle to determine the boundary values of pressure in order for the Kleiser–Schumann derivative boundary condition $\boldsymbol{\nabla}\cdot\boldsymbol{u}|_S = 0$ to be satisfied. The latter, after introducing the Fourier representation for plane channel problems and using the velocity boundary condition $\boldsymbol{u}|_S = 0$, assumes the form

$$\boldsymbol{\nabla}\cdot\boldsymbol{u}|_S = 0 \;\rightarrow\; Du(\pm a) = 0,$$

where $u(x; k_y, k_z)$ is the Fourier component of the velocity component normal to the channel wall. Instead of using Green identity to derive conditions to be imposed on the pressure, the solution (p, u) is expressed according to the following linear combination:

$$\begin{bmatrix} p \\ u \end{bmatrix} = \begin{bmatrix} p_0 \\ u_0 \end{bmatrix} + \sum_{j=1,2} \lambda_j \begin{bmatrix} p_j \\ u_j \end{bmatrix}. \tag{5.34}$$

The solution (p, u) must satisfy the system of equations

$$\begin{aligned} (-D^2 + \kappa^2)p &= -Dg + \kappa_y g_y + \kappa_z g_z, & Du(\pm a) &= 0, \\ (-D^2 + \kappa_\gamma^2)u &= -Dp + g, & u(\pm a) &= 0, \end{aligned}$$

which are coupled together due to the presence of two pairs of boundary conditions for the single unknown u. Now, the three components of (p, u) in expansion (5.34) are characterized as solution to the following three problems: for (p_0, u_0)

$$\begin{aligned} (-D^2 + \kappa^2)p_0 &= -Dg + \kappa_y g_y + \kappa_z g_z, & p_0(\pm a) &= 0; \\ (-D^2 + \kappa_\gamma^2)u_0 &= -Dp_0 + g, & u_0(\pm a) &= 0; \end{aligned}$$

and for (p_j, u_j), $j = 1, 2$,

$$\begin{aligned} (-D^2 + \kappa^2)p_j &= 0, & p_j(\pm a) &\neq 0; \\ (-D^2 + \kappa_\gamma^2)u_j &= -Dp_j, & u_j(\pm a) &= 0. \end{aligned}$$

All these problems contain only independent boundary value problems for second-order equations that can be solved in sequence. In particular, the boundary conditions for p_j can be chosen, for instance, in the form $p_1(-a) = 1$, $p_1(a) = 0$, and $p_2(-a) = 0$, $p_2(a) = 1$. By imposing that the linear combination (5.34) be solution to the coupled system for (p, u) leads to the linear system of two equations $\widetilde{B}\lambda = \widetilde{\beta}$ where

$$\widetilde{B} = \begin{bmatrix} u_1'(-a) & u_2'(-a) \\ u_1'(+a) & u_2'(+a) \end{bmatrix}, \qquad \widetilde{\beta} = -\begin{bmatrix} u_0'(-a) \\ u_0'(+a) \end{bmatrix}, \tag{5.35}$$

the prime denoting the derivative with respect to x.

This influence matrix method has been implemented by Kleiser and Schumann to compute plane channel flows by means of a Chebyshev spectral approximation to the Navier–Stokes equations. Such a technique has been subsequently extended with success to problems with nonperiodic boundary conditions in two dimensions by Le Quéré and Alziary de Roquefort (1982) and to three-dimensional problems within cylindrical walls with axial periodicity by Marcus (1984) and with top and bottom rigid walls by Le Quéré and Pécheux (1989). A detailed analysis of the influence matrix method and of its approximation by means of Legendre polynomials has been conducted by Canuto and Sacchi Landriani (1986). The influence matrix method has also been employed for solving Chebyshev approximations to the nonprimitive variable equations by Ehrenstein and Peyret (1989).

5.5.3 Numerical comparison

To compare the influence matrix method with the method using the integral conditions for pressure the modal equations of a two-dimensional problem are condidered. In this case the number of second-order equations to be solved for each mode is $6+1=7$ for the influence matrix method against $2+1=3$ for the integral condition method. The analogous counting for the three-dimensional problem would be $6+2=8$ equations against $2+2=4$.

The Chebyshev approximation to the equations (5.32) and (5.33) or to the similar equations for the unknowns p_j and u_j in (5.34) is set up exactly as in section 2.11 for the nonprimitive variable problem. The primitive variable equations in a plane channel are such that the matrix of the linear systems for all the unknowns can be transformed into the bordered quasi-pentadiagonal form which allows for the efficient UL factorization algorithm described in section 2.11.4.

The performances of two methods are compared by considering a simple analytical solution defined as follows:

$$u(x) = e^{\alpha x} \qquad p(x) = e^{\beta x}$$

so that, by virtue of the incompressibility condition (in two dimension) $v(x) = \kappa^{-1}\alpha u(x)$. The solution u is modified by adding also a component which is solution to the homogeneous equation, to give

$$u(x) = e^{\alpha x} + C_1\chi_\gamma^{(1)}(x) + C_2\chi_\gamma^{(2)}(x),$$

where C_1 and C_2 are arbitrary constants. From such u and p the source functions g and g_y of the velocity equations are determined analytically and then the two methods are employed to compute the approximate solution.

In Table 5.1 the relative errors for the three unknowns p, u and v are reported for the following values of the parameters:

$$\kappa = 2, \qquad \gamma = 1, \qquad \alpha = 2, \qquad \beta = 3, \qquad C_1 = 100, \qquad C_2 = 100.$$

For reference, Table 5.2 contains the numerical error of the solutions obtained imposing the boundary values of p evaluated from the exact solution, for the same number N of Chebyshev polynomials, namely, $0 \leq N \leq 20$.

The comparison of the numerical results indicates that the two methods provide almost identical solutions for the velocity components. On the contrary, the pressure is calculated with a substantially lower accuracy by the influence matrix method than by the integral condition method. This is in accordance with the results obtained for the nonprimitive variable equations and shown in section 2.11. In both cases, methods which avoid the direct imposition of global constraints *via* their integration kernels and which require to solve more equations than those of the original problem can be numerically unstable: in fact,

	Influence matrix method			Integral condition method		
N	rerr(p)	rerr(u)	rerr(v)	rerr(p)	rerr(u)	rerr(v)
6	0.70(−00)	0.68(−03)	0.71(−03)	0.25(−02)	0.54(−03)	0.55(−03)
8	0.23(−01)	0.86(−05)	0.94(−05)	0.65(−04)	0.77(−05)	0.79(−05)
10	0.21(−03)	0.81(−07)	0.90(−07)	0.10(−05)	0.75(−07)	0.77(−07)
12	0.15(−05)	0.58(−09)	0.65(−09)	0.13(−07)	0.54(−09)	0.55(−09)
14	0.83(−08)	0.32(−11)	0.36(−11)	0.13(−09)	0.30(−11)	0.30(−11)
16	0.34(−10)	0.13(−13)	0.15(−13)	0.10(−11)	0.13(−13)	0.13(−13)
18	0.22(−11)	0.14(−14)	0.15(−14)	0.69(−14)	0.13(−14)	0.12(−14)
20	0.26(−11)	0.18(−14)	0.16(−14)	0.30(−14)	0.15(−14)	0.12(−14)

Table 5.1:

	Pressure boundary conditions		
N	rerr(p)	rerr(u)	rerr(v)
6	0.25 (−02)	0.54 (−03)	0.55 (−03)
8	0.62 (−04)	0.77 (−05)	0.79 (−05)
10	0.10 (−05)	0.75 (−07)	0.77 (−07)
12	0.13 (−07)	0.54 (−09)	0.55 (−09)
14	0.13 (−09)	0.30 (−11)	0.30 (−11)
16	0.10 (−11)	0.13 (−13)	0.13 (−13)
18	0.72 (−14)	0.13 (−14)	0.12 (−14)
20	0.30 (−14)	0.15 (−14)	0.13 (−14)

Table 5.2:

they provide numerical results for the integrally conditioned unknown which are sensitive to the specific form of the solution to the additional equations, although the problem for the constrained variable is completely independent.

To sum up, the methods which enforce the integral conditions by completing the second-order operator for the pressure or vorticity represent a fortunate example in which conceptual simplicity, computational efficiency and numerical stability can be simultaneously achieved.

5.6 Direct Stokes solver

In this section we describe a very important direct method introduced by Glowinski and Pironneau for the solution of the time-discretized primitive variable equations (5.1) by means of finite elements. The present exposition is rather sketchy, the purpose being that of showing the structural similarity of this method with the method using the integral conditions described previously. For more details, the interested reader is referred to the original papers (Glowinski and Pironneau 1978a-b), to Glowinski (1984) and also to the extended report by Bristeau *et al.* (1987) which contains several applications to the calculation of realistic flows in two and three dimensions.

Assuming for simplicity, as usual, Dirichlet boundary conditions for the velocity along the entire boundary, consider the system

$$\begin{aligned}(-\nabla^2+\gamma)\boldsymbol{u}+\boldsymbol{\nabla}p&=\boldsymbol{g},\\ \boldsymbol{\nabla}\cdot\boldsymbol{u}&=0,\\ \boldsymbol{u}|_S&=\boldsymbol{b},\end{aligned}\tag{5.36}$$

which defines the so-called Stokes problem. Strictly speaking a Stokes problem would require $\gamma=0$, but in the computational literature this denomination applies often also to time-discretized versions of the linearized incompressible equations.

The principle leading to the direct Stokes solver is to introduce, aside to the Poisson equation $-\nabla^2 p=-\boldsymbol{\nabla}\cdot\boldsymbol{g}$ derived from the momentum equation thank to the incompressibility, another Poisson for a scalar unknown ψ of the form $-\nabla^2\psi=\boldsymbol{\nabla}\cdot\boldsymbol{u}$ which is supplemented by a homogeneous Dirichlet condition $\psi|_S=0$. Applying the operator $(-\nabla^2+\gamma)$ to this equation, taking into account the divergence of the momentum equation and the fact that the pressure satisfies the Poisson equation, it is not diffcult to verify that ψ must be solution to the fourth-order elliptic equation

$$(-\nabla^2+\gamma)\nabla^2\psi=0.\tag{5.37}$$

Therefore, the equation $-\nabla^2\psi=\boldsymbol{\nabla}\cdot\boldsymbol{u}$ means that one can assure $\boldsymbol{\nabla}\cdot\boldsymbol{u}=0$ by making $\psi=0$, while the solution of the fourth-order equation above will be identically zero provided that $\psi|_S=0$ and $(\partial\psi/\partial n)|_S=0$. It follows that the equations and the boundary condition of the Stokes problem (5.36) will be satisfied by p and $\boldsymbol{u}$ solution of the system

$$\begin{aligned}-\nabla^2 p&=-\boldsymbol{\nabla}\cdot\boldsymbol{g},\\ (-\nabla^2+\gamma)\boldsymbol{u}&=-\boldsymbol{\nabla}p+\boldsymbol{g}, &\qquad \boldsymbol{u}|_S&=\boldsymbol{b},\\ -\nabla^2\psi&=\boldsymbol{\nabla}\cdot\boldsymbol{u}, &\qquad \psi|_S&=0,\end{aligned}\tag{5.38}$$

provided that the auxiliary unknown ψ satisfies also the Neumann boundary condition

$$\left.\frac{\partial\psi}{\partial n}\right|_S = 0. \tag{5.39}$$

Note that the last boundary condition represents in some sense an equivalent substitute for the laking boundary condition for the pressure. Stated in other terms, the system above (5.38)–(5.39), due to the lack of an explicit boundary condition for p, constitutes a set of equations coupled together by the imposition of two boundary conditions on ψ. To determine the boundary value of p assuring that incompressibility is satisfied, consider the system of equations

$$\begin{aligned} -\nabla^2 p_\lambda &= -\boldsymbol{\nabla}\cdot\boldsymbol{g}, & p_\lambda|_S &= \lambda;\\ (-\nabla^2+\gamma)\boldsymbol{u}_\lambda &= -\boldsymbol{\nabla}p_\lambda + \boldsymbol{g}, & \boldsymbol{u}_\lambda|_S &= \boldsymbol{b};\\ -\nabla^2\psi_\lambda &= \boldsymbol{\nabla}\cdot\boldsymbol{u}_\lambda, & \psi_\lambda|_S &= 0; \end{aligned} \tag{5.40}$$

and determine λ so that $(\partial\psi_\lambda/\partial n)|_S = 0$. The latter condition is then formulated in a variational (integral) form writing

$$-\oint \frac{\partial\psi_\lambda}{\partial n}\mu\, dS = 0, \tag{5.41}$$

for any function μ defined on S (the minus sign has been introduced to maintain the original notation of the method).

It should be remarked that the choice of the specific form of the elliptic operator in the equation for ψ is arbitrary. For instance, any operator of the type $(-\nabla^2+\delta)$, $\delta > 0$, would fit the needs of the construction. For computational purposes the operator $-\nabla^2$ is the most convenient because it is the same operator of the pressure equation, whose matrix of coefficients is already available in the computer memory.

Using (scalar) Green identity for the operator ∇^2 to transform the surface integral into volume integrals and taking into account the equations of the λ-system (5.40) and of an analogous μ-system, the integral on the left-hand side of the variational equation can be written in the form

$$-\oint \frac{\partial\psi_\lambda}{\partial n}\mu\, dS = \int (\gamma\boldsymbol{u}_\lambda\cdot\boldsymbol{u}_\mu + \boldsymbol{\nabla}\times\boldsymbol{u}_\lambda\cdot\boldsymbol{\nabla}\times\boldsymbol{u}_\mu + \boldsymbol{\nabla}\cdot\boldsymbol{u}_\lambda\ \boldsymbol{\nabla}\cdot\boldsymbol{u}_\mu)\, dV, \tag{5.42}$$

which shows the symmetric character of the integral. The expression (5.42) is however not convenient for computational purposes since it would require to determine and store the solutions $\boldsymbol{u}_\lambda$ for every linearly independent function λ. It is more convenient to decompose the solution $(p, \boldsymbol{u}, \psi)$ as follows:

$$\begin{bmatrix} p(\boldsymbol{x})\\ \boldsymbol{u}(\boldsymbol{x})\\ \psi(\boldsymbol{x}) \end{bmatrix} = \begin{bmatrix} p^0(\boldsymbol{x})\\ \boldsymbol{u}^0(\boldsymbol{x})\\ \psi^0(\boldsymbol{x}) \end{bmatrix} + \oint \begin{bmatrix} p'(\boldsymbol{x};\boldsymbol{\sigma}')\\ \boldsymbol{u}'(\boldsymbol{x};\boldsymbol{\sigma}')\\ \psi'(\boldsymbol{x};\boldsymbol{\sigma}') \end{bmatrix} \lambda(\boldsymbol{\sigma}')\, dS(\boldsymbol{\sigma}'), \tag{5.43}$$

where the primed functions are the solutions, for any $\boldsymbol{\sigma}' \in S$, to the three elliptic problems

$$-\nabla^2 p' = 0, \qquad p'|_S = \delta^{(2)}(\boldsymbol{s} - \boldsymbol{\sigma}'); \tag{5.44a}$$

$$(-\nabla^2 + \gamma)\boldsymbol{u}' = -\boldsymbol{\nabla} p', \qquad \boldsymbol{u}'|_S = 0; \tag{5.44b}$$

$$-\nabla^2 \psi' = \boldsymbol{\nabla}\cdot\boldsymbol{u}', \qquad \psi'|_S = 0; \tag{5.44c}$$

whereas the functions with superscript 0 are the solution to

$$-\nabla^2 p^0 = -\boldsymbol{\nabla}\cdot\boldsymbol{g}, \qquad p^0|_S = 0; \tag{5.45a}$$

$$(-\nabla^2 + \gamma)\boldsymbol{u}^0 = -\boldsymbol{\nabla} p^0 + \boldsymbol{g}, \qquad \boldsymbol{u}^0|_S = \boldsymbol{b}; \tag{5.45b}$$

$$-\nabla^2 \psi^0 = \boldsymbol{\nabla}\cdot\boldsymbol{u}^0, \qquad \psi^0|_S = 0. \tag{5.45c}$$

Furthermore, instead of using the functions μ defined only on S, one introduces the auxiliary scalar functions $w(\boldsymbol{x};\boldsymbol{\sigma})$ which are defined over the domain V according to

$$w(\boldsymbol{x};\boldsymbol{\sigma}) \text{ arbitrary in } V, \quad w(\boldsymbol{x};\boldsymbol{\sigma})|_S = \delta^{(2)}(\boldsymbol{s} - \boldsymbol{\sigma}). \tag{5.46}$$

Using the solutions of the two sets of problems above and the auxiliary volumic functions $w(\boldsymbol{x};\boldsymbol{\sigma})$, the variational boundary problem (5.41) gives the following linear problem

$$A\lambda = \beta, \tag{5.47}$$

where A and β are expressed by

$$A(\boldsymbol{\sigma},\boldsymbol{\sigma}') = -\int (\boldsymbol{\nabla}\psi' + \boldsymbol{u}')\cdot\boldsymbol{\nabla} w\, dV, \tag{5.48}$$

$$\beta(\boldsymbol{\sigma}) = +\int (\boldsymbol{\nabla}\psi^0 + \boldsymbol{u}^0)\cdot\boldsymbol{\nabla} w\, dV. \tag{5.49}$$

In practice, the functions $w(\boldsymbol{x};\boldsymbol{\sigma})$ are taken different from zero only in a narrow shell along the boundary, so that the volume integrals above can be evaluated much more efficiently than by means of their symmetric expressions.

The advantage of the present method over the decomposition scheme described in section 5.4 is that the linear operator A is symmetric. Therefore, in the discrete case, the corresponding linear system of equations can be solved iteratively by means of the efficient conjugate gradient method. This avoids the explicit determination of the full matrix A_h, which would become prohibitively large in "real-life" finite element calculations, such as those considered by Bristeau *et al.* (1987).

5.7 General boundary conditions

The attention has been restricted so far to boundary conditions imposing the value of velocity on the entire boundary. In this section we discuss different kinds of boundary conditions for the time-discretized Navier–Stokes equations and examine the complete set of conditions to be considered when the problem is written as a set of uncoupled elliptic equations or solved by means of the direct Stokes solver. As a general principle, a total of only three scalar boundary conditions can be prescribed for three-dimensional problems (two conditions for two-dimensional problems) since the incompressibility condition implies another relationship between the variables on the boundary. The present analysis is an elaboration of the formulation provided by Pironneau (1986) for the Stokes problem.

Let us suppose that the velocity $\boldsymbol{u}$ be prescribed only on a portion S_0 of the boundary S, according to the specification

$$\boldsymbol{u}|_{S_0} = \boldsymbol{b},$$

whereas different kinds of boundary conditions for velocity and pressure are specified on the remaining parts of S, which will be denoted by S_j, $j = 1, 2, \ldots, 6$. Let us suppose that on S_1, S_2 and S_3 one condition for pressure and only two scalar conditions for velocity are prescribed, in the following form:

$$\begin{aligned} p|_{S_1} &= q, & \boldsymbol{n}\times\boldsymbol{u}|_{S_1} &= \boldsymbol{n}\times\boldsymbol{b}; \\ p|_{S_2} &= q, & \boldsymbol{n}\times\boldsymbol{\nabla}\times\boldsymbol{u}|_{S_2} &= \boldsymbol{n}\times\boldsymbol{z}; \\ \boldsymbol{n}\cdot\boldsymbol{\nabla}p|_{S_3} &= r, & \boldsymbol{n}\times\boldsymbol{u}|_{S_3} &= \boldsymbol{n}\times\boldsymbol{b}; \end{aligned}$$

where q, r, $\boldsymbol{n}\times\boldsymbol{b}$ and $\boldsymbol{n}\times\boldsymbol{z}$ are defined on their respective portions S_j of S. Introducing the Poisson equation for pressure to satisfy incompressibility requires to consider the Kleiser–Schumann boundary condition

$$\boldsymbol{\nabla}\cdot\boldsymbol{u}|_{S_0\cup S_1\cup S_2\cup S_3} = 0.$$

(The condition $\boldsymbol{\nabla}\cdot\boldsymbol{u}|_{S_4\cup S_5\cup S_6} = 0$ is satisfied automatically, as it will be explained later.) Thus, on S_0 there is one redundant boundary condition for velocity whereas on S_1, S_2 and S_3 the correct number of conditions for both pressure and velocity is obtained. As a consequence, integral conditions for pressure have to be introduced only for S_0 while the derivative boundary condition $\boldsymbol{\nabla}\cdot\boldsymbol{u}|_{S_j} = 0$, $j = 1, 2, 3$, can be used together with the two velocity boundary conditions already specified on these parts of S. We have therefore the following complete sets of boundary conditions for pressure and velocity:

$$p|_{S_1} = q; \qquad \boldsymbol{n}\times\boldsymbol{u}|_{S_1} = \boldsymbol{n}\times\boldsymbol{b}, \qquad \boldsymbol{\nabla}\cdot\boldsymbol{u}|_{S_1} = 0;$$

$$p|_{S_2} = q; \qquad \boldsymbol{n}\times\boldsymbol{\nabla}\times\boldsymbol{u}|_{S_2} = \boldsymbol{n}\times\boldsymbol{z}, \qquad \boldsymbol{\nabla}\cdot\boldsymbol{u}|_{S_2} = 0;$$

$$\boldsymbol{n}\cdot\boldsymbol{\nabla}p|_{S_3} = r; \qquad \boldsymbol{n}\times\boldsymbol{u}|_{S_3} = \boldsymbol{n}\times\boldsymbol{b}, \qquad \boldsymbol{\nabla}\cdot\boldsymbol{u}|_{S_3} = 0.$$

Concerning the boundary conditions satisfied by the metaharmonic fields $\boldsymbol{\chi}_\gamma$ to be used in the integral conditions which determine $\lambda = p|_{S_0}$, one has evidently $\boldsymbol{n}\times\boldsymbol{\chi}_\gamma|_{S_0} = 0$ and the other conditions on S_j, $j = 1, 2, \ldots, 6$, can be easily deduced from the corresponding boundary conditions for $\boldsymbol{u}$.

Let us now consider the part S_4 of S on which three scalar conditions for $\boldsymbol{u}$ are prescribed, one imposing the normal component of velocity and the other two conditions fixing the two tangential components of vorticity, namely,

$$\boldsymbol{n}\cdot\boldsymbol{u}|_{S_4} = \boldsymbol{n}\cdot\boldsymbol{b}, \qquad \boldsymbol{n}\times\boldsymbol{\nabla}\times\boldsymbol{u}|_{S_4} = \boldsymbol{n}\times\boldsymbol{z},$$

where $\boldsymbol{n}\cdot\boldsymbol{b}$ and $\boldsymbol{n}\times\boldsymbol{z}$ are specified on S_4. These boundary conditions and the governing equations imply a Neumann boundary condition for the pressure. In fact, using vector identity $-\nabla^2\boldsymbol{\alpha} = \boldsymbol{\nabla}\times\boldsymbol{\nabla}\times\boldsymbol{\alpha} - \boldsymbol{\nabla}(\boldsymbol{\nabla}\cdot\boldsymbol{\alpha})$ and the continuity equation, the momentum equation $(-\nabla^2 + \gamma)\boldsymbol{u} + \boldsymbol{\nabla}p = \boldsymbol{g}$ can be written as

$$\boldsymbol{\nabla}\times\boldsymbol{\nabla}\times\boldsymbol{u} + \gamma\boldsymbol{u} + \boldsymbol{\nabla}p = \boldsymbol{g}.$$

Since the tangential components of $\boldsymbol{\nabla}\times\boldsymbol{u}$ are prescribed on S_4, one can take the normal component of the momentum equation written in this form. Solving with respect to the pressure term and using the boundary conditions prescribed on S_4, one obtains

$$\boldsymbol{n}\cdot\boldsymbol{\nabla}p|_{S_4} = -\boldsymbol{n}\cdot\boldsymbol{\nabla}_S\times\boldsymbol{z} - \gamma\,\boldsymbol{n}\cdot\boldsymbol{b} + \boldsymbol{n}\cdot\boldsymbol{g}|_{S_4},$$

where $\boldsymbol{\nabla}_S\times$ denotes the surface differential operator defined in section 3.4.1. Conversely, the imposition of this pressure boundary condition guarantees the fulfillment of the boundary condition $\boldsymbol{\nabla}\cdot\boldsymbol{u}|_{S_4} = 0$. In fact, by the previous vector identity, the normal component of the momentum equation on S_4 implies that $\boldsymbol{n}\cdot\boldsymbol{\nabla}(\boldsymbol{\nabla}\cdot\boldsymbol{u})|_{S_4} = 0$, and therefore $\boldsymbol{\nabla}\cdot\boldsymbol{u} = 0$, since the pressure Poisson equation is satisfied and the boundary condition $\boldsymbol{\nabla}\cdot\boldsymbol{u}|_{S_0\cup S_1\cup S_2\cup S_3} = 0$ is imposed.

Next, let us examine the portion S_5 where derivative boundary conditions for both the pressure and velocity are prescribed according to:

$$\boldsymbol{n}\cdot\boldsymbol{\nabla}p|_{S_5} = r, \qquad \boldsymbol{n}\times\boldsymbol{\nabla}\times\boldsymbol{u}|_{S_5} = \boldsymbol{n}\times\boldsymbol{z},$$

where r and $\boldsymbol{n}\times\boldsymbol{z}$ are defined on S_5. By the same argument used to derive the pressure boundary condition on S_4, a boundary condition for the normal component of velocity is easily obtained, provided that $\gamma > 0$. In this case, the complete set of conditions to be used in the uncoupled formulation or in the direct Stokes solver is

$$\boldsymbol{n}\cdot\boldsymbol{\nabla}p|_{S_5} = r; \quad \gamma\,\boldsymbol{n}\cdot\boldsymbol{u}|_{S_5} = -\boldsymbol{n}\cdot\boldsymbol{\nabla}_S\times\boldsymbol{z} - r + \boldsymbol{n}\cdot\boldsymbol{g}|_{S_5}, \quad \boldsymbol{n}\times\boldsymbol{\nabla}\times\boldsymbol{u}|_{S_5} = \boldsymbol{n}\times\boldsymbol{z}.$$

If $\gamma = 0$ (steady-state problem), it is possible to impose the (independent) boundary condition $\boldsymbol{n}\cdot\boldsymbol{u}|_{S_5} = \boldsymbol{n}\cdot\boldsymbol{b}$ together with the derivative conditions for both pressure and velocity, but the data r and $\boldsymbol{n}\times\boldsymbol{z}$ of the latter must satisfy the following compatibility condition with the source term $\boldsymbol{g}$ of the momentum equation: $\boldsymbol{n}\cdot\boldsymbol{\nabla}_S\times\boldsymbol{z} + r = \boldsymbol{n}\cdot\boldsymbol{g}|_{S_5}$, at each point $\in S_5$.

Finally, if the following set of *four* scalar boundary conditions is prescribed on S_6

$$\boldsymbol{n}\cdot\boldsymbol{\nabla}p|_{S_6} = r, \qquad \boldsymbol{n}\cdot\boldsymbol{u}|_{S_6} = \boldsymbol{n}\cdot\boldsymbol{b}, \qquad \boldsymbol{n}\times\boldsymbol{\nabla}\times\boldsymbol{u}|_{S_6} = \boldsymbol{n}\times\boldsymbol{z},$$

then, as shown by Pironneau (1986), all the boundary data r, $\boldsymbol{n}\cdot\boldsymbol{b}$ and $\boldsymbol{n}\times\boldsymbol{z}$ defined on S_6 must be compatible with the boundary value of the normal component of the source term $\boldsymbol{g}$ according to the relationship:

$$\boldsymbol{n}\cdot\boldsymbol{\nabla}_S\times\boldsymbol{z} + \gamma\,\boldsymbol{n}\cdot\boldsymbol{b} + r = \boldsymbol{n}\cdot\boldsymbol{g}|_{S_6}.$$

5.8 Extension to compressible equations

The direct Stokes solver has been generalized to the equations governing compressible flows by Bristeau *et al.* (1987). In this section we briefly describe such a generalization and from it derive an uncoupled method using integral conditions for the solution of the same equations.

Let us consider the compressible Navier–Stokes equations for an ideal gas. After the time discretization and a convenient linearization of advection and other nonlinear terms, the compressible equations provide the following system of equations

$$\begin{aligned} \boldsymbol{\nabla}\cdot\boldsymbol{u} + \alpha q &= f, \\ (-\nu\nabla^2 + \alpha)\boldsymbol{u} + \beta\boldsymbol{\nabla}q &= \boldsymbol{g}, \\ (-\kappa\nabla^2 + \alpha)T &= h. \end{aligned} \tag{5.50}$$

The variable q represents the logarithm of the density of the fluid and T denotes its absolute temperature, while f, $\boldsymbol{g}$ and h are known source terms of the three equations. The quantities α, β, ν and κ are all positive constants whose value may depend on the properties of the fluid and on the time-step size Δt, as well as on the time-discretization scheme itself. For simplicity, let us assume that the system (5.50) is supplemented by Dirichlet conditions for the velocity and the temperature, namely,

$$\boldsymbol{u}|_S = \boldsymbol{b} \qquad \text{and} \qquad T|_S = c, \tag{5.51}$$

where $\boldsymbol{b}$ and c are the value of velocity and temperature prescribed on S. The problem for the temperature is uncoupled from the other two equations, so that we can concentrate on the latter which constitute a coupled system for the two unknows $\boldsymbol{u}$ and q, supplemented by the boundary condition $\boldsymbol{u}|_S = \boldsymbol{b}$, namely,

$$\begin{aligned} \boldsymbol{\nabla}\cdot\boldsymbol{u} + \alpha q &= f, \\ (-\nu\nabla^2 + \alpha)\boldsymbol{u} + \beta\boldsymbol{\nabla} q &= \boldsymbol{g}, \\ \boldsymbol{u}|_S &= \boldsymbol{b}. \end{aligned} \tag{5.52}$$

This problem is usually referred to as *generalized Stokes problem*, the generalization being intended from the incompressible equations to the compressible ones.

A possible method for solving problem (5.52) consists in eliminating the scalar unknown q from the second equation by means of the first equation, to give

$$\Big(-\nu\nabla^2 - \tfrac{1}{\alpha}\boldsymbol{\nabla}(\boldsymbol{\nabla}\cdot\ \ldots) + \alpha\Big)\boldsymbol{u} = \boldsymbol{g} - \tfrac{\beta}{\alpha}\boldsymbol{\nabla} f.$$

For $\nu > 0$ and $\alpha > 0$ the operator so obtained is symmetric definite positive. However, due to the presence of the parameter α at both the numerator and denominator in the operator coefficients, the condition number of the discrete counterpart of the operator can become unfavourable as α becomes small. Under those circumstances, it is preferable to envisage a method in which the scalar variable q is retained in the formulation and a convenient equation for q is explicitly admitted aside the momentum equation.

5.8.1 Generalized Stokes solver

Taking the divergence of the momentum equation of the generalized Stokes problem and substituting $\boldsymbol{\nabla}\cdot\boldsymbol{u}$ from the equation of continuity gives the elliptic equation for q

$$(-\xi\nabla^2 + \alpha^2)q = (-\nu\nabla^2 + \alpha)f - \boldsymbol{\nabla}\cdot\boldsymbol{g},$$

where the parameter ξ is defined by

$$\xi = \alpha\nu + \beta.$$

Now, similarly to the incompressible situation, an auxiliary scalar unknown ψ is introduced through the following Dirichlet problem

$$(-\xi\nabla^2 + \alpha^2)\psi = \alpha q + \boldsymbol{\nabla}\cdot\boldsymbol{u} - f, \qquad \psi|_S = 0,$$

the source term of which contains all the terms of the continuity equation of the original Stokes problem. Applying the operator $(-\nu\nabla^2+\alpha)$ to the equation for ψ, using the divergence of the momentum equation and the fact that q satisfies the elliptic equation above, one can see that ψ satisfies also the fourth-order elliptic equation

$$(-\xi\nabla^2+\alpha^2)(-\nu\nabla^2+\alpha)\psi=0. \tag{5.53}$$

It follows that the equations and the boundary condition of the generalized Stokes problem (5.52) will be satisfied by q and $\boldsymbol{u}$ solution to the system

$$\begin{aligned}(-\xi\nabla^2+\alpha^2)q &= (-\nu\nabla^2+\alpha)f-\boldsymbol{\nabla}\cdot\boldsymbol{g}, & & \\ (-\nu\nabla^2+\alpha\ )\boldsymbol{u} &= -\beta\boldsymbol{\nabla}q+\boldsymbol{g}, & \boldsymbol{u}|_S&=\boldsymbol{b}, \\ (-\xi\nabla^2+\alpha^2)\psi &= \alpha q+\boldsymbol{\nabla}\cdot\boldsymbol{u}-f, & \psi|_S&=0,\end{aligned} \tag{5.54}$$

provided that the auxiliary unknown ψ satisfies also the Neumann boundary condition $(\partial\psi/\partial n)|_S=0$. Such a condition is then expressed in a variational form writing

$$-\oint\frac{\partial\psi}{\partial n}\mu\,dS=0, \tag{5.55}$$

for any function μ defined on S. Introducing the following system of uncoupled λ-equations

$$\begin{aligned}(-\xi\nabla^2+\alpha^2)q_\lambda &= (-\nu\nabla^2+\alpha)f-\boldsymbol{\nabla}\cdot\boldsymbol{g}, & q_\lambda|_S&=\lambda; \\ (-\nu\nabla^2+\alpha\ )\boldsymbol{u}_\lambda &= -\beta\boldsymbol{\nabla}q_\lambda+\boldsymbol{g}, & \boldsymbol{u}_\lambda|_S&=\boldsymbol{b}; \\ (-\xi\nabla^2+\alpha^2)\psi_\lambda &= \alpha q_\lambda+\boldsymbol{\nabla}\cdot\boldsymbol{u}_\lambda-f, & \psi_\lambda|_S&=0;\end{aligned} \tag{5.56}$$

by means of Green identity one can show that

$$\begin{aligned}-\oint\frac{\partial\psi_\lambda}{\partial n}\mu\,dS=\int\Big[&\tfrac{\alpha}{\xi}q_\lambda\,q_\mu+\tfrac{\alpha}{\xi\beta}\boldsymbol{u}_\lambda\cdot\boldsymbol{u}_\mu \\ &+\tfrac{\nu}{\xi\beta}(\boldsymbol{\nabla}\times\boldsymbol{u}_\lambda\cdot\boldsymbol{\nabla}\times\boldsymbol{u}_\mu+\boldsymbol{\nabla}\cdot\boldsymbol{u}_\lambda\ \boldsymbol{\nabla}\cdot\boldsymbol{u}_\mu)\Big]\,dV.\end{aligned} \tag{5.57}$$

Thus, the unknown boundary value λ of q can be shown to be the solution of a linear problem of the type

$$A\lambda=\beta, \tag{5.58}$$

with the operator A symmetric. In particular, the quantities A and β of such a linear problem can be expressed in the form

$$A(\boldsymbol{\sigma},\boldsymbol{\sigma}')=-\int\Big[\tfrac{\alpha}{\xi}(\alpha\psi'-q')w+(\boldsymbol{\nabla}\psi'+\tfrac{1}{\xi}\boldsymbol{u}')\cdot\boldsymbol{\nabla}w\Big]\,dV, \tag{5.59}$$

$$\beta(\boldsymbol{\sigma})=+\int\Big[\tfrac{\alpha}{\xi}(\alpha\psi^0-q^0)w+(\boldsymbol{\nabla}\psi^0+\tfrac{1}{\xi}\boldsymbol{u}^0)\cdot\boldsymbol{\nabla}w\Big]\,dV, \tag{5.60}$$

where the auxiliary scalar functions $w(\boldsymbol{x};\boldsymbol{\sigma})$ are the same defined in (5.46) for the incompressible problem. The primed functions appearing in the expression of A are the solutions to the uncoupled elliptic problems

$$(-\xi\nabla^2+\alpha^2)q' = 0, \qquad q'|_S = \delta^{(2)}(\boldsymbol{s}-\boldsymbol{\sigma}'); \quad (5.61a)$$

$$(-\nu\nabla^2+\alpha\ )\boldsymbol{u}' = -\beta\boldsymbol{\nabla}q', \qquad \boldsymbol{u}'|_S = 0; \quad (5.61b)$$

$$(-\xi\nabla^2+\alpha^2)\psi' = \alpha q' + \boldsymbol{\nabla}\cdot\boldsymbol{u}', \qquad \psi'|_S = 0. \quad (5.61c)$$

Similarly, the functions with superscript 0 occurring in the expression of β are the solutions to the problems

$$(-\xi\nabla^2+\alpha^2)q^0 = (-\nu\nabla^2+\alpha)f - \boldsymbol{\nabla}\cdot\boldsymbol{g}, \qquad q^0|_S = 0; \quad (5.62a)$$

$$(-\nu\nabla^2+\alpha\ )\boldsymbol{u}^0 = -\beta\boldsymbol{\nabla}q^0 + \boldsymbol{g}, \qquad \boldsymbol{u}^0|_S = \boldsymbol{b}; \quad (5.62b)$$

$$(-\xi\nabla^2+\alpha^2)\psi^0 = \alpha q^0 + \boldsymbol{\nabla}\cdot\boldsymbol{u}^0 - f, \qquad \psi^0|_S = 0. \quad (5.62c)$$

For more general boundary conditions, the analysis developed in section 5.7 can be easily adapted to the generalized Stokes problem. Another kind of boundary condition suitable for dealing with truncated infinite domains has been considered by Quarteroni, Sacchi Landriani and Valli (1991) and by Rogé (1990).

5.8.2 Integral conditions

The direct solver for the generalized Stokes problem can be used as a template to derive another solution method which imposes integral conditions on the "density" q to uncouple the determination of this variable from the evaluation of velocity. The method starts by taking the divergence of the momentum equation and using the continuity equation to derive an elliptic equation for q. Thus, the equations of the generalized Stokes problem (5.52) are replaced by the following system of two elliptic equations

$$(-\xi\nabla^2+\alpha^2)q = (-\nu\nabla^2+\alpha)f - \boldsymbol{\nabla}\cdot\boldsymbol{g},$$

$$(-\nu\nabla^2+\alpha\ )\boldsymbol{u} = -\beta\boldsymbol{\nabla}q + \boldsymbol{g}.$$

The fulfillment of the equation for q does not guarantee that the equation of continuity is satisfied. In fact, taking the divergence of the second equation and using the first one to eliminate q gives

$$(-\xi\nabla^2+\alpha^2)(\boldsymbol{\nabla}\cdot\boldsymbol{u}+\alpha q - f) = 0,$$

which means that $(\boldsymbol{\nabla}\cdot\boldsymbol{u}+\alpha q-f)$ is only metaharmonic and not necessarily zero. On the other hand, imposing the additional boundary condition $(\boldsymbol{\nabla}\cdot\boldsymbol{u}+\alpha q)|_S =$

$f|_S$ assures that the continuity equation is identically satisfied. Therefore the generalized Stokes problem (5.52) is equivalent to the problem

$$(-\xi\nabla^2+\alpha^2)q = (-\nu\nabla^2+\alpha)f - \boldsymbol{\nabla\cdot g},$$

$$(-\nu\nabla^2+\alpha\ )\boldsymbol{u} = -\beta\boldsymbol{\nabla}q+\boldsymbol{g},$$

$$\boldsymbol{u}|_S = \boldsymbol{b},$$

$$(\boldsymbol{\nabla\cdot u}+\alpha q)|_S = f|_S,$$

which represents a system of two elliptic equations coupled together by the second of the boundary conditions. To uncouple the equation for q from that for $\boldsymbol{u}$ it is necessary to consider the vector Green identity for the operator $(-\nu\nabla^2+\alpha)$ and introduce the vector fields $\boldsymbol{\chi}_{\alpha/\nu}$ solution to the following metaharmonic problem

$$(-\nu\nabla^2+\alpha)\boldsymbol{\chi}_{\alpha/\nu}=0, \qquad \boldsymbol{n}\times\boldsymbol{\chi}_{\alpha/\nu}|_S=0, \quad \boldsymbol{n}\cdot\boldsymbol{\chi}_{\alpha/\nu}|_S\neq 0.$$

Using the momentum equation and the two boundary conditions of the equivalent formulation of the generalized Stokes problem, it is not difficult to show that q must satisfy the integral condition

$$\int\Big(-\xi\boldsymbol{\nabla}q\boldsymbol{\cdot\chi}_{\alpha/\nu} - \alpha\nu q\,\boldsymbol{\nabla\cdot\chi}_{\alpha/\nu}\Big)\,dV = -\int\boldsymbol{g\cdot\chi}_{\alpha/\nu}\,dV$$

$$+\ \nu\oint\Big(\boldsymbol{n}\times\boldsymbol{b}\boldsymbol{\cdot\nabla}\times\boldsymbol{\chi}_{\alpha/\nu}+\boldsymbol{n\cdot b}\ \boldsymbol{\nabla\cdot\chi}_{\alpha/\nu} - f|_S\ \boldsymbol{n\cdot\chi}_{\alpha/\nu}\Big)\,dS,$$

for any $\boldsymbol{\chi}_{\alpha/\nu}$ solution to the problem above. Thus, the integral conditions for compressible flows involve the gradient of the "density" variable q as well as the variable q itself. Moreover, the boundary term contains an additional contribution, with respect to the incompressible equations, which results from the source term f of the continuity equation. To implement these generalized integral conditions in a computational scheme one can finally resort to a decomposition scheme similar to that described in section 5.4.

Chapter 6

Evolutionary pressure–velocity equations

6.1 Introduction

There is a fundamental difference between the integral conditions for the vorticity and those for the pressure considered in the preceding chapter. In fact, while the former are formulated for the governing equations with a continuous time dependence, the latter refer to a time-discretized version of the problem. It seems therefore legitimate to ask whether pressure integral conditions can be established also for the unsteady primitive variable Navier–Stokes equations, or for any linearized version of them, before the time discretization is introduced.

The elucidation of this aspect is very important in order to gain a better theoretical understanding of the influence of solid boundaries on the pressure field and on its dynamics in incompressible viscous flows. In fact, as well known, the time-dependent incompressible equations imply that the pressure is governed by a simple a Poisson equation but they do not provide any apparent indication about which kind of conditions should be associated with this elliptic equation. Since solid boundaries surely affect the pressure field, uncovering the precise form of the conditions satisfied by the pressure in transient problems under no-slip conditions is a necessary prerequisite to realize the rôle played by the boundaries on the dynamic of pressure when both viscosity and incompressibity are in action. The issue has been the object of many speculations since long ago, but it has proven so far to be rather elusive.

Only recently, the appropriate conditions supplementing the Poisson equation for pressure have been recongnized and the general expressions describing the dynamics of this variable in bounded incompressible viscous flows have been discovered, in the particular case of the unsteady Stokes problem (Quartapelle and Valz-Gris 1990). It has been found that the same theorem of orthogonal projection considered in the previous chapters can be used effectively to derive also the conditions supplementing the Poisson equation for pressure in transient flows. However, the problem being now characterized by a continuous time dependence, the theorem leads to integral conditions more general than those established so far. In fact, it turns out that the conditions for the time-dependent problem are integral *both in space and in time.* This results means that the description of the dynamics of the pressure field in the presence of solid boundaries is "conditioned" by the *space-time* integral conditions supplementing the elliptic, "nonevolutionary," equation for the pressure—a rather surprising and counterintuitive result. In other words, for this kind of motion of an incompressible viscous fluid, pressure dynamics is found to be "hidden" in the conditions instead of being governed by an evolutionary equation.

The aim of the present chapter is to derive these space-time integral conditions for the pressure valid for the unsteady Stokes problem and to show that the ensuing dynamics of the pressure field is described by an infinite system of integral equations. The content of the chapter is expected to contribute to elucidate the structural difficulties encountered in the development of numerical methods for the solution of the primitive–variable Navier–Stokes equations under no-slip velocity conditions. In particular, we hope that understanding the very peculiar nature of the evolution of the pressure in the presence of solid boundaries can be helpful to fully appreciate the advantages as well as to realize some subtleties of the fractional-step method that will be described in the next chapter.

6.2 Unsteady Stokes problem with continuous time

Let us consider the unsteady Stokes problem, that is, the equations governing the motion of an incompressible viscous fluid with negligible inertial effects, namely,

$$\frac{\partial \boldsymbol{u}}{\partial t} = -\boldsymbol{\nabla} P + \nu \nabla^2 \boldsymbol{u} + \boldsymbol{f}, \tag{6.1}$$

$$\boldsymbol{\nabla} \cdot \boldsymbol{u} = 0. \tag{6.2}$$

The momentum equation has been obtained by dropping the nonlinear advection term $(\boldsymbol{u} \cdot \boldsymbol{\nabla})\boldsymbol{u}$ and adding a source term $\boldsymbol{f}(\boldsymbol{x}, t)$ which represents an external body force. As for the Navier–Stokes problem introduced in the first chapter, equations (6.1)–(6.2) are supplemented by the boundary and initial conditions:

$$\boldsymbol{u}\big|_S = \boldsymbol{b}(\boldsymbol{x}_S, t) \qquad \text{and} \qquad \boldsymbol{u}\big|_{t=0} = \boldsymbol{u}_0(\boldsymbol{x}), \tag{6.3}$$

where S is the boundary of the fluid domain V, which is assumed to be sufficiently smooth, and $\boldsymbol{x}_S \in S$. As usual, the velocity $\boldsymbol{b}$ of the fluid prescribed on the boundary and the initial velocity field $\boldsymbol{u}_0(\boldsymbol{x})$ are supposed to satisfy the conditions

$$\oint \boldsymbol{n} \cdot \boldsymbol{b} \, dS = 0 \qquad \text{and} \qquad \boldsymbol{\nabla} \cdot \boldsymbol{u}_0 = 0, \tag{6.4}$$

respectively. Finally, the boundary and initial data are assumed to satisfy the following compatibility condition

$$\boldsymbol{n} \cdot \boldsymbol{b}|_{t=0} = \boldsymbol{n} \cdot \boldsymbol{u}_0|_S, \tag{6.5}$$

where $\boldsymbol{n} \cdot \boldsymbol{b}(\boldsymbol{x}_S, t)$ is taken to be a continuous function of time as $t \to 0^+$.

Equations (6.1)–(6.5) define an initial boundary value linear problem which has a unique solution $(\boldsymbol{u}(\boldsymbol{x}, t), P(\boldsymbol{x}, t))$, the pressure P being of course defined up to an arbitrary additive function of time.

This problem can be formulated in a different, but equivalent, way, by introducing a Poisson equation for the pressure. In fact, taking the divergence of the momentum equation (6.1) and assuming that the incompressibility condition (6.2) is satisfied, one obtains the equation $\nabla^2 P = \boldsymbol{\nabla} \cdot \boldsymbol{f}$. However, as already noted for the time-discretized problem, this equation is only a necessary condition for the solenoidal character of the velocity field. The condition required to obtain an equivalent statement of the set (6.1)–(6.3) is obtained by the following very simple argument due to Kleiser and Schumann (1980).

Let $\boldsymbol{v}(\boldsymbol{x})$ be a velocity field solution of (6.1), with P solution of the aforementioned Poisson equation, and satisfying the boundary and initial conditions in (6.3), under the assumption $\boldsymbol{\nabla} \cdot \boldsymbol{u}_0 = 0$ in (6.4). To verify whether $\boldsymbol{v}$ is solenoidal, one takes the divergence of the momentum equation for $\boldsymbol{v}$ and obtains the following heat equation

$$\frac{\partial D}{\partial t} = \nu \nabla^2 D,$$

governing the scalar variable $D = \boldsymbol{\nabla} \cdot \boldsymbol{v}$. By the initial condition $\boldsymbol{v}|_{t=0} = \boldsymbol{u}_0$ and by the assumption $\boldsymbol{\nabla} \cdot \boldsymbol{u}_0 = 0$, D satisfies the homogeneous initial condition $D|_{t=0} = 0$. Therefore D will be identically zero for any $t > 0$ if and only if the heat equation is supplemented also by the homogeneous boundary condition $D|_S = 0$ for all $t > 0$. It follows that the problem (6.1)–(6.3) admits the following equivalent formulation

$$\begin{aligned}
&\frac{\partial \boldsymbol{u}}{\partial t} = -\boldsymbol{\nabla} P + \nu \nabla^2 \boldsymbol{u} + \boldsymbol{f}, \\
&-\nabla^2 P = -\boldsymbol{\nabla} \cdot \boldsymbol{f}, \\
&\boldsymbol{u}|_S = \boldsymbol{b}(\boldsymbol{x}_S, t), \qquad \boldsymbol{\nabla} \cdot \boldsymbol{u}|_S = 0, \\
&\boldsymbol{u}|_{t=0} = \boldsymbol{u}_0(\boldsymbol{x}),
\end{aligned} \tag{6.6}$$

with the same conditions (6.4) and (6.5) on the boundary and initial data. It is to be noted that the equation $\nabla^2 P = \boldsymbol{\nabla}\cdot \boldsymbol{f}$ alone allows to determine P only up to a function which is harmonic in V and depends on time. The pressure field is nevertheless unique (disregarding of course the arbitrariness by the additive "constant" $C = C(t)$) because P must be chosen so that all the remaining equations and conditions for $\boldsymbol{u}$ in problem (6.6) be *compatible*.

6.3 Space-time integral conditions for pressure

The pressure field P solution to problem (6.6) can be characterized uniquely, in a form which is completely independent of the velocity field $\boldsymbol{u}$, except for the value of $\boldsymbol{u}$ prescribed on the boundary and at the initial time. One notices that problem (6.6) implies that

$$-\boldsymbol{\nabla}P + \boldsymbol{f} = \left(-\nu\nabla^2 + \tfrac{\partial}{\partial t}\right)\boldsymbol{u}$$

with the velocity field $\boldsymbol{u}$ such that

$$\boldsymbol{u}|_S = \boldsymbol{b}(\boldsymbol{x}_S, t), \qquad \boldsymbol{\nabla}\cdot \boldsymbol{u}|_S = 0 \qquad \text{and} \qquad \boldsymbol{u}|_{t=0} = \boldsymbol{u}_0(\boldsymbol{x}).$$

Consider first the problem with homogeneous conditions. Then $-\boldsymbol{\nabla}P + \boldsymbol{f}$ belongs to the range of the diffusion operator $\left(-\nu\nabla^2 + \frac{\partial}{\partial t}\right)$ acting on (time-dependent) vector fields which satisfy the homogeneous version of the considered boundary and initial conditions. Consider the orthogonality theorem

$$\overline{\mathcal{R}(A)} = \mathcal{N}(A^{\dagger})^{\perp}, \tag{6.7}$$

where $\mathcal{R}(A)$ and $\mathcal{N}(A)$ denote the range (or image) and null space (or kernel) of a linear operator A, whose range is a subspace of $\boldsymbol{L}^2(V \times [0,T])$; moreover, $A^{\dagger}$ is the operator adjoint to A, the bar indicates the closure and $\perp$ the orthogonal complement. Taking the operator A in this theorem as the diffusion operator $\left(-\nu\nabla^2 + \frac{\partial}{\partial t}\right)$ supplemented by the aforementioned conditions, it follows that $-\boldsymbol{\nabla}P + \boldsymbol{f}$ must belong to the orthogonal complement, in the space $\boldsymbol{L}^2(V\times[0,T])$, of the null space $\mathcal{N}\left(\left(-\nu\nabla^2 + \frac{\partial}{\partial t}\right)^{\dagger}\right)$. The adjoint operator $\left(-\nu\nabla^2 + \frac{\partial}{\partial t}\right)^{\dagger}$ and its respective conditions are identified by the (space-time) Green identity for the original operator $\left(-\nu\nabla^2 + \frac{\partial}{\partial t}\right)$, which reads

$$\begin{aligned}
&\int_0^T\!\!\int \left[\boldsymbol{v}\cdot\left(-\nu\nabla^2 + \tfrac{\partial}{\partial t}\right)\boldsymbol{u} - \boldsymbol{u}\cdot\left(-\nu\nabla^2 - \tfrac{\partial}{\partial t}\right)\boldsymbol{v}\right] dV\,dt \\
&\quad = \nu\int_0^T\!\!\oint \left[\boldsymbol{n}\times\boldsymbol{u}\cdot\boldsymbol{\nabla}\times\boldsymbol{v} - \boldsymbol{n}\times\boldsymbol{v}\cdot\boldsymbol{\nabla}\times\boldsymbol{u} + \boldsymbol{n}\cdot\boldsymbol{u}\,\boldsymbol{\nabla}\cdot\boldsymbol{v} - \boldsymbol{n}\cdot\boldsymbol{v}\,\boldsymbol{\nabla}\cdot\boldsymbol{u}\right] dS\,dt \\
&\qquad + \int \left[(\boldsymbol{u}\cdot\boldsymbol{v})|_{t=T} - (\boldsymbol{u}\cdot\boldsymbol{v})|_{t=0}\right] dV.
\end{aligned} \tag{6.8}$$

Therefore, as far as the adjoint operator is concerned, Green identity gives

$$\left(-\nu\nabla^2+\tfrac{\partial}{\partial t}\right)^\dagger=\left(-\nu\nabla^2-\tfrac{\partial}{\partial t}\right), \tag{6.9}$$

which is the *antidiffusion* operator, *i.e.*, the operator of diffusion with a backward time evolution. On the other hand, the quantities $\boldsymbol{n}\times\nabla\times\boldsymbol{u}|_S$ and $\boldsymbol{u}|_{t=T}$ being unknown, the conditions to be associated with $\left(-\nu\nabla^2-\frac{\partial}{\partial t}\right)$ are the vanishing of the tangential components of the vector field and the vanishing of the field itself at the final time $t=T$. Thus the space $\mathcal{N}\left(-\nu\nabla^2-\frac{\partial}{\partial t}\right)$ is the space of the fields $\boldsymbol{\chi}(\boldsymbol{x},t;T)$ which are solution to the antidiffusion vector equation:

$$\left(-\nu\nabla^2-\tfrac{\partial}{\partial t}\right)\boldsymbol{\chi}(\boldsymbol{x},t;T)=0, \tag{6.10}$$

supplemented by the following boundary and *final* conditions

$$\boldsymbol{n}\times\boldsymbol{\chi}|_S=0,\quad 0<t<T,\qquad\text{and}\qquad \boldsymbol{\chi}|_{t=T}=0,\quad \forall\boldsymbol{x}\in V. \tag{6.10a}$$

Since problem (6.10)–(6.10a) does not contain any incompressibility condition, the data specifying the boundary values and the final values of $\boldsymbol{\chi}$ are not required to satisfy any compatibility condition similar to those in (6.4) and (6.5) for the original (Navier–) Stokes problem. Thus, to construct the linear space $\{\boldsymbol{\chi}(\boldsymbol{x},t;T)\}$ one takes $\boldsymbol{n}\cdot\boldsymbol{\chi}(\boldsymbol{x},t;T)|_S\neq 0$ for any $t\leq T$ and, in particular, one can take $\boldsymbol{n}\cdot\boldsymbol{\chi}(\boldsymbol{x}_S,t;T)|_{t=T}\neq 0$ even though the values of such a nonhomogeneous boundary condition at the final time are different from the boundary values of the homogeneous final condition in (6.10a).

Comparing the *boundary* conditions for the time-dependent fields $\boldsymbol{\chi}(\boldsymbol{x},t;T)$ with those for the fields $\boldsymbol{\chi}_\gamma(\boldsymbol{x})$ of the time-discretized equations (see chapter 5) one notices that they coincide. The discretized problem is however much simpler since the differential operator $(-\nabla^2+\gamma)$ is equal to its adjoint (disregarding their respective boundary conditions, of course): any aspect associated with the time evolution is condensed in the presence of the constant γ in the operator $(-\nabla^2+\gamma)$.

Using the vector fields $\boldsymbol{\chi}$ solution to problem (6.10)–(6.10a), the orthogonality theorem (6.7) gives the following *space-time* integral conditions for $-\nabla P+\boldsymbol{f}$

$$\int_0^T\!\!\int(-\nabla P+\boldsymbol{f})\cdot\boldsymbol{\chi}\,dV\,dt=0. \tag{6.11}$$

For the more general situation of nonhomogeneous conditions, the integral conditions fix the orthogonal projection of $-\nabla P+\boldsymbol{f}$ on the linear space $\{\boldsymbol{\chi}(\boldsymbol{x},t;T)\}$ according to the relationship:

$$\begin{aligned}&\int_0^T\!\!\int(-\nabla P+\boldsymbol{f})\cdot\boldsymbol{\chi}\,dV\,dt \qquad (6.12)\\ &\quad=\nu\int_0^T\!\!\oint(\boldsymbol{n}\times\boldsymbol{b}\cdot\nabla\times\boldsymbol{\chi}+\boldsymbol{n}\cdot\boldsymbol{b}\,\nabla\cdot\boldsymbol{\chi})\,dS\,dt-\int\boldsymbol{u}_0\cdot\boldsymbol{\chi}|_{t=0}\,dV,\end{aligned}$$

as it can be readily verified by means of Green identity (6.8) with $\boldsymbol{v} = \boldsymbol{\chi}$.

The conclusion is that, in order that the equations and conditions for $\boldsymbol{u}$ in problem (6.6) be compatible, the gradient of the pressure field must satisfy the integral conditions (6.12) for any time-dependent vector field $\boldsymbol{\chi}(\boldsymbol{x}, t; T)$ solution to problem (6.10)–(6.10a). In other words, the incompressible problem (6.6) implies that the pressure is characterized by an independent problem given by

$$-\nabla^2 P = -\boldsymbol{\nabla}\cdot \boldsymbol{f}(\boldsymbol{x}, t), \tag{6.13}$$

$$\begin{aligned} -\int_0^t \int \boldsymbol{\nabla} P \cdot \boldsymbol{\chi}\, dV\, dt' &= -\int_0^t \int \boldsymbol{f}\cdot \boldsymbol{\chi}\, dV\, dt' \\ &+\nu \int_0^t \oint (\boldsymbol{n}\times\boldsymbol{b}\cdot \boldsymbol{\nabla}\times\boldsymbol{\chi} + \boldsymbol{n}\cdot\boldsymbol{b}\, \boldsymbol{\nabla}\cdot \boldsymbol{\chi})\, dS\, dt' - \int \boldsymbol{u}_0 \cdot \boldsymbol{\chi}|_{t'=0}\, dV, \end{aligned} \tag{6.13a}$$

$$\forall \boldsymbol{\chi} \in \mathcal{N}\big(-\nu\nabla^2 - \tfrac{\partial}{\partial t'}\big), \qquad \boldsymbol{n}\times\boldsymbol{\chi}|_S = 0, \qquad \boldsymbol{\chi}|_{t'=t} = 0. \tag{6.13b}$$

Here the variable for the time integration has been indicated by t' and the upper limit of the integrals by t, so that the symbol $\boldsymbol{\chi}$ means now $\boldsymbol{\chi}(\boldsymbol{x}, t'; t)$. The solution of problem (6.13)–(6.13a) allows the subsequent determination of the velocity $\boldsymbol{u}(\boldsymbol{x}, t)$ by solving the following parabolic equation supplemented by boundary and initial conditions

$$\begin{gathered} \big(-\nu\nabla^2 + \tfrac{\partial}{\partial t}\big)\boldsymbol{u} = -\boldsymbol{\nabla} P + \boldsymbol{f}, \\ \boldsymbol{u}|_S = \boldsymbol{b} \quad \text{and} \quad \boldsymbol{u}|_{t=0} = \boldsymbol{u}_0, \end{gathered} \tag{6.14}$$

without requiring to impose the incompressibility condition explicitly, since it will be satisfied automatically for any $t > 0$.

6.4 Drag on a sphere in nonuniform motion

The equation governing the pressure field is purely spatial. The actual evolution of this variable, letting aside the variation caused by the possible time dependence of the body force $\boldsymbol{f}(\boldsymbol{x}, t)$, is determined by the integral conditions (6.13a). Thus, by virtue of the orthogonality theorem, the dynamics of the pressure in incompressible viscous flows, under no-slip velocity conditions, is discovered to be governed by the conditions of space-time integral character, supplementing the purely spatial Poisson equation satisfied by the pressure. This situation is to be contrasted with what has been found to occurr to the vorticity, a variable governed by a dynamical equation and subject to integral conditions of a purely spatial character.

An important characteristic of the entire problem for the determination of pressure is that it is not possible to formulate an instantaneous elliptic problem so as to obtain the pressure field $P(\boldsymbol{x}, t)$ at any given instant of time in terms of the data of the Stokes problem. In fact, the temporal content of the pressure integral conditions is such that one can set up a pressure problem only for a whole time interval, but not for a single time instant. However counterintuitive, this limitation is in accordance with the fact that the determination of the pressure from the original momentum equation (6.1) would require to know the instantaneous value of the acceleration of the fluid and the latter can be determined only by solving the complete initial boundary value problem defined by (6.1)–(6.3). An almost evident corollary of this result is that the initial pressure field cannot be characterized by a Poisson problem with boundary conditions derived from the data of the unsteady Stokes problem by means of a local evaluation process: in other words, the very concept of an initial pressure field in incompressible viscous problems with no-slip boundaries appears to be excluded by the presence of the space-time integral conditions, unless a higher degree of regularity is assumed in the solution.

Another related consequence of the temporal character of the integral conditions for pressure is that they imply a *memory* aspect[1] in this variable. The idea of a *hereditary* character in unsteady Stokes flows is not new. It has been put in evidence originally by Basset (1887) and Boussinesq (1885) who have calculated the force experienced by a sphere moving with a nonzero acceleration in a viscous incompressible fluid at rest at a large distance from the body. Assuming that the motion of the sphere with respect to the stationary fluid is given by the function $U(t)$, it is possible to evaluate the expression for the force $F(t)$ acting at time t on a sphere of radius a (for the simplest and most general derivation see Picciati 1907, see also Landau and Lifshitz [26], p. 96–97)

$$F(t) = 6\pi a\mu U(t) + \tfrac{2}{3}\pi a^3 \rho U'(t)$$

$$+ 6a^2\sqrt{\pi\mu\rho}\int_0^t \frac{U'(\tau)\, d\tau}{\sqrt{t-\tau}}$$

$$- \frac{4a\sqrt{\pi\mu\rho}}{\sqrt{t}} \int_a^\infty \left[\frac{d}{dr}\left(\frac{d\psi_0}{dr} + \frac{\psi_0}{r}\right)\right] e^{-\frac{(r-a)^2}{4\nu t}}\, dr,$$

where ρ is the density of the fluid, μ the coefficient of dynamic viscosity ($\nu = \mu/\rho$). The function $\psi_0(r)$ is related to the stream function $\Psi_0(r, \theta)$ of the initial velocity field $\boldsymbol{u}_0(r, \theta)$, which is assumed of course axisymmetric, through the relation $\Psi_0(r, \theta) = \psi_0(r) \sin^2\theta$, so that

$$u_{r,0}(r, \theta) = \frac{1}{r^2 \sin\theta} \frac{\partial \Psi_0(r, \theta)}{\partial \theta} = \frac{2}{r^2}\psi_0(r) \cos\theta,$$

$$u_{\theta,0}(r, \theta) = -\frac{1}{r \sin\theta} \frac{\partial \Psi_0(r, \theta)}{\partial r} = -\frac{1}{r}\frac{d\psi_0(r)}{dr} \sin\theta.$$

[1]This circumstance has been pointed to our attention by Jean-Luc Achard.

There are four contributions to the instantaneous value of the force $F(t)$ experienced by the sphere. The first term, proportional to the velocity of the sphere, provides a viscous contribution to the force, given by the well-known Stokes law for steady (slow) motion, the value of velocity being taken at the considered instant of time. This term is the only one that remains if the sphere reaches a steady motion in the fluid, whereas the other three terms are due to the unsteady character of the problem.

The second term is proportional to the acceleration of the sphere, with a coefficient equal to half the mass of fluid displaced by the sphere. It provides an inviscid unsteady contribution to the force, which corresponds to the resistance encountered by a sphere in nonuniform translational motion in an ideal fluid.

The third term, discovered by Basset and Boussinesq, takes into account the history of the sphere acceleration from the initial time $t = 0$ to the actual time t. The presence of this term shows that the force acting on the sphere at time t depends on its earlier motion and not only on its instantaneous velocity and acceleration. The structure of the integrand is such that the acceleration $U'(\tau)$ at the older times has a more attenuated influence. The fourth and last term describes instead the influence at time t due to the initial velocity field $\boldsymbol{u}_0$ throughout the entire fluid domain: the temporal coefficient within the spatial integral makes the influence of $\boldsymbol{u}_0$ to decrease as the distance from the sphere increases and the time elapses.

One can observe that the two integrals, respectively in time and in space, appearing in the unsteady force formula above, have their counterparts in the conditions for pressure in problem (6.13). Similarly, in both cases the complete solution of the problem is achieved only in terms of the data $\boldsymbol{b}(\boldsymbol{x}_S, t)$, $\boldsymbol{u}_0(\boldsymbol{x})$ and $\boldsymbol{f}(\boldsymbol{x}, t)$. Remarkably enough, the formulation (6.13)–(6.14) and, in particular, the uncoupled formulation of the pressure problem (6.13) reveal that the hereditary aspect associated with the pressure in incompressible viscous flows with boundaries can be given a geometrical interpretation in terms of a general orthogonality principle within the context of Hilbert space theory. In this sense, the orthogonality principle underlying the various kinds of integral conditions discussed in the present study provides a simple and unifying geometrical picture of the effect of no-slip velocity conditions at solid boundaries in incompressible viscous flows, whenever the Navier–Stokes equations are reduced to a system of second-order parabolic and elliptic equations.

6.5 Pressure dynamics in incompressible viscous flows with boundaries

A consequence of the integral conditions (6.13a) is the possibility of describing the evolution of the pressure field by means of a system of Volterra integral equations. This can be demonstrated by considering a basis $\{\chi_i\}$ of the space $\mathcal{N}\big(-\nu\nabla^2 - \frac{\partial}{\partial t'}\big)$, defined as follows

$$\boldsymbol{n}\cdot\boldsymbol{\chi}_i(\boldsymbol{x},t';t)|_S = g_i(\boldsymbol{x}_S)\,k(t-t'), \tag{6.15}$$

where $\{g_i(\boldsymbol{x}_S)\}$ is any basis of $L^2(S)$ and $k(t)$ is an arbitrary function, defined for $t \geq 0$. The field P solution to the equation (6.13) at any time can be decomposed in its harmonic and nonharmonic components according to the relationship

$$P(\boldsymbol{x},t) = \sum_i p_i(t)\,h_i(\boldsymbol{x}) + P^{[\boldsymbol{f}]}(\boldsymbol{x},t). \tag{6.16}$$

Here $\{h_i(\boldsymbol{x})\}$ represents any basis of the subspace of $L^2(V)$ consisting of the harmonic functions [it is possible, for instance, to take $\boldsymbol{n}\cdot\boldsymbol{\nabla}h_i|_S = g_i(\boldsymbol{x}_S)$], whereas $P^{[\boldsymbol{f}]}(\boldsymbol{x},t)$ is the solution, at any time, of the following Neumann problem

$$-\nabla^2 P^{[\boldsymbol{f}]} = -\boldsymbol{\nabla}\cdot\boldsymbol{f}(\boldsymbol{x},t), \qquad \boldsymbol{n}\cdot\boldsymbol{\nabla}P^{[\boldsymbol{f}]}|_S = |S|^{-1}\oint \boldsymbol{n}\cdot\boldsymbol{f}\,dS. \tag{6.17}$$

Here $|S|$ denotes the area of the surface S and the (constant) quantity on the right-hand side of the Neumann condition has been introduced to satisfy the solvability condition for the Neumann problem.

By introducing expression (6.16) into the integral conditions (6.13a), one obtains the following problem for the functions of time $p_i(t)$, $i = 1, 2, \ldots$,

$$\sum_j \int_0^t K_{ij}(t',t)\,p_j(t')\,dt' = c_i(t), \quad t > 0, \tag{6.18}$$

where

$$K_{ij}(t',t) = -\int \boldsymbol{\nabla}h_j(\boldsymbol{x})\cdot\boldsymbol{\chi}_i(\boldsymbol{x},t';t)\,dV, \tag{6.19}$$

and

$$\begin{aligned} c_i(t) = \nu \int_0^t \oint (\boldsymbol{n}\times\boldsymbol{b}\cdot\boldsymbol{\nabla}\times\boldsymbol{\chi}_i + \boldsymbol{n}\cdot\boldsymbol{b}\,\boldsymbol{\nabla}\cdot\boldsymbol{\chi}_i)\,dS\,dt' \\ - \int \boldsymbol{u}_0\cdot\boldsymbol{\chi}_i|_{t'=0}\,dV + \int_0^t \int \big(\boldsymbol{\nabla}P^{[\boldsymbol{f}]} - \boldsymbol{f}\big)\cdot\boldsymbol{\chi}_i\,dV\,dt' \end{aligned} \tag{6.20}$$

Equation (6.18) represents an infinite system of Volterra integral equations of *the first kind*. We notice that the functions $K_{ij}(t',t)$ depend only on the difference $t-t'$, while their explicit form depends on the representation chosen for the functions $g_i(\boldsymbol{x}_S)$ and $k(t)$. Figure 6.1 contains a schematic picture of the procedure of performing the operation of orthogonal projection underlying the fulfillment of the space-time integral conditions and of the associated Volterra integral equations. It is important to remind that, in the case of primitive variable equations with a continuous time dependence considered here, the geometrical operations refer to points and vectors which represent not a field at a given instant of time, but the entire evolution of the field over a finite time interval, *e.g.*, $0 < t' < t$, with $t > 0$. Thus, while the overall geometrical picture is very similar to those introduced in the previous chapters, its correct interpretation require to extend the concept of orthogonality to the space-time domain.

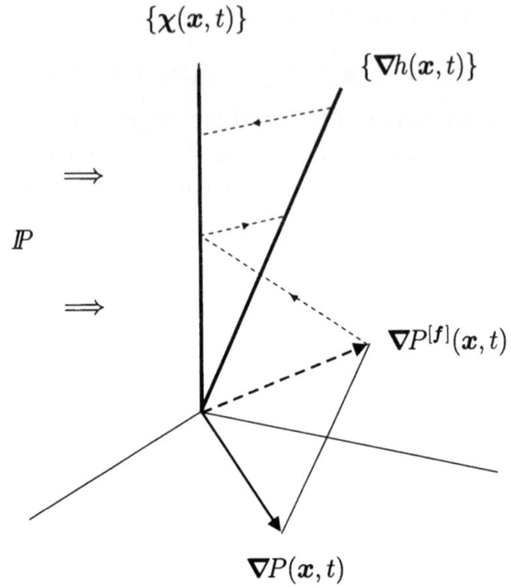

Figure 6.1: Schematic of the construction of the pressure field in the unsteady Stokes problem with a continuous time dependence. Each point represents the entire evolution over the time interval $0 \leq t \leq T$. (P denotes the pressure field of the incompressible flow and h its harmonic component.)

6.6 Comments

The uncoupling of the pressure from the velocity in the unsteady Stokes problem has been obtained at the cost of introducing the nonstandard problem (6.13), whose solution implies the simultaneous determination of $P(\boldsymbol{x}, t)$ at different instants of time. However, this uncoupled formulation relies upon Volterra integral equations. The latter, once recast in a discretized form to give a numerical method, are characterized by very favourable stability properties. Therefore, the uncoupled formulation is expected to be a convenient starting point for approximations of spectral type both in time and in space of the unsteady Stokes problem formulated as a system of second-order elliptic and parabolic equations. In this formulation there is no need to satisfy the conditions of compatibility of nonlocal type which are encountered when the Poisson equation for pressure is supplemented by the boundary condition consisting in the trace of the momentum equation on S: these conditions are very difficult to verify in practice while their fulfillement has been found to be essential for the application of space-time spectral methods.

When the full nonlinear Navier–Stokes equations for the primitive variables are considered, the uncoupling of the pressure from the velocity demonstrated in this chapter is no more possible, due to the presence of the inertial term $(\boldsymbol{u} \cdot \boldsymbol{\nabla})\boldsymbol{u}$ [cf. the equation (6.1) with $\boldsymbol{f} = -(\boldsymbol{u} \cdot \boldsymbol{\nabla})\boldsymbol{u}$]. In this general situation, the hereditary aspect associated with the integral conditions supplementing the equation $-\nabla^2 P = \boldsymbol{\nabla} \cdot [(\boldsymbol{u} \cdot \boldsymbol{\nabla})\boldsymbol{u}]$ is "entangled" with the nonlinear character of the momentum equation. This remark represents a further confirmation, obtained form a different viewpoint, of the well-known fact that flows governed by the Navier–Stokes equations in the presence of solid walls are characterized by dynamical aspects intrinsically different with respect to the fluid motions in unbounded domains or in bounded regions with spatially periodic boundary conditions.

Perhaps, of not minor interest is the fact that the space-time integral conditions for pressure can be generalized to the complete Navier–Stokes equations governing compressible flows. Generalization of this kind have been considered by Bristeau, Glowinski, Périaux *et al.*, and have been briefly presented in the previous chapter. Similar ideas can be pursued starting from the formulation that has been presented in this chapter. The subject of the present study being limited to incompressible problems, the results of such an investigation will be described elsewhere.

Chapter 7

Fractional-step projection method

7.1 Introduction

Dulcis in fundo, the last method to be described for the solution of the incompressible Navier–Stokes is the fractional-step projection method introduced by Chorin (1968a-b and 1969) and Temam (1969a-b, see also [35, p. 395]). This method is the most frequently used for solving the primitive variable Navier–Stokes equations and has been the first numerical scheme enabling a cost-effective solution of three-dimensional time-dependent problems. Since its initial appearance, the fractional-step method has been repeatedly considered in CFD literature and several implementations have been proposed using spatial discretizations based on finite difference, finite elements or spectral methods. A general description of the method with the details of its most relevant variants can be found in [35], [14] and [6].

The aim of the present chapter is to provide an elementary introduction to the fractional-step projection method and to describe its unique features in comparison with all other schemes for the numerical solution of the incompressible Navier–Stokes equations. In particular, we intend to emphasize the (abstract) geometrical characteristics of the method and to explore its relationship with the nonfractional-step method using the integral conditions for the pressure described in chapter 5. It will be shown that the orthogonal projection involved in the fractional-step method serves a completely different purpose with respect to the projection operation we have introduced in order to satisfy conditions of integral type. In spite of that, for problems involving solid boundaries, a sort of geometrical correspondence is found to exist in the treatment of the boundary conditions in the fractional-step and nonfractional-step methods.

A very peculiar feature of the projection method is that the boundary conditions to be imposed in the incompressible phase of the calculation are different depending on the form assumed by the equations and on the kind of spatial discretization adopted. We will address this issue and describe a version of the fractional-step method in which a Poisson equation for the pressure is used in place of the incompressibility condition. The boundary condition which supplements such an equation will be derived and discussed.

Then, we will consider the fractional-step projection method when the equations enforcing the incompressibility are left in the original grad-div form. In particular, a finite element implementation of such a form of the projection method is described, giving the explicit expression of the discrete counterpart of the projection operator for a representative computational molecule of the equations for two-dimensional problems. This finite element projection method is based on quadrilateral elements with a bilinear interpolation of the velocity and a piecewise constant approximation of the pressure (Donea *et al.* 1982). The method has been proved to be effective in the solution of problems with solid boundaries without being plagued by the well known difficulty of the spurious modes for the pressure.

7.2 Ladyzhenskaya theorem

Following Chorin (1969a), let the Navier–Stokes equations, describing the motion of an incompressible viscous fluid, be written in the form

$$\begin{aligned} \frac{\partial \boldsymbol{u}}{\partial t} + \boldsymbol{\nabla} P &= -(\boldsymbol{u} \cdot \boldsymbol{\nabla})\boldsymbol{u} + \nu \nabla^2 \boldsymbol{u}, \\ \boldsymbol{\nabla} \cdot \boldsymbol{u} &= 0, \end{aligned} \tag{7.1}$$

where $P = p/\rho$ is the pressure (per unit density of the fluid). Basic to the derivation of the fractional-step method is a theorem of orthogonal decomposition due to Ladyzhenskaya [24], see also [9, p. 37]. This decomposition is one instance of the general Hodge orthogonal decomposition which has been mentioned in chapter 3 (section 3.5.3). Here it is discussed in its simplest form to see the precise consequences of the various assumptions.

Theorem 7.1. *Any vector field $\boldsymbol{v}$ defined in V admits the unique orthogonal decomposition*

$$\boldsymbol{v} = \boldsymbol{w} + \boldsymbol{\nabla}\varphi, \tag{7.2}$$

where $\boldsymbol{w}$ is a solenoidal field with zero normal component on the boundary S of V.

Proof. First we establish the orthogonality of the decomposition, namely,

$$\int \boldsymbol{w} \cdot \boldsymbol{\nabla}\varphi \, dV = 0.$$

In fact, by the identity $\boldsymbol{\nabla}\cdot(\boldsymbol{w}\varphi) = (\boldsymbol{\nabla}\cdot\boldsymbol{w})\,\varphi + \boldsymbol{w}\cdot\boldsymbol{\nabla}\varphi$, the condition $\boldsymbol{\nabla}\cdot\boldsymbol{w} = 0$ and the divergence theorem give

$$\int \boldsymbol{w}\cdot\boldsymbol{\nabla}\varphi\, dV = \int \boldsymbol{\nabla}\cdot(\boldsymbol{w}\,\varphi)\, dV = \oint \boldsymbol{n}\cdot\boldsymbol{w}\,\varphi\, dS = 0,$$

since $\boldsymbol{n}\cdot\boldsymbol{w}|_S = 0$.

The orthogonality is used to prove uniqueness. Suppose that $\boldsymbol{u} = \boldsymbol{w}_1 + \boldsymbol{\nabla}\varphi_1 = \boldsymbol{w}_2 + \boldsymbol{\nabla}\varphi_2$. Then

$$0 = \boldsymbol{w}_1 - \boldsymbol{w}_2 + \boldsymbol{\nabla}(\varphi_1 - \varphi_2).$$

Taking the scalar product with $\boldsymbol{w}_1 - \boldsymbol{w}_2$ and integrating on V, we get

$$\begin{aligned} 0 &= \int \Big[|\boldsymbol{w}_1 - \boldsymbol{w}_2|^2 + (\boldsymbol{w}_1 - \boldsymbol{w}_2)\cdot\boldsymbol{\nabla}(\varphi_1 - \varphi_2)\Big]\, dV \\ &= \int |\boldsymbol{w}_1 - \boldsymbol{w}_2|^2\, dV, \end{aligned}$$

by the orthogonality of the decomposition. It follows that $\boldsymbol{w}_1 = \boldsymbol{w}_2$ and so $\boldsymbol{\nabla}\varphi_1 = \boldsymbol{\nabla}\varphi_2$, that is, $\varphi_1 = \varphi_2 + \text{constant}$.

If $\boldsymbol{v} = \boldsymbol{w} + \boldsymbol{\nabla}\varphi$, we notice that $\boldsymbol{\nabla}\cdot\boldsymbol{v} = \nabla^2\varphi$ and $\boldsymbol{n}\cdot\boldsymbol{v}|_S = \boldsymbol{n}\cdot\boldsymbol{\nabla}\varphi|_S$. This remark allows to prove existence. In fact, given $\boldsymbol{v}$ let φ be defined by the solution to the Neumann problem

$$\nabla^2\varphi = \boldsymbol{\nabla}\cdot\boldsymbol{v}, \qquad \boldsymbol{n}\cdot\boldsymbol{\nabla}\varphi|_S = \boldsymbol{n}\cdot\boldsymbol{v}|_S.$$

The solution φ to this problem exists and is defined up to an arbitrary additive constant because the solvability condition of the Neumann problem, namely, $\int \boldsymbol{\nabla}\cdot\boldsymbol{v}\, dV = \oint \boldsymbol{n}\cdot\boldsymbol{v}\, dS$, is satisfied by the divergence theorem. This completes the proof. □

Let us now assume that the normal component of the boundary condition for the velocity $\boldsymbol{u}$, solution of the equations (7.1), is homogeneous:

$$\boldsymbol{n}\cdot\boldsymbol{u}|_S = 0.$$

In this case it is natural to introduce the operator of orthogonal projection of vector fields onto the space

$$\boldsymbol{J}_0^0(V) = \Big\{\boldsymbol{w} \in \boldsymbol{L}^2(V), \quad \boldsymbol{\nabla}\cdot\boldsymbol{w} = 0, \quad \boldsymbol{n}\cdot\boldsymbol{w}|_S = 0\Big\}.$$

The space $\boldsymbol{J}_0^0(V)$ is a closed subspace of $\boldsymbol{L}^2(V)$ and to simplify the notation it will be indicated also as $\boldsymbol{J}_0$. The operator of orthogonal projection onto $\boldsymbol{J}_0$ is denoted by $\mathcal{P}$, or, more explicitly,

$$\mathcal{P} = \mathcal{P}_{\boldsymbol{J}_0}.$$

(To distinguish the projection operators occurring in the fractional-step method from those associated with the imposition of the integral conditions a different notation is used.)

Taking the time derivative of $\nabla \cdot \boldsymbol{u} = 0$ gives $\nabla \cdot (\partial \boldsymbol{u}/\partial t) = 0$, so that the orthogonal decomposition (7.2) allows to write the momentum equation in (7.1), under homogeneous boundary condition for the normal component, in the following "projected" form

$$\frac{\partial \boldsymbol{u}}{\partial t} = \mathcal{P}[-(\boldsymbol{u} \cdot \nabla)\boldsymbol{u} + \nu \nabla^2 \boldsymbol{u}]. \tag{7.3}$$

Equation (7.3) means that the use of the orthogonal projection operator $\mathcal{P}$ eliminates the pressure variable from the problem. It should be noted that the vector fields $\boldsymbol{w}$ of the projection space $\boldsymbol{J}_0$ are subject to a boundary condition which prescribes only the normal component of $\boldsymbol{w}$. This fact is very important, especially in view of finite element Galerkin approximations of the equations after their time-discretization. The boundary condition for the normal component of the velocity is an essential condition in the weak variational formulation of the equations used to satisfy the incompressibility by means of the orthogonal projection operator. In other terms, the complete boundary condition for the velocity, which involves both the normal and the tangential components, are not the appropriate condition for supplementing the equations of the projection phase of the calculation.

Therefore, to make the orthogonal projection operator $\mathcal{P}$ viable for solving the viscous incompressible equations, one has to *split* the viscosity term from the treatment of the incompressibility, *i.e.*, from the projective part of the method which determines the solenoidal (component of the) velocity field. This leads to the requirement that the equations be discretized in time by a *fractional-step* method which separates the effect of the viscous diffusion from the treatment of the incompressibility condition. It must be noted that this requirement is due to the noncommutativity of the projection $\mathcal{P}$ with the Laplace operator appearing in the viscous term when solid boundaries are present, and can be ultimately traced back to the complicated space-time character of the conditions for the uncoupled pressure equation in incompressible flows discussed in chapter 6.

7.3 Fractional-step projection method

The typical form of the time-discretized equations of the projection method consists in two distinct steps. First, an *intermediate* velocity field $\boldsymbol{u}^{n+1/2}$, not satisfying the condition of incompressibility, is calculated as solution of a discretized version of the momentum equation with the pressure term omitted. Considering for instance and for simplicity a fully explicit scheme, one has the problem

$$\begin{aligned} &\frac{\boldsymbol{u}^{n+1/2} - \boldsymbol{u}^n}{\Delta t} = -(\boldsymbol{u}^n \cdot \nabla)\boldsymbol{u}^n + \nu \nabla^2 \boldsymbol{u}^n, \\ &\boldsymbol{u}^{n+1/2}|_S = \boldsymbol{b}^{n+1}. \end{aligned} \tag{7.4}$$

Then, the intermediate field $\boldsymbol{u}^{n+1/2}$ is decomposed into the sum of a solenoidal velocity field $\boldsymbol{u}^{n+1}$ and the gradient of a scalar function proportional to the unknown pressure, namely, $\boldsymbol{\nabla} P^{n+1}$, according to the following equations and boundary condition:

$$\frac{\boldsymbol{u}^{n+1}-\boldsymbol{u}^{n+1/2}}{\Delta t} = -\boldsymbol{\nabla} P^{n+1},$$

$$\boldsymbol{\nabla}\cdot \boldsymbol{u}^{n+1} = 0, \tag{7.5}$$

$$\boldsymbol{n}\cdot \boldsymbol{u}^{n+1}|_S = \boldsymbol{n}\cdot \boldsymbol{b}^{n+1}.$$

The specification of the boundary condition only for the normal component of velocity is an essential aspect of the second half-step problem (7.5), and is a consequence of the definition of the space $\boldsymbol{J}_0$ involved in the orthogonal decomposition of Theorem 7.1. Stated in physical terms, this boundary condition means that the equations of the second half-step define an *inviscid* flow problem.

It is interesting to note that when the fractional-step method (7.4)–(7.5) is used to calculate stationary flows, the steady-state numerical solution depends on the value of the time step Δt. In fact, consider the first equation in (7.5) and observe that, at convergence, the solution $\boldsymbol{u}^{n+1}$ would be independent of Δt only if, for n suitably large, $\boldsymbol{u}^{n+1}$ is coincident with the auxiliary field $\boldsymbol{u}^{n+1/2}$. But this cannot be the case, except possibly in the absence of no-slip boundaries. This behaviour is an intrinsic characteristic of the fractional-step method—a method without any steady-state counterpart.

7.3.1 Homogeneous boundary condition

We note that the equation involving the pressure in (7.5) can also be written in the form

$$\boldsymbol{u}^{n+1/2} = \boldsymbol{u}^{n+1} + \Delta t\, \boldsymbol{\nabla} P^{n+1}. \tag{7.6}$$

In the particular situation of homogeneous boundary value for the normal component, namely, $\boldsymbol{n}\cdot\boldsymbol{b} = 0$, this equation is nothing but the orthogonal decomposition (7.2) for the intermediate velocity field $\boldsymbol{u}^{n+1/2}$, since $\boldsymbol{\nabla}\cdot\boldsymbol{u}^{n+1} = 0$ and $\boldsymbol{n}\cdot\boldsymbol{u}^{n+1}|_S = 0$ by (7.5). Therefore, in this case, one can express the final velocity and pressure fields as

$$\boldsymbol{u}^{n+1} = \mathcal{P}\boldsymbol{u}^{n+1/2} \qquad \text{and} \qquad \Delta t\, \boldsymbol{\nabla} P^{n+1} = (\mathcal{I}-\mathcal{P})\boldsymbol{u}^{n+1/2}, \tag{7.7}$$

a construction which is depicted in Figure 7.1.

7.3.2 Nonhomogeneous boundary condition

The situation is different when the boundary condition for the velocity is such that $\boldsymbol{n}\cdot\boldsymbol{b}^{n+1}|_S \neq 0$, but a geometrical interpretation of the construction can

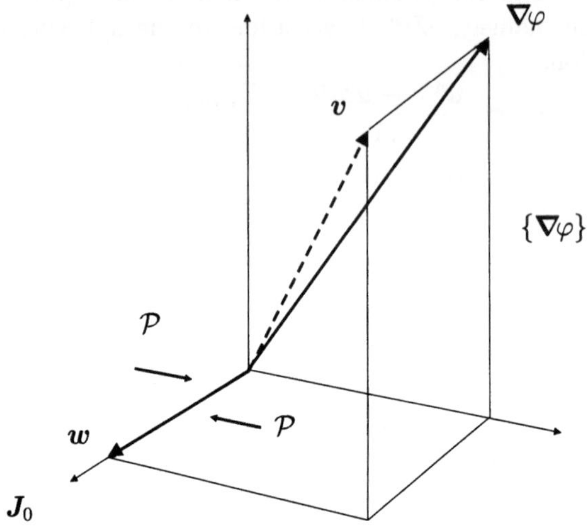

Figure 7.1: Construction of the solenoidal velocity field in the fractional-step projection method with homogeneous boundary conditions.

still be given. To this purpose, an orthogonal decomposition of the space of the gradient fields is required (Temam [35, p. 16]).

Theorem 7.2. *Any vector field which is the gradient of a scalar function φ defined in V admits the unique orthogonal decomposition*

$$\boldsymbol{\nabla}\varphi = \boldsymbol{\nabla}\varphi_0 + \boldsymbol{\nabla}h, \tag{7.8}$$

where the function φ_0 vanishes on the boundary S of V and the function h is harmonic in V.

Proof. First of all, we establish the orthogonality of the decomposition (see Figure 7.2):

$$\int \boldsymbol{\nabla}\varphi_0 \cdot \boldsymbol{\nabla}h \, dV = 0.$$

In fact, by the identity $\boldsymbol{\nabla}\cdot(\varphi_0 \boldsymbol{\nabla}h) = \boldsymbol{\nabla}\varphi_0 \cdot \boldsymbol{\nabla}h + \varphi_0 \nabla^2 h$ and the divergence theorem, we get

$$\begin{aligned}\int \boldsymbol{\nabla}\varphi_0 \cdot \boldsymbol{\nabla}h \, dV &= -\int \varphi_0 \nabla^2 h \, dV + \int \boldsymbol{\nabla}\cdot(\varphi_0 \boldsymbol{\nabla}h) \, dV \\ &= \oint \varphi_0 \, \boldsymbol{n} \cdot \boldsymbol{\nabla}h \, dS = 0,\end{aligned}$$

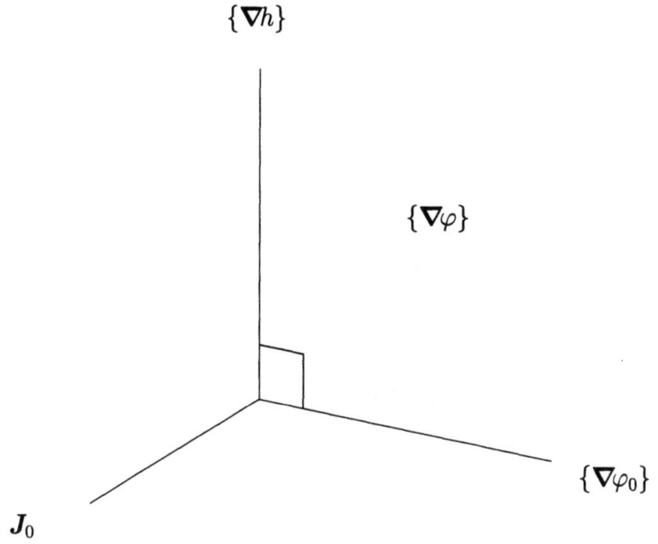

Figure 7.2: Fractional-step projection method. Orthogonal decomposition of the space of the gradient fields, for analyzing the imposition of nonhomogeneous boundary condition on the normal velocity.

since h is harmonic and $\varphi_0|_S = 0$.

The orthogonality is used to prove uniqueness. Suppose that $\nabla\varphi = \nabla\varphi_{01} + \nabla h_1 = \nabla\varphi_{02} + \nabla h_2$. Then

$$0 = \nabla(\varphi_{01} - \varphi_{02}) + \nabla(h_1 - h_2).$$

Taking the scalar product with $\nabla(h_1 - h_2)$ and integrating on V, we get

$$\begin{aligned} 0 &= \int \left[\nabla(h_1 - h_2)\cdot\nabla(\varphi_{01} - \varphi_{02}) + |\nabla(h_1 - h_2)|^2\right] dV \\ &= \int |\nabla(h_1 - h_2)|^2\, dV \end{aligned}$$

by the orthogonality of $\{\nabla h_j\}$ and $\{\nabla\varphi_{0j}\}$. It follows that $\nabla(h_1 - h_2) = 0$, that is, $h_1 = h_2 + \text{constant}$. Therefore also $\nabla(\varphi_1 - \varphi_2) = 0$, which means $\varphi_1 = \varphi_2 + \text{constant}$.

If $\nabla\varphi = \nabla\varphi_0 + \nabla h$, we notice that $\varphi|_S = \varphi_0|_S + h|_S + \text{const} = h|_S + \text{const}$, which allows to prove the existence of the decomposition. In fact, given φ let h be the harmonic function satisfying the boundary condition $h|_S = \varphi|_S$. This h exists and is unique. With this choice of h we define φ_0 as solution of the Dirichlet problem

$$\nabla^2\varphi_0 = \nabla^2\varphi, \qquad \varphi_0|_S = 0,$$

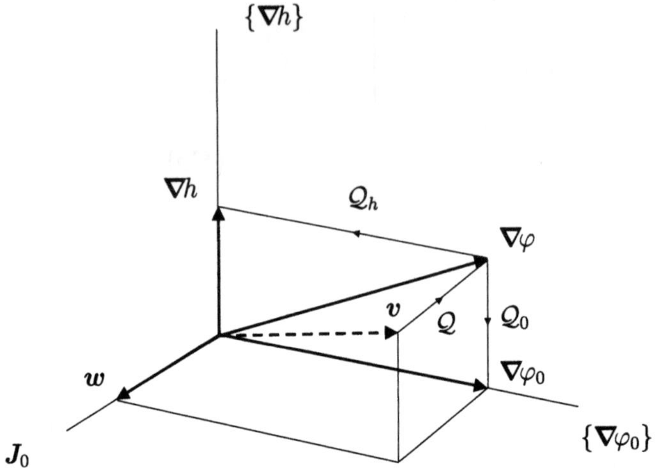

Figure 7.3: Fractional-step projection method. Orthogonal projection operators in the presence of boundaries.

which is again uniquely defined and has the desired properties. □

By virtue of Theorems 7.1 and 7.2, any vector field $\boldsymbol{v}$ can be decomposed as follows

$$\boldsymbol{v} = \boldsymbol{w} + \boldsymbol{\nabla}\varphi = \boldsymbol{w} + \boldsymbol{\nabla}h + \boldsymbol{\nabla}\varphi_0, \tag{7.9}$$

where the functions φ_0 and h are determined uniquely from $\boldsymbol{v}$, by solving (in sequence) the two elliptic problems

$$\nabla^2\varphi_0 = \boldsymbol{\nabla}\cdot\boldsymbol{v}, \qquad \varphi_0|_S = 0;$$

$$\nabla^2 h = 0, \qquad \boldsymbol{n}\cdot\boldsymbol{\nabla}h|_S = \boldsymbol{n}\cdot(\boldsymbol{v} - \boldsymbol{\nabla}\varphi_0)|_S.$$

The solvability condition of the Neumann problem is satisfied since, by the divergence theorem, it results

$$\oint \boldsymbol{n}\cdot(\boldsymbol{v} - \boldsymbol{\nabla}\varphi_0)\, dS = \int \boldsymbol{\nabla}\cdot(\boldsymbol{v} - \boldsymbol{\nabla}\varphi_0)\, dV = \int (\boldsymbol{\nabla}\cdot\boldsymbol{v} - \nabla^2\varphi_0)\, dV = 0.$$

Introducing the "complementary" projection operator $\mathcal{Q} = \mathcal{I} - \mathcal{P}$, one has

$$\mathcal{Q} = \mathcal{Q}_0 + \mathcal{Q}_h,$$

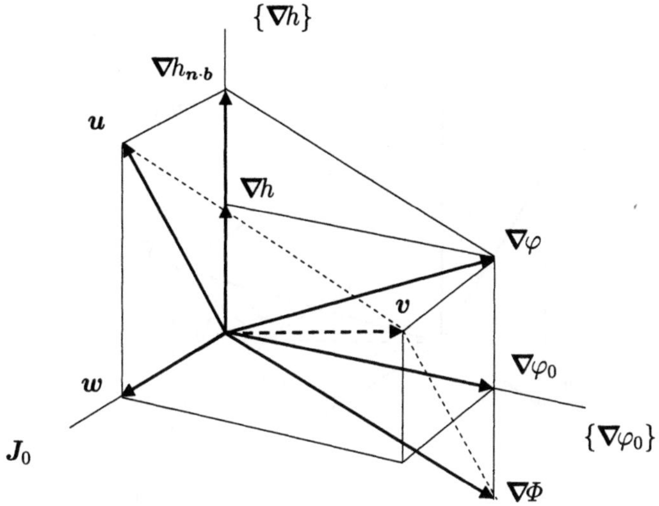

Figure 7.4: Construction of the solenoidal velocity field satisfying a nonhomogeneous boundary condition for the normal component.

where $\mathcal{Q}_0$ and $\mathcal{Q}_h$ denote the operators of orthogonal projection onto the linear subspaces $\{\boldsymbol{\nabla}\varphi_0 \,|\, \varphi_0|_S = 0\}$ and $\{\boldsymbol{\nabla} h \,|\, \nabla^2 h = 0\}$, respectively (cf. Figure 7.3).

Suppose now that $\boldsymbol{v}$ is a given vector field. Let $\boldsymbol{u}$ denote the solenoidal component of $\boldsymbol{v}$ which assumes a prescribed boundary value for the normal component, namely, $\boldsymbol{n}\cdot\boldsymbol{u}|_S = \boldsymbol{n}\cdot\boldsymbol{b}$, with $\boldsymbol{n}\cdot\boldsymbol{b}$ given. Such a solenoidal part $\boldsymbol{u}$ of $\boldsymbol{v}$ can be obtained from the following decomposition:

$$\boldsymbol{v} = \boldsymbol{w} + \boldsymbol{\nabla} h_{\boldsymbol{n}\cdot\boldsymbol{b}} + \boldsymbol{\nabla}(h - h_{\boldsymbol{n}\cdot\boldsymbol{b}}) + \boldsymbol{\nabla}\varphi_0$$

where $h_{\boldsymbol{n}\cdot\boldsymbol{b}}$ denotes the harmonic function sastisfying the Neumann condition $\boldsymbol{n}\cdot\boldsymbol{\nabla} h_{\boldsymbol{n}\cdot\boldsymbol{b}}|_S = \boldsymbol{n}\cdot\boldsymbol{b}$, a boundary condition which is admissible since $\boldsymbol{b}$ satisfies $\oint \boldsymbol{n}\cdot\boldsymbol{b}\, dS = 0$. It follows that $\boldsymbol{u} = \boldsymbol{w} + \boldsymbol{\nabla} h_{\boldsymbol{n}\cdot\boldsymbol{b}}$ and therefore, defining $\Phi = h - h_{\boldsymbol{n}\cdot\boldsymbol{b}} + \varphi_0$, the decomposition above assumes the form

$$\boldsymbol{v} = \boldsymbol{u} + \boldsymbol{\nabla}\Phi.$$

The geometric representation of such a construction, which is required for a nonhomogeneous velocity boundary condition, is depicted in Figure 7.4.

Actually, Figure 7.4 does not tell the whole story, because the space $\{\boldsymbol{\nabla} h\}$ is not one-dimensional, so that, in general, the "vectors" representing the fields

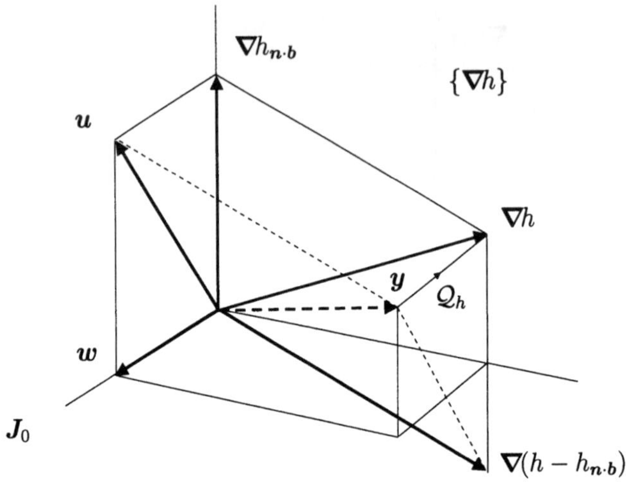

Figure 7.5: Detailed illustration of the construction of the solenoidal velocity field satisfying nonhomogeneous normal boundary condition $[\boldsymbol{y} = (\mathcal{P} + \mathcal{Q}_h)\boldsymbol{v}]$.

∇h and $\nabla h_{n \cdot b}$ point to different directions. Thus, the illustration of Figure 7.5, where the space $\{\nabla \varphi_0\}$ has been suppressed while the space $\{\nabla h\}$ spans a vertical plane and $\boldsymbol{y} = (\mathcal{P} + \mathcal{Q}_h)\boldsymbol{v}$, would provide a more appropriate description of the construction required for nonhomogeneous boundary conditions. In any case, the velocity field $\boldsymbol{u}$ is obtained from the expression

$$\boldsymbol{u}^{n+1} = \mathcal{P}\boldsymbol{u}^{n+1/2} + \nabla h_{n \cdot b}.$$

This construction reveals that in the fractional-step method the projection out of the subspace $\{\nabla \varphi_0\}$ is associated with the fulfillment of the incompressibility condition at internal points of the fluid domain, whereas the projection out of the subspace $\{\nabla h\}$ has a connection with the imposition of the boundary condition on the velocity. In the particular situation of absence of boundaries or with spatially periodic boundary conditions, the second subspace disappears and the construction of the fractional-step method simplifies substantially, as depicted in Figure 7.6. Notice that the projection $\mathcal{P}$ is always to be performed in the fractional-step method even in the absence of no-slip boundaries, whereas the operator of orthogonal projection $I\!P$ of methods using the integral conditions disappears when no-slip boundary are excluded.

In problems with no-slip velocity conditions, the subspace $\{\nabla h\}$ involved by

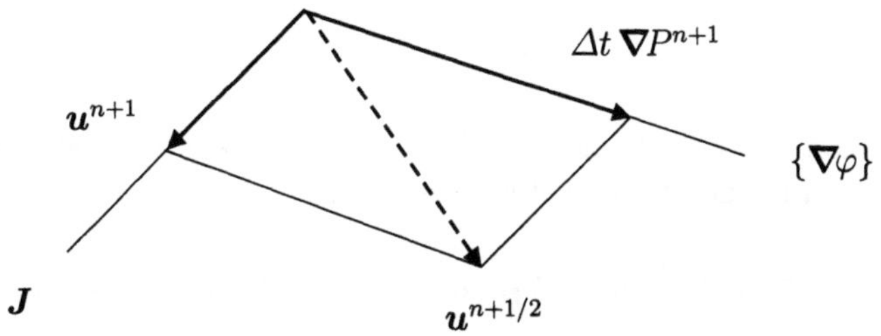

Figure 7.6: Schematic representation of the fractional-step projection method in the absence of boundaries.

the imposition of the boundary condition in the fractional-step method remind us the analogous subspace $\{\boldsymbol{\nabla} p'\}$ encountered in the method using the integral conditions for pressure (cf. section 5.5 and Figure 5.1). This correspondence is somewhat surprising, because the two spaces serve quite different purposes. In the nonfractional-step method, the subspace $\{\boldsymbol{\nabla} p'\}$ [observe, however, that the projection is done onto a different subspace, *i.e.*, $\{\boldsymbol{\chi}_\gamma\}$] is involved in the fulfillment of the pressure integral conditions which enforce simultaneously the incompressibility and the no-slip conditions, so that it is basically related to the viscous character of the equations. On the contrary, in the fractional-step projection method the subspace $\{\boldsymbol{\nabla} h\}$ is used to satisfy a boundary condition whose character is essentially inviscid.

7.4 Poisson equation for pressure

A possible method for formulating the equations (7.5) of the second half-step consists in deriving a Poisson equation for the pressure P^{n+1}. In fact, by taking the divergence of the first equation of problem (7.5) and using the second one, we obtain $\nabla^2 P^{n+1} = (\Delta t)^{-1}\boldsymbol{\nabla}\cdot\boldsymbol{u}^{n+1/2}$. The boundary condition for this equation is obtained by taking the normal component on the boundary of the (inviscid) momentum equation in (7.5). By the boundary conditions imposed in the first half-step $\boldsymbol{u}^{n+1/2}|_S = \boldsymbol{b}^{n+1}$ and in the second half-step $\boldsymbol{n}\cdot\boldsymbol{u}^{n+1}|_S = \boldsymbol{n}\cdot\boldsymbol{b}^{n+1}$, we have $\boldsymbol{n}\cdot\boldsymbol{\nabla} P^{n+1}|_S = 0$. Therefore, the pressure P^{n+1} must be solution of the Neumann problem

$$\begin{aligned} &-\nabla^2 P^{n+1} = -(\Delta t)^{-1}\boldsymbol{\nabla}\cdot\boldsymbol{u}^{n+1/2}, \\ &\boldsymbol{n}\cdot\boldsymbol{\nabla} P^{n+1}|_S = 0. \end{aligned} \tag{7.10}$$

The condition of solvability of this Neumann problem is

$$\int \nabla\cdot \boldsymbol{u}^{n+1/2}\, dV = 0.$$

This condition is satisfied since the divergence theorem and the boundary condition imposed in the first half-step give

$$\int \nabla\cdot \boldsymbol{u}^{n+1/2}\, dV = \oint \boldsymbol{n}\cdot \boldsymbol{u}^{n+1/2}\, dS = \oint \boldsymbol{n}\cdot \boldsymbol{b}^{n+1}\, dS = 0,$$

the last passage being justified since $\boldsymbol{b}$ satisfies $\oint \boldsymbol{n}\cdot\boldsymbol{b}\, dS = 0$ at any time $t > 0$.

It is important to note that the boundary condition for P^{n+1} in problem (7.10) has been obtained *after* a time-discretization process of *fractional-step* type. As a consequence, it is not correct to attempt to reinterpret this condition as something pertaining to the pressure variable appearing in the original problem before the time discretization process. Similarly, it is not legitimate to attempt to infer from the original momentum equation in (7.1) any boundary condition for supplementing the Poisson pressure equation occurring in the second half-step (7.10) of the fractional-step method. Another obstacle to such an attempt comes from the Laplace operator of the viscous term in the momentum equation: the presence of such second-order operator prevents taking the normal component of the equation on the boundary, unless an unnecessary increase of the regularity of the solution is assumed.

Once the pressure field P^{n+1} has been determined, the final solenoidal velocity field is obtained from the equations

$$\begin{aligned} \boldsymbol{u}^{n+1} &= \boldsymbol{u}^{n+1/2} - \Delta t\, \nabla P^{n+1}, \\ \boldsymbol{n}\cdot \boldsymbol{u}^{n+1}\big|_S &= \boldsymbol{n}\cdot \boldsymbol{b}^{n+1}. \end{aligned} \tag{7.11}$$

It is very important to note that the final velocity $\boldsymbol{u}^{n+1}$ so obtained satisfies only the normal component of the boundary condition, and thus that, in general, it will be

$$\boldsymbol{n}\times\boldsymbol{u}^{n+1}\big|_S \neq \boldsymbol{n}\times\boldsymbol{b}^{n+1}.$$

On the other hand, the tangential components of the velocity boundary condition are imposed on the intermediate velocity $\boldsymbol{u}^{n+1/2}$, so that the violation indicated above will be small enough to be acceptable. It can be also remarked that the failure to satisfy the tangential components of the velocity boundary condition is of the same kind of the discontinuity between the boundary and initial data occurring at time $t = 0$ in problems with an impulsive initial condition, discussed in chapter 1 (section 1.2).

7.4.1 On higher-order methods

The violation of the velocity boundary condition for the tangential components could make it appear desirable to introduce the viscosity term into the equations of the incompressibility step, in order to satisfy the "entire" boundary condition $\boldsymbol{u}^{n+1}|_S = \boldsymbol{b}^{n+1}$. Let us analyze the situation, considering the case of a Crank–Nicolson scheme applied only to the viscous diffusion term, for the moment being. The problem for the first half-step would become

$$\frac{\boldsymbol{u}^{n+1/2} - \boldsymbol{u}^n}{\Delta t} = -(\boldsymbol{u}^n \cdot \boldsymbol{\nabla})\boldsymbol{u}^n + \tfrac{1}{2}\nu\nabla^2\boldsymbol{u}^n, \qquad \boldsymbol{u}^{n+1/2}|_S = \boldsymbol{b}^{n+1}, \tag{7.12}$$

whereas the problem for the second half-step would read

$$\begin{aligned} &\frac{\boldsymbol{u}^{n+1} - \boldsymbol{u}^{n+1/2}}{\Delta t} = -\boldsymbol{\nabla}P^{n+1} + \tfrac{1}{2}\nu\nabla^2\boldsymbol{u}^{n+1}, \qquad \boldsymbol{u}^{n+1}|_S = \boldsymbol{b}^{n+1}, \\ &\boldsymbol{\nabla}\cdot \boldsymbol{u}^{n+1} = 0. \end{aligned} \tag{7.13}$$

By substituting $\boldsymbol{u}^{n+1/2}$ obtained from equation (7.12) into the first equation of (7.13), one verifies the second-order accuracy in the treatment of the viscous term. The boundary condition for $\boldsymbol{u}^{n+1}$ in the incompressible step now imposes both the normal and the tangential components of velocity. Thus, the character of the complete set (7.13) is indeed that of a Stokes problem (for the time-discretized equations, of course) so that this second-order method can be put hardly under the banner of the projection methods. Nevertheless, let us examine how the incompressibility condition is satisfied in the second half-step of the second-order method under examination.

Taking the divergence of the momentum equation in the second half-step problem (7.13) and using the incompressibility condition give the following Poisson equation P^{n+1}:

$$-\nabla^2 P^{n+1} = -(\Delta t)^{-1}\boldsymbol{\nabla}\cdot \boldsymbol{u}^{n+1/2}.$$

Hovewer, the fact that P^{n+1} satisfies this equation is not sufficient to guarantee the solution $\boldsymbol{u}^{n+1}$ of the momentum equation to be solenoidal. In fact, let $\boldsymbol{v}^{n+1}$ be a solution of the momentum equation in (7.13) with P^{n+1} a solution of the Poisson equation above. Then, taking the divergence of the momentum equation gives

$$(\Delta t)^{-1}\boldsymbol{\nabla}\cdot \boldsymbol{v}^{n+1} = \tfrac{1}{2}\nu\nabla^2\boldsymbol{\nabla}\cdot \boldsymbol{v}^{n+1},$$

that is, after dividing by $\nu/2$,

$$(-\nabla^2 + 2\gamma)\boldsymbol{\nabla}\cdot \boldsymbol{v}^{n+1} = 0,$$

where $\gamma = 1/(\nu\Delta t)$. Therefore, by repeating Kleiser–Schumann argument, in order that the equation $\boldsymbol{\nabla}\cdot \boldsymbol{v}^{n+1} = 0$ be satisfied, the boundary condition

$\boldsymbol{\nabla}\cdot\boldsymbol{v}^{n+1}|_S = 0$ must be imposed in addition to $\boldsymbol{v}^{n+1}|_S = \boldsymbol{b}^{n+1}$. Thus, as anticipated, if the fractional-step method is formulated with the viscous term included in the incompressible phase and using the Poisson equation for pressure, the boundary conditions for the velocity are faced with exactly the same difficulty encountered by nonfractional-step methods using a Poisson equation for pressure. As a consequence, the same results obtained in chapter 5 apply to the present second-order fractional-step method and the pressure P^{n+1} turns out to be subject to conditions of an integral character previously discussed. In the present situation, one has to project the vector field

$$-2\nu^{-1}\boldsymbol{\nabla}P^{n+1} + 2\gamma\,\boldsymbol{u}^{n+1/2}\,\Big[= (-\nabla^2 + 2\gamma)\boldsymbol{u}^{n+1}\,\Big]$$

onto the linear space

$$\big\{\boldsymbol{\chi}_{2\gamma}\,|\,(-\nabla^2 + 2\gamma)\boldsymbol{\chi}_{2\gamma} = 0, \quad \boldsymbol{n}\times\boldsymbol{\chi}_{2\gamma}|_S = 0\big\}.$$

Things do not change if one adopts a second-order treatment also for the pressure term. The equations of the two fractional steps would change in the following form:

$$\frac{\boldsymbol{u}^{n+1/2} - \boldsymbol{u}^n}{\Delta t} = -(\boldsymbol{u}^n\cdot\boldsymbol{\nabla})\boldsymbol{u}^n - \tfrac{1}{2}\boldsymbol{\nabla}P^n + \tfrac{1}{2}\nu\nabla^2\boldsymbol{u}^n, \qquad \boldsymbol{u}^{n+1/2}|_S = \boldsymbol{b}^{n+1}, \quad (7.14)$$

and

$$\frac{\boldsymbol{u}^{n+1} - \boldsymbol{u}^{n+1/2}}{\Delta t} = -\tfrac{1}{2}\boldsymbol{\nabla}P^{n+1} + \tfrac{1}{2}\nu\nabla^2\boldsymbol{u}^{n+1}, \qquad \boldsymbol{u}^{n+1}|_S = \boldsymbol{b}^{n+1},$$
$$\boldsymbol{\nabla}\cdot\boldsymbol{u}^{n+1} = 0. \quad (7.15)$$

Disregarding the very subtle difficulty of evaluating the initial pressure field (cf. the comment at the end of the second paragraph of section 6.4), there are only two differences with respect to the equations (7.12)–(7.13): firstly, the Poisson equation becomes

$$-\nabla^2 P^{n+1} = -2(\Delta t)^{-1}\boldsymbol{\nabla}\cdot\boldsymbol{u}^{n+1/2}$$

and, secondly, the vector field to be projected is

$$-\nu^{-1}\boldsymbol{\nabla}P^{n+1} + 2\gamma\,\boldsymbol{u}^{n+1/2}\,\Big[= (-\nabla^2 + 2\gamma)\boldsymbol{u}^{n+1}\,\Big].$$

Now, the following question arises spontaneously: If treating the viscous term within the incompressible step leads to the same mathematical structure typical of nonfractional-step methods, why not to employ the latter directly? We raise this question not to prevent the use of the fractional-step projection method, which remains the most elegant and powerful method for incompressible viscous flow problems and perhaps the only method to be recommended without any reservation, but to keep the reader's attention alert with respect to those implementations of the method which risk to contradict its basic premises when no-slip boundaries are present.

7.5 A finite element projection method

Another method for implementing the fractional-step method is to introduce the spatial discretization directly in equations (7.4)–(7.5) without deriving the Poisson equation for the pressure. This is the most frequently used approach and has been considered using finite differences, finite elements and spectral approximations. Of course, the detailed expression of the fully discrete equations depends on the method which is adopted for the spatial approximation. In particular, for the equations of the second half-step, there are substantial differences in the boundary conditions to be imposed according to the kind of spatial discretization which is employed.

Basically, one has to distinguish whether the discrete representation is such that the pressure variable is *defined* or not on the boundary. If the pressure is defined on the boundary, an additional boundary condition is required to determine the boundary value of this variable. The boundary condition appropriate for this case is the condition $\nabla \cdot \boldsymbol{u}^{n+1}|_S = 0$, originally introduced by Chorin in his finite difference method with the velocity and pressure defined at the same points of the mesh. The same derivative boundary condition has also been used by Ku, Hirsh and Taylor (1987) to implement the fractional-step projection method by means of a Chebyshev spectral approximation.

The second alternative is when the pressure variable is not defined on the boundary S. This occurs, for instance, with finite element approximations representing the pressure field by means of piece-wise polynomials which are discontinous at interelement boundaries, so that speaking of the value of P^{n+1} on S has no meaning. In such a case, no boundary condition for the pressure is required at all.

Therefore, we can list (at least) four different kinds of boundary conditions which are appropriate for being imposed on the pressure in direct methods for the numerical calculation of incompressible viscous flows by finite elements:

1. integral conditions for the pressure in nonfractional-step (uncoupled) methods or in fractional-step methods with a simultaneous treatment of viscosity and incompressibility and using a Poisson equation for pressure;

2. homogeneous Neumann condition for the pressure in fractional-step methods still using an explicit derived Poisson equation for pressure, but with the viscosity and the incompressibility left split from each other;

3. boundary condition $\nabla \cdot \boldsymbol{u}^{n+1}|_S = 0$ in fractional-step methods left in the starting grad-div form and with the discrete pressure unknown defined on the boundary;

4. no boundary condition at all in fractional-step methods left in the starting grad-div form but with the discrete pressure variable not defined on the boundary.

Perhaps, a hierarchical description of all these different conditions for the pressure in incompressible viscous problems with no-slip boundaries can help to remind the various possible alternatives.

- Nonfractional-step methods, uncoupled formulations
  ⟶ integral conditions
- Fractional-step methods:
  - with a Poisson equation for pressure:
    * viscosity recoupled with incompressibility
      ⟶ integral conditions
    * viscosity left split from incompressibility
      ⟶ Neumann homogeneous boundary condition
  - without a Poisson equation for pressure (*i.e.* grad-div form):
    * the discrete pressure is defined on the boundary
      ⟶ Chorin's boundary condition $\boldsymbol{\nabla}\cdot \boldsymbol{u}^{n+1}|_S = 0$
    * the discrete pressure is *not* defined on the boundary
      ⟶ no boundary condition for pressure is required

In the following we intend to describe a finite element version of the fractional-step projection method based on interpolations of different order for the velocity and pressure and with discontinuous approximation of the latter (Donea *et al.* 1982). An important characteristic of this numerical method for the incompressible Navier–Stokes equations is the absence the parasitic modes of the pressure (the so-called pressure chequerboarding) which are instead found in other finite element methods for two-dimensional problems. Two different methods have been proposed: the first method is based on a bilinear interpolation for the velocity and piecewise constant approximation for the pressure on four-noded quadrilateral elements, while the second method uses biquadratic velocity interpolation and a piece-wise bilinear but discontinuous pressure approximation on nine-noded quadrilaters (Donea *et al.* 1982). Here, a detailed account of the matrix structure will be given only of the finite element projection method which uses four-noded quadrilateral elements.

7.5.1 Variational formulation

To construct a finite element fractional-step method, a weak variational formulation of the problems (7.4) and (7.5) is necessary. To this purpose, the (time-discretized) governing equations are multipled by suitable weighting functions and integrated over V. Let $\boldsymbol{H}^1(V)$ denote the Sobolev space of vector-valued functions defined on V which are square integrable and have square integrable first derivatives. Furthermore, let $L^2(V)$ be the Hilbert space of scalar square integrable functions defined on V. Multiplying the momentum equation in (7.4) by the weighting function $\boldsymbol{v} \in \boldsymbol{H}^1(V)$ with $\boldsymbol{v}|_S = 0$, integrating over V and using the integration by parts for the second-order derivative term, one obtains the following variational problem for the intermediate velocity $\boldsymbol{u}^{n+1/2}$

$$\begin{aligned}
&\left(\boldsymbol{v}, \frac{\boldsymbol{u}^{n+1/2} - \boldsymbol{u}^n}{\Delta t}\right) = -(\boldsymbol{v}, (\boldsymbol{u}^n \cdot \boldsymbol{\nabla})\boldsymbol{u}^n) - \nu((\boldsymbol{v}, \boldsymbol{u}^n)), \quad \forall \boldsymbol{v} \in \boldsymbol{H}^1(V),\ \boldsymbol{v}|_S = 0, \\
&\boldsymbol{u}^{n+1/2}|_S = \boldsymbol{b}^{n+1},
\end{aligned} \tag{7.16}$$

where $((\cdot, \cdot))$ is defined by

$$((\boldsymbol{v}, \boldsymbol{u})) = \sum_{i=1,2,3} (\boldsymbol{\nabla} v_i, \boldsymbol{\nabla} u_i),$$

assuming to consider the equations in three dimensions. In a similarly vein, the variational form of the problem for the end-of-step velocity $\boldsymbol{u}^{n+1}$ is obtained by multiplying the momentum equation in (7.5) by the weighting function $\boldsymbol{v} \in \boldsymbol{H}^1(V)$ with $\boldsymbol{n} \cdot \boldsymbol{v}|_S = 0$ and the continuity equation by the weighting function $Q \in L^2(V)$, and integrating over V, to give, after integrating by parts the pressure gradient term,

$$\begin{aligned}
&\left(\boldsymbol{v}, \frac{\boldsymbol{u}^{n+1} - \boldsymbol{u}^{n+1/2}}{\Delta t}\right) - (\boldsymbol{\nabla} \cdot \boldsymbol{v}, P^{n+1}) = 0, \quad \forall \boldsymbol{v} \in \boldsymbol{H}^1(V),\ \boldsymbol{n} \cdot \boldsymbol{v}|_S = 0, \\
&(Q, \boldsymbol{\nabla} \cdot \boldsymbol{u}^{n+1}) = 0, \qquad \forall Q \in L^2(V), \\
&\boldsymbol{n} \cdot \boldsymbol{u}^{n+1}|_S = \boldsymbol{n} \cdot \boldsymbol{b}^{n+1}.
\end{aligned} \tag{7.17}$$

Note that $\boldsymbol{u} \in \boldsymbol{H}^1(V)$ means that the interpolation of the velocity are continuous across interelement boundaries, whereas $P \in L^2(V)$ means that the pressure interpolation can be discontinuous. Actually, the condition for $\boldsymbol{u}$ is not necessary, since the optimal setting for this variable would be to take $\boldsymbol{v} \in \boldsymbol{L}^2(V)$ with $\boldsymbol{\nabla} \cdot \boldsymbol{v} = 0$ and $\boldsymbol{n} \cdot \boldsymbol{v}|_S = 0$. The space for the velocity considered in (7.17) is however also correct since it is dense in the optimal one. We emphasize once more the fact that the essential boundary condition for the second half-step problem involves only the normal component of the velocity. We also note that, in the first half-step, other time integration schemes different from the fully explicit one employed here can be considered without affecting the second half-step equations; an implicit scheme will be described in section 7.5.5.

7.5.2 Finite element equations

Let us now consider the finite element approximation of the weak variational problems (7.16) and (7.17) obtained by using quadrilateral elements with a bilinear interpolation for the velocity and a piecewise constant approximation for the pressure. The approximate unknowns $\boldsymbol{u}_h$ and P_h are expanded using this finite dimensional representation and the weighting functions $\boldsymbol{v}_h$ and Q_h for the approximate problem are chosen in the same finite dimensional spaces of $\boldsymbol{u}_h$ and P_h. The resulting Galerkin finite element equations of the first half-step problem (7.16) are written in the standard matrix form

$$\begin{aligned} &\mathbf{M}(\mathbf{U}^{n+1/2} - \mathbf{U}^n)/\Delta t = [-\mathbf{A}(\mathbf{U}^n) + \mathbf{K}]\mathbf{U}^n, \\ &\mathbf{U}^{n+1/2}|_S = \mathbf{B}^{n+1}, \end{aligned} \qquad (7.18)$$

and those of the second half-step problem (7.17) as

$$\begin{aligned} &\mathbf{M}(\mathbf{U}^{n+1} - \mathbf{U}^{n+1/2})/\Delta t - \mathbf{C}\mathbf{P}^{n+1} = \mathbf{0}, \\ &\mathbf{C}^{\mathrm{T}}\mathbf{U}^{n+1} = \mathbf{0}, \\ &\boldsymbol{n}\cdot\mathbf{U}^{n+1}|_S = \boldsymbol{n}\cdot\mathbf{B}^{n+1}. \end{aligned} \qquad (7.19)$$

In these equations, $\mathbf{U} = (\mathbf{U}_i, i = 1, 2)$ denotes the vector of the nodal values of velocity, $\mathbf{P}$ is the vector of the element values of pressure, $\mathbf{M}$ is the consistent mass matrix, $\mathbf{K}$ the stiffness matrix and $\mathbf{A}(\mathbf{U})$ the matrix corresponding to the advection operator $(\boldsymbol{u}\cdot\boldsymbol{\nabla})$. Furthermore, $\mathbf{C}$ and $\mathbf{C}^{\mathrm{T}}$ are the rectangular matrices representing the gradient and divergence operators for the considered spatial discretization, and are the transpose of each other. Finally, $\mathbf{B}$ denotes the vector of the nodal values of the velocity $\boldsymbol{b}$ prescribed on S.

The boundary condition of problem (7.19) is such that the discrete counterpart $\boldsymbol{v}_h$ of the weighting function $\boldsymbol{v}$ must be chosen so that $\boldsymbol{n}_h\cdot\boldsymbol{v}_h|_{S_h} = 0$. The normal $\boldsymbol{n}_h$ to the boundary at a node of the spatially discretized boundary S_h is defined simply as the direction bisecting the angle formed by two sides of elements on the boundary. Thus, in two dimensions each node of the boundary gives only one equation instead of two, for the system of the finite element equations (7.19) of the incompressible step; in other terms, the matrix $\mathbf{C}$ has two rows for every internal node but only one row for each boundary node. Accordingly, the transposed matrix $\mathbf{C}^{\mathrm{T}}$ has only one column for each node on the boundary. (In three dimensions each boundary node will contribute two equations as opposed to three equations for every internal node.) We note that in the original implementation of this method the imposition of the boundary condition for the normal component of velocity was achieved by a sligthly different procedure, which has been subsequently modified in the form just described for the equations in 2D and 3D (Donea and Giuliani, private communication 1988).

To derive the discrete counterpart $\boldsymbol{\mathcal{P}}$ of the projection operator $\mathcal{P}$ in terms of the finite element matrices, the first equation in (7.19) is solved with respect to $\mathbf{U}^{n+1}$, to give

$$\mathbf{U}^{n+1} = \mathbf{U}^{n+1/2} + \Delta t\, \mathbf{M}^{-1}\mathbf{C}\mathbf{P}^{n+1}. \tag{7.20}$$

This equation is multiplied by the matrix $\mathbf{C}^{\mathrm{T}}$, which, by virtue of the equation $\mathbf{C}^{\mathrm{T}}\mathbf{U}^{n+1} = \mathbf{0}$, gives

$$\mathbf{C}^{\mathrm{T}}\mathbf{M}^{-1}\mathbf{C}\mathbf{P}^{n+1} = -\,\mathbf{C}^{\mathrm{T}}\mathbf{U}^{n+1/2}/\Delta t. \tag{7.21}$$

Notice that this equation is *not* the discretized counterpart of a Poisson equation, because there is no Poisson equation in the present grad-div formulation of the projection method. After eliminating the indeterminacy of $\mathbf{P}$ up to an additive constant by fixing arbitrarily the value of pressure value in a given element, the solution $\mathbf{P}^{n+1}$ of the previous finite element equation can be expressed in the form

$$\mathbf{P}^{n+1} = -\left(\mathbf{C}^{\mathrm{T}}\mathbf{M}^{-1}\mathbf{C}\right)^{-1}\mathbf{C}^{\mathrm{T}}\mathbf{U}^{n+1/2}/\Delta t. \tag{7.22}$$

If now the pressure $\mathbf{P}^{n+1}$ so obtained is substituted into equation (7.20), one obtains

$$\begin{aligned}\mathbf{U}^{n+1} &= \mathbf{U}^{n+1/2} - \mathbf{M}^{-1}\mathbf{C}\left(\mathbf{C}^{\mathrm{T}}\mathbf{M}^{-1}\mathbf{C}\right)^{-1}\mathbf{C}^{\mathrm{T}}\mathbf{U}^{n+1/2}\\ &= \left[\boldsymbol{\mathcal{I}} - \mathbf{M}^{-1}\mathbf{C}\left(\mathbf{C}^{\mathrm{T}}\mathbf{M}^{-1}\mathbf{C}\right)^{-1}\mathbf{C}^{\mathrm{T}}\right]\mathbf{U}^{n+1/2}\\ &= (\boldsymbol{\mathcal{I}} - \boldsymbol{\mathcal{Q}})\mathbf{U}^{n+1/2} = \boldsymbol{\mathcal{P}}\mathbf{U}^{n+1/2}.\end{aligned} \tag{7.23}$$

Here, we have introduced the operator

$$\boldsymbol{\mathcal{Q}} = \mathbf{M}^{-1}\mathbf{C}\left(\mathbf{C}^{\mathrm{T}}\mathbf{M}^{-1}\mathbf{C}\right)^{-1}\mathbf{C}^{\mathrm{T}} \tag{7.24}$$

which is idempotent, namely $\boldsymbol{\mathcal{Q}}^2 = \boldsymbol{\mathcal{Q}}$, since

$$\begin{aligned}&\left[\mathbf{M}^{-1}\mathbf{C}\left(\mathbf{C}^{\mathrm{T}}\mathbf{M}^{-1}\mathbf{C}\right)^{-1}\mathbf{C}^{\mathrm{T}}\right]^2\\ &\quad= \mathbf{M}^{-1}\mathbf{C}\left(\mathbf{C}^{\mathrm{T}}\mathbf{M}^{-1}\mathbf{C}\right)^{-1}\mathbf{C}^{\mathrm{T}}\,\mathbf{M}^{-1}\mathbf{C}\left(\mathbf{C}^{\mathrm{T}}\mathbf{M}^{-1}\mathbf{C}\right)^{-1}\mathbf{C}^{\mathrm{T}}\\ &\quad= \mathbf{M}^{-1}\mathbf{C}\left(\mathbf{C}^{\mathrm{T}}\mathbf{M}^{-1}\mathbf{C}\right)^{-1}\mathbf{C}^{\mathrm{T}}.\end{aligned}$$

Therefore, the operator $\boldsymbol{\mathcal{Q}}$ represents the discrete counterpart of the projection operator $\mathcal{Q} = \mathcal{I} - \mathcal{P}$. Equation (7.24) achieves the formal elimination of the pressure variable from the discrete equations, as well as the fulfillment of the discrete counterpart of the incompressibility condition through the "filtering" action of the operator $\boldsymbol{\mathcal{P}} = \boldsymbol{\mathcal{I}} - \boldsymbol{\mathcal{Q}}$. However, due to the very large dimension of the matrix operator $\boldsymbol{\mathcal{P}}$, it is more convenient, for reasons of computational

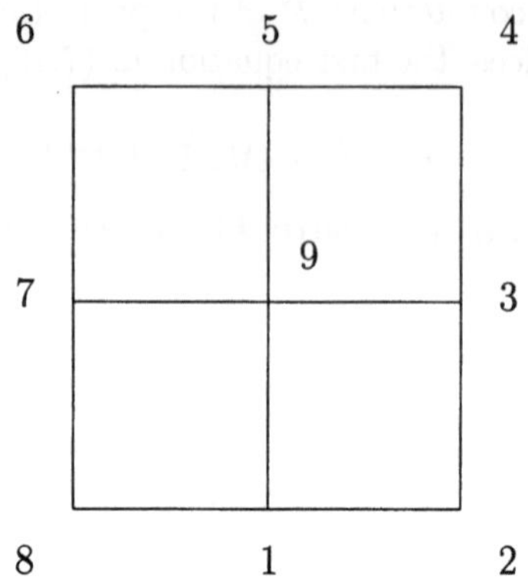

Figure 7.7: Node numbering of the computational molecule containing four quadrilateral elements with a bilinear interpolation of the variable.

efficiency, to retain the pressure in the numerical scheme and to simulate the projection of (7.23) by the following procedure.

Firstly, evaluate the right-hand side of equation (7.21) by multiplying the known vector $\mathbf{U}^{n+1/2}$ by the matrix $\mathbf{C}^{\mathrm{T}}$. Then, calculate the solution $\mathbf{P}^{n+1}$ of the linear system (7.21) with a number of equations equal to the number of the degrees of freedom of the pressure (minus one). Finally, determine the vector $\mathbf{U}^{n+1}$ of the end-of-step velocity by means of equation (7.20). This procedure can also be regarded as a block Gaussian elimination scheme which exploits the special structure of the original linear system (7.19). The matrices $\mathbf{C}$ and $\mathbf{C}^{\mathrm{T}}$ need not to be stored since the usual element-by-element multiplication technique can be used to compute $\mathbf{C}^{\mathrm{T}}\mathbf{U}^{n+1/2}$ and $\mathbf{C}\mathbf{P}^{n+1}$, and also to generate the matrix $\mathbf{C}^{\mathrm{T}}\mathbf{M}^{-1}\mathbf{C}$ once and for all at the beginning of the computation.

7.5.3 Discretized projection operator

To investigate the precise form of all these matrix equations for the second half-step, we consider a patch of four equal square elements, each of size h, with the nodal numbering indicated in Figure 7.7. By standard calculation, the consistent mass matrix $\mathbf{M}_s$ for a *scalar* equation is easily found to be

$$\mathbf{M}_s = \left(\frac{h}{6}\right)^2 \begin{pmatrix} 8 & 2 & 1 & & & & 1 & 2 & 4 \\ 2 & 4 & 2 & & & & & & 1 \\ 1 & 2 & 8 & 2 & 1 & & & & 4 \\ & & 2 & 4 & 2 & & & & 1 \\ & & 1 & 2 & 8 & 2 & 1 & & 4 \\ & & & & 2 & 4 & 2 & & 1 \\ 1 & & & & 1 & 2 & 8 & 2 & 4 \\ 2 & & & & & & 2 & 4 & 1 \\ 4 & 1 & 4 & 1 & 4 & 1 & 4 & 1 & 16 \end{pmatrix}$$

Of course, the mass matrix $\mathbf{M}_s$ valid for the scalar problem is related to the mass matrix $\mathbf{M}$ appearing in the vector problems (7.18) and (7.19) by means of the block-diagonal construction

$$\mathbf{M} = \begin{pmatrix} \mathbf{M}_s & \mathbf{0} \\ \mathbf{0} & \mathbf{M}_s \end{pmatrix}.$$

For reasons to be examined in a while, it is convenient to introduce the so-called diagonally lumped matrix $\mathbf{D}_s$ obtained from $\mathbf{M}_s$ by summing all the entries of a single row and placing the result on the diagonal. From $\mathbf{M}_s$ one easily obtains the diagonal matrix

$$\mathbf{D}_s = \left(\frac{h}{2}\right)^2 \mathrm{diag}(2, 1, 2, 1, 2, 1, 2, 1, 4),$$

whose inverse is evidently

$$\mathbf{D}_s^{-1} = \frac{1}{h^2}\,\mathrm{diag}(2, 4, 2, 4, 2, 4, 2, 4, 1).$$

The inverse of $\mathbf{M}_s$ can be determined by first obtaining the $\mathbf{LDL}^{\mathrm{T}}$ factorization and then evaluating the inverse $\mathbf{M}_s^{-1} = (\mathbf{L}^{\mathrm{T}})^{-1}\mathbf{D}^{-1}\mathbf{L}^{-1}$, to give

$$\mathbf{M}_s^{-1} = \left(\frac{1}{2h}\right)^2 \begin{pmatrix} 28 & -14 & 4 & -2 & 4 & -2 & 4 & -14 & -8 \\ -14 & 49 & -14 & 7 & -2 & 1 & -2 & 7 & 4 \\ 4 & -14 & 28 & -14 & 4 & -2 & 4 & -2 & -8 \\ -2 & 7 & -14 & 49 & -14 & 7 & -2 & 1 & 4 \\ 4 & -2 & 4 & -14 & 28 & -14 & 4 & -2 & -8 \\ -2 & 1 & -2 & 7 & -14 & 49 & -14 & 7 & 4 \\ 4 & -2 & 4 & -2 & 4 & -14 & 28 & -14 & -8 \\ -14 & 7 & -2 & 1 & -2 & 7 & -14 & 49 & 4 \\ -8 & 4 & -8 & 4 & -8 & 4 & -8 & 4 & 16 \end{pmatrix}$$

To perform the analysis of the discrete counterpart of the incompressibility condition we consider now a *computational molecule* consisting of four elements,

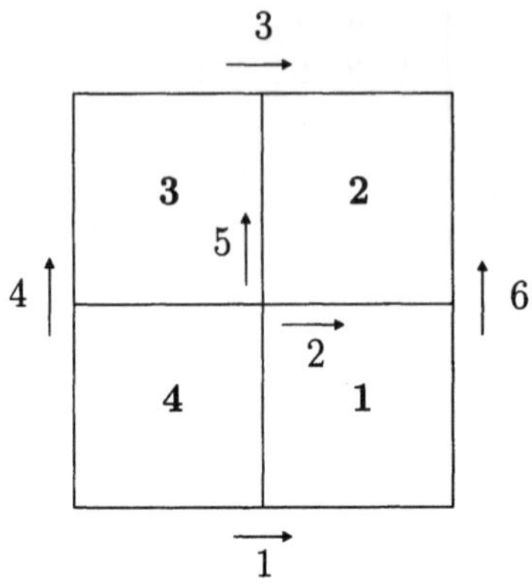

Figure 7.8: Node and element numbering of the computational molecule containing four quadrilateral elements: bilinear interpolation of velocity and constant approximation of pressure.

assuming that the external sides constitute the boundary of the computational domain. Thus, we concentrate our attention on the six degrees of freedom for the velocity indicated in Figure 7.8 and eliminate the nodal values of velocity components normal to the boundary. As a matter of fact, a complete discussion would require to include the additional four degrees of freedom corresponding to the tangential velocity at the corners, *i.e.*, the four velocity components whose directions form an angle of 45° with element sides. Since this would increase the total number of velocity degrees of freedom to ten, the corresponding analysis is omitted here, to simplify the exposition.

With the numbering of the velocity and pressure degrees of freedom shown in Figure 7.8, the matrix $\mathbf{C}$ representing the gradient operator is

$$\mathbf{C} = \frac{h}{2} \begin{pmatrix} -1 & 0 & 0 & 1 \\ -1 & -1 & 1 & 1 \\ 0 & -1 & 1 & 0 \\ 0 & 0 & -1 & 1 \\ 1 & -1 & -1 & 1 \\ 1 & -1 & 0 & 0 \end{pmatrix}$$

The submatrices of $\mathbf{M}_s$ and $\mathbf{M}_s^{-1}$ corresponding to the three degrees of freedom for each vector component of velocity are

$$\mathbf{M}_s = \left(\frac{h}{3}\right)^2 \begin{pmatrix} 2 & 1 & 0 \\ 1 & 4 & 1 \\ 0 & 1 & 2 \end{pmatrix}$$

and

$$\mathbf{M}_s^{-1} = \frac{1}{h^2} \begin{pmatrix} 7 & -2 & 1 \\ -2 & 4 & -2 \\ 1 & -2 & 7 \end{pmatrix}$$

respectively.

The matrix $\mathbf{C}^{\mathrm{T}}\mathbf{M}^{-1}\mathbf{C}$ for the considered computational molecule is easily calculated to be

$$\mathbf{C}^{\mathrm{T}}\mathbf{M}^{-1}\mathbf{C} = \frac{1}{2} \begin{pmatrix} 7 & -3 & -1 & -3 \\ -3 & 7 & -3 & -1 \\ -1 & -3 & 7 & -3 \\ -3 & -1 & -3 & 7 \end{pmatrix}$$

It is singular, as requested, since $\mathbf{P}$ is defined up to an additive constant. To fix the solution $\mathbf{P}^{n+1}$ uniquely, an arbitrary value of the pressure must be imposed in one element, which means that the 4×4 matrix is replaced by the following reduced version of order 3

$$\mathbf{C}^{\mathrm{T}}\mathbf{M}^{-1}\mathbf{C} = \frac{1}{2} \begin{pmatrix} 7 & -3 & -1 \\ -3 & 7 & -3 \\ -1 & -3 & 7 \end{pmatrix}$$

The inverse of the latter is immediately obtained as

$$\left(\mathbf{C}^{\mathrm{T}}\mathbf{M}^{-1}\mathbf{C}\right)^{-1} = \frac{1}{12} \begin{pmatrix} 5 & 3 & 2 \\ 3 & 6 & 3 \\ 2 & 3 & 5 \end{pmatrix}$$

A simple calculation gives

$$\mathbf{M}^{-1}\mathbf{C}\left(\mathbf{C}^{\mathrm{T}}\mathbf{M}^{-1}\mathbf{C}\right)^{-1} = \frac{1}{2h} \begin{pmatrix} -2 & -1 & -1 \\ -1 & -1 & 0 \\ 0 & -1 & 1 \\ -1 & -1 & -2 \\ 0 & -1 & -1 \\ 1 & -1 & 0 \end{pmatrix}$$

Multiplying $\mathbf{C}^{\mathrm{T}}$ on the left by this matrix, one finally obtains the square matrix of the projection operator in the form

$$\mathcal{Q} = \frac{1}{4} \begin{pmatrix} 2 & 2 & 0 & 1 & 0 & -1 \\ 1 & 2 & 1 & 0 & 0 & 0 \\ 0 & 2 & 2 & -1 & 0 & 1 \\ 1 & 0 & -1 & 2 & 2 & 0 \\ 0 & 0 & 0 & 1 & 2 & 1 \\ -1 & 0 & 1 & 0 & 2 & 2 \end{pmatrix}$$

It is not difficult to verify that this matrix is idempotent, namely, $\mathcal{Q}^2 = \mathcal{Q}$.

7.5.4 Diagonalization of the mass matrix

The computational cost of the projection step of the finite element method discussed so far can be reduced by replacing the consistent mass matrix $\mathbf{M}$ by some approximate mass matrix which appears in diagonal form. We discuss here the effect of replacing the consistent mass matrix $\mathbf{M}_s$ by its diagonal approximation $\mathbf{D}_s$ which is obtained by adding all terms of each row of $\mathbf{M}_s$ and placing the results on the diagonal. For the computational molecule considered in Figure 7.8, one obtains

$$\mathbf{D}_s = \frac{h^2}{4}\begin{pmatrix} 2 & 0 & 0 \\ 0 & 4 & 0 \\ 0 & 0 & 2 \end{pmatrix}$$

with its inverse

$$\mathbf{D}_s^{-1} = \frac{1}{h^2}\begin{pmatrix} 2 & 0 & 0 \\ 0 & 1 & 0 \\ 0 & 0 & 2 \end{pmatrix}$$

Using the matrix $\mathbf{C}$ calculated previously, it is easy to obtain the matrix of the linear system for the pressure within the diagonal mass matrix approximation

$$\mathbf{C}^{\mathrm{T}}\mathbf{D}^{-1}\mathbf{C} = \frac{1}{2}\begin{pmatrix} 3 & -1 & -1 & -1 \\ -1 & 3 & -1 & -1 \\ -1 & -1 & 3 & -1 \\ -1 & -1 & -1 & 3 \end{pmatrix}$$

This matrix is different from the one obtained using the consistent mass matrix representation, but is still singular. After eliminating the singularity by reducing the order of the matrix by one, the matrix becomes

$$\mathbf{C}^{\mathrm{T}}\mathbf{D}^{-1}\mathbf{C} = \frac{1}{2}\begin{pmatrix} 3 & -1 & -1 \\ -1 & 3 & -1 \\ -1 & -1 & 3 \end{pmatrix}$$

and can be readily inverted to give

$$\left(\mathbf{C}^{\mathrm{T}}\mathbf{D}^{-1}\mathbf{C}\right)^{-1} = \frac{1}{2}\begin{pmatrix} 2 & 1 & 1 \\ 1 & 2 & 1 \\ 1 & 1 & 2 \end{pmatrix}$$

Multiplying this matrix by $\mathbf{D}^{-1}\mathbf{C}$ gives the rectangular matrix

$$\mathbf{D}^{-1}\mathbf{C}\left(\mathbf{C}^{\mathrm{T}}\mathbf{D}^{-1}\mathbf{C}\right)^{-1} = \frac{1}{2h}\begin{pmatrix} -2 & -1 & -1 \\ -1 & -1 & 0 \\ 0 & -1 & 1 \\ -1 & -1 & -2 \\ 0 & -1 & -1 \\ 1 & -1 & 0 \end{pmatrix}$$

which is found to be concident with the one calculated using the consistent mass matrix, namely,

$$\mathbf{D}^{-1}\mathbf{C}\big(\mathbf{C}^{\mathrm{T}}\mathbf{D}^{-1}\mathbf{C}\big)^{-1} = \mathbf{M}^{-1}\mathbf{C}\big(\mathbf{C}^{\mathrm{T}}\mathbf{M}^{-1}\mathbf{C}\big)^{-1}.$$

This calculation shows that, at least for a uniform mesh of bilinear elements, the replacement of the consistent mass matrix $\mathbf{M}$ by its diagonalized counterpart $\mathbf{D}$ does not modify the matrix $\mathcal{Q}$ and henceforth neither the operation of orthogonal projection which provides the discretely solenoidal velocity field. By contrast, the diagonal approximation modifies the matrix of the linear system governing the pressure. Therefore, if the diagonal mass matrix $\mathbf{D}$ is employed in the computational procedure described at the end of section 7.5.2, one actually calculates an approximate pressure $\mathbf{P}_{\mathbf{D}}^{n+1}$ (a "diagonalized pressure" we could say) which is related to the correct pressure $\mathbf{P}^{n+1}$ of the considered finite element representation by the obvious relationship

$$\mathbf{C}^{\mathrm{T}}\mathbf{D}^{-1}\mathbf{C}\mathbf{P}_{\mathbf{D}}^{n+1} = \mathbf{C}^{\mathrm{T}}\mathbf{M}^{-1}\mathbf{C}\mathbf{P}^{n+1}.$$

Therefore, the correct pressure $\mathbf{P}^{n+1}$ can be obtained from $\mathbf{P}_{\mathbf{D}}^{n+1}$ at any stage of the calculation in an independent step of post-processing, requiring the "consistently accurate" matrix $\mathbf{C}^{\mathrm{T}}\mathbf{M}^{-1}\mathbf{C}$.

7.5.5 Taylor–Galerkin scheme for advection-diffusion

In this section we examine the problem of solving the advection-diffusion equation (7.16) numerically. For simplicity of the exposition, an explicit scheme has been considered so far, but it can become inefficient for very small viscosity. In these situations, it becomes necessary to resort to an implicit integration scheme. Two approaches are possible for obtaining time-integration schemes for evolutionary problems. A first method consists in discretizing the entire equation, with all its terms beeing taken into account simultaneously according to the same time-integration scheme. For finite element approximations the approach most appropriate for transient equations containing advection terms is the Taylor–Galerkin method introduced by Donea (1984). In particular, we will describe a two-level second-order accurate scheme of this class which has been proposed for obtaining time-accurate solutions of advection-diffusion problems (Donea *et al.* 1984).

Another possible approach consists in tackling advection and viscous diffusion separately, using the time integration scheme most suitable for each term. For the diffusion the Crank–Nicolson scheme can be used, whereas for the advection several options are possible within the class of the Taylor–Galerkin schemes. Third- and fourth-order Taylor–Galerkin schemes for pure advection problems will be described in the next chapter discussing the numerical solution of the incompressible Euler equations by finite elements.

Let us consider the nonlinear advection-diffusion equation

$$\boldsymbol{u}_t + (\boldsymbol{u} \cdot \boldsymbol{\nabla})\boldsymbol{u} = \nu \nabla^2 \boldsymbol{u},$$

with no pressure term in it, as occurring in the first half-step equation of the fractional-step method. Assuming that the advection velocity $\boldsymbol{u}$ can be represented by a "frozen" velocity field $\boldsymbol{a}(\boldsymbol{x})$, the following linearized version of the advection-diffusion equation is obtained:

$$\boldsymbol{u}_t + (\boldsymbol{a} \cdot \boldsymbol{\nabla})\boldsymbol{u} = \nu \nabla^2 \boldsymbol{u}. \tag{7.25}$$

Letting the spatial variable continuous, this linear equation can be discretized in time by means of the following Taylor series expansion in the time step Δt, up to the second order:

$$\boldsymbol{u}^{n+1} = \boldsymbol{u}^n + \Delta t\, \boldsymbol{u}_t^n + \tfrac{1}{2}(\Delta t)^2 \boldsymbol{u}_{tt}^n + \mathrm{O}[(\Delta t)^3].$$

While the first-order time derivative is provided directly by (7.25), the second-order derivative can be obtained by differentiating the governing partial differential equation, namely,

$$\begin{aligned} \boldsymbol{u}_{tt}^n &= -(\boldsymbol{a} \cdot \boldsymbol{\nabla})\boldsymbol{u}_t^n + \nu \nabla^2 \boldsymbol{u}_t^n \\ &= \left[-\boldsymbol{a} \cdot \boldsymbol{\nabla} + \nu \nabla^2\right] \frac{\boldsymbol{u}^{n+1} - \boldsymbol{u}^n}{\Delta t}. \end{aligned}$$

We note that the right-hand side has been left in mixed spatial–temporal form, because the elimination of $\boldsymbol{u}_t$ through (7.25) would introduce third-order spatial derivatives, which would preclude the use of finite elements with only C^0-continuity for the spatial discretization. For the same reason the Taylor series expansion has been limited to the second-order time derivative. This is in constrast to the situation of the pure advection equation ($\nu = 0$) for which the third-order time derivative can be retained to obtain schemes of third-order temporal accuracy, as it will be discussed in the next chapter.

The substitution of $\boldsymbol{u}_t^n$ and $\boldsymbol{u}_{tt}^n$ in the Taylor expansion yields the semi-discrete equation

$$\left[1 - \tfrac{1}{2}\Delta t(-\boldsymbol{a} \cdot \boldsymbol{\nabla} + \nu \nabla^2)\right] \frac{\boldsymbol{u}^{n+1} - \boldsymbol{u}^n}{\Delta t} = -(\boldsymbol{a} \cdot \boldsymbol{\nabla})\boldsymbol{u}^n + \nu \nabla^2 \boldsymbol{u}^n. \tag{7.26}$$

It is immediate to recognize that this scheme represents the Crank–Nicolson time-stepping method written in *incremental form*.

To analyze the numerical properties of this schemes combined with a finite-element-based spatial discretization, a scalar and one-dimensional version of the problem is considered, namely,

$$\left[1 - \tfrac{1}{2}\Delta t(-a\partial_x + \nu \partial_x^2)\right] \frac{u^{n+1} - u^n}{\Delta t} = (-a\partial_x + \nu \partial_x^2)u^n. \tag{7.27}$$

The fully discrete equations are obtained by introducing a local approximation of the unknown $u^n(x)$, according to the expansion

$$U^n(x) = \sum_j U_j^n v_j(x),$$

where $v_j(x)$ are the basis functions. By the standard Galerkin formulation, the expansion is substituted into the semi-discrete equation and the resulting equation is made orthogonal to the space of the basis functions. After integrating by parts the terms with the second-order spatial derivative, one obtains

$$\left[1 + \tfrac{1}{6}\delta^2 - \tfrac{1}{2}(-c\Delta_0 + d\delta^2)\right]\left(U^{n+1} - U^n\right) = (-c\Delta_0 + d\delta^2)U^n \tag{7.28}$$

where $c = a\Delta t/h$ is the Courant number and $d = \nu\Delta t/h^2$ is the diffusion number. Here Δ_0 and δ^2 are standard notations for the central difference operators, namely,

$$(\Delta_0 U)_j = \tfrac{1}{2}(U_{j+1} - U_{j-1}),$$
$$(\delta^2 U)_j = U_{j-1} - 2U_j + U_{j+1}.$$

This Taylor–Galerkin scheme of second-order accuracy has an amplification in one time step of

$$G_{\mathrm{TG}}(\xi, c, d) = 1 + \frac{-\mathrm{i}c\sin\xi - 4d\sin^2\frac{1}{2}\xi}{1 - (\frac{2}{3} - 2d)\sin^2\frac{1}{2}\xi + \frac{1}{2}\mathrm{i}c\sin\xi}$$
$$= 1 - \mathrm{i}c\xi - \left(d + \tfrac{1}{2}c^2\right)\xi^2 + \mathrm{i}c\left(d + \tfrac{1}{4}c^2\right)\xi^3 + \cdots \quad \text{as } \xi \to 0,$$

where $\xi = kh$ is the dimensionless wave number (k is the dimensional wave number of the Fourier mode $e^{\mathrm{i}kx}$).

The corresponding quantity for the differental equation is $G_{\mathrm{ex}} = e^{-\delta_{\mathrm{ex}} + \mathrm{i}\phi_{\mathrm{ex}}}$ where $\delta_{\mathrm{ex}} = d\xi^2$ and $\phi_{\mathrm{ex}} = c\xi$ are the exact damping and phase speed, respectively. To evaluate the accuracy of the scheme beyond the asymptotic limit, we introduce the damping δ_{num} and phase speed ϕ_{num} of the fully discrete equations through the relation $G_{\mathrm{TG}} = e^{\delta_{\mathrm{num}} + \mathrm{i}\phi_{\mathrm{num}}}$.

In Tables 7.1 and 7.2 the relative phase error $(\phi_{\mathrm{num}} - \phi_{\mathrm{ex}})/\phi_{\mathrm{ex}}$ and the damping ratio $\delta_{\mathrm{num}}/\delta_{\mathrm{ex}}$ for the scheme are compared with those of the Crank–Nicolson finite difference scheme (CN–FD) and the explicit Euler–Galerkin finite element scheme (EG–FE).

From Table 7.1 one notes that the Taylor–Galerkin scheme shows a substantial improvement in phase speed with respect to the implicit finite difference scheme and the Euler finite element method. The values of the damping ratio in Table 7.2 indicate that the Taylor–Galerkin method reproduces very well the physical diffusion in the region of accurate spatial resolution ($0 < \xi < \pi/4$) and

			Relative phase error			
c	d	ξ	CN–FD	EG–FE	TG–FE	TG–3p
		$\frac{1}{4}\pi$	−0.1010	−0.0223	−0.0040	−0.0032
	0.05	$\frac{1}{2}\pi$	−0.3639	−0.0800	−0.0497	−0.0545
		$\frac{3}{4}\pi$	−0.6982	−0.0562	−0.2829	−0.3475
		$\frac{1}{4}\pi$	−0.1004	0.0571	−0.0032	−0.0023
0.2	0.1	$\frac{1}{2}\pi$	−0.3592	0.2888	−0.0311	−0.0247
		$\frac{3}{4}\pi$	−0.6915	1.1360	−0.1888	−0.2118
		$\frac{1}{4}\pi$	−0.0981	—	−0.0001	0.0004
	0.2	$\frac{1}{2}\pi$	−0.3395	—	0.0375	0.0375
		$\frac{3}{4}\pi$	−0.6613	—	0.6122	0.6122
		$\frac{1}{4}\pi$	−0.1087	—	−0.0145	−0.0071
	0.05	$\frac{1}{2}\pi$	−0.3748	—	−0.0822	−0.0620
		$\frac{3}{4}\pi$	−0.7009	—	−0.3187	−0.3727
		$\frac{1}{4}\pi$	−0.1082	—	−0.0137	−0.0074
0.5	0.1	$\frac{1}{2}\pi$	−0.3705	—	−0.0696	−0.0370
		$\frac{3}{4}\pi$	−0.6945	—	−0.2510	−0.2652
		$\frac{1}{4}\pi$	−0.1060	—	−0.0108	−0.0074
	0.2	$\frac{1}{2}\pi$	−0.3528	—	−0.0162	0.0176
		$\frac{3}{4}\pi$	−0.6662	—	0.1263	0.1547

Table 7.1:

that it is overdiffusive in the high-frequency region where the phase-speed error is maximum. The Crank–Nicolson finite difference scheme is instead systematically underdiffusive, in particular for short wavelengths. Finally, the Euler finite element method has a rather limited stability interval and does not exhibit enough damping in the region of accuracy. Moreover, in the limit $\xi \to 0$ the damping ration for Euler scheme is found to be $1 - c^2/(2d)$ instead of one. Therefore the diffusion of signals with long wavelengths is not correctly reproduced by this scheme unless $c^2/(2d) \to 0$, a condition which implies the use of exceedingly small time steps in advection dominated advection–diffusion problems.

			Diffusion ratio			
c	d	ξ	CN–FD	EG–FE	TG–FE	TG–3p
0.2	0.05	$\frac{1}{4}\pi$	0.945	0.650	1.046	1.041
		$\frac{1}{2}\pi$	0.803	0.841	1.191	1.115
		$\frac{3}{4}\pi$	0.613	1.351	1.406	1.174
	0.1	$\frac{1}{4}\pi$	0.945	0.863	1.046	1.045
		$\frac{1}{2}\pi$	0.805	1.104	1.197	1.166
		$\frac{3}{4}\pi$	0.617	1.704	1.458	1.301
	0.2	$\frac{1}{4}\pi$	0.946	—	1.047	1.048
		$\frac{1}{2}\pi$	0.812	—	1.222	1.220
		$\frac{3}{4}\pi$	0.636	—	1.726	1.655
0.5	0.05	$\frac{1}{4}\pi$	0.920	—	1.014	1.027
		$\frac{1}{2}\pi$	0.763	—	1.067	0.764
		$\frac{3}{4}\pi$	0.597	—	1.227	0.817
	0.1	$\frac{1}{4}\pi$	0.921	—	1.014	1.032
		$\frac{1}{2}\pi$	0.765	—	1.069	0.976
		$\frac{3}{4}\pi$	0.601	—	1.242	1.025
	0.2	$\frac{1}{4}\pi$	0.921	—	1.015	1.033
		$\frac{1}{2}\pi$	0.770	—	1.080	1.095
		$\frac{3}{4}\pi$	0.617	—	1.248	1.174

Table 7.2:

Iterative method for the Taylor–Galerkin equations

The Taylor–Galerkin equations above require to solve at each time step a system of algebraic equations of the form

$$\mathbf{M}(\mathbf{a})\mathbf{V} = \mathbf{F},$$

where the vector $\mathbf{V} = \mathbf{U}^{n+1} - \mathbf{U}^n$ is the unknown of the incremental nodal values and

$$\mathbf{M}(\mathbf{a}) = \mathbf{M} - \tfrac{1}{2}\Delta t[-\mathbf{A}(\mathbf{a}) + \nu\mathbf{K}],$$
$$\mathbf{F} = \mathbf{F}(\mathbf{U}^n) = \Delta t[-\mathbf{A}(\mathbf{a}) + \nu\mathbf{K}]\mathbf{U}^n.$$

For the Navier–Stokes equations $\mathbf{a} = \boldsymbol{u}^n$ and therefore the matrix of the linear system changes at each time level. This makes the use of iterative methods for the approximate solution of the system attractive. On the other hand, for the present Taylor–Galerkin scheme it is essential that the approximate solution retains the consistent character of the "mass" matrix $\mathbf{M}(\mathbf{a})$ which results from the Crank–Nicolson time discretization of the advection-diffusion equation. This can be achieved as follows. Let us rewrite the equation in the simpler form

$$\mathbf{MV} = \mathbf{F}$$

and consider the identity

$$\mathbf{M} = \mathbf{D} + (\mathbf{M} - \mathbf{D}),$$

where $\mathbf{D}$ is the *lumped diagonal* matrix obtained from $\mathbf{M}$ by the well known row-sum technique,

$$\mathrm{D}_{ii} = \sum_j \mathrm{M}_{ij}, \qquad \mathrm{D}_{ij} = 0, \quad j \neq i.$$

Since $\mathbf{D}$ is diagonal and has positive entries, it is possible to write

$$\mathbf{M} = \mathbf{D}^{1/2}(\mathbf{I} + \mathbf{X})\mathbf{D}^{1/2}$$

where

$$\mathbf{X} = \mathbf{D}^{-1/2}(\mathbf{M} - \mathbf{D})\mathbf{D}^{-1/2}.$$

Then, under the assumption $\|\mathbf{X}\| \leq 1$, the inverse of $\mathbf{M}$ written in the form above can be expressed by the following series

$$\mathbf{M}^{-1} = \mathbf{D}^{-1/2}(\mathbf{I} - \mathbf{X} + \mathbf{X}^2 - \mathbf{X}^3 + \cdots)\mathbf{D}^{-1/2}.$$

Truncating the series after the term $\mathbf{X}$ gives the following two-term approximation of the inverse of $\mathbf{M}$,

$$\mathbf{M}^{-1}(2) = 2\mathbf{D}^{-1}\big(\mathbf{I} - \tfrac{1}{2}\mathbf{MD}^{-1}\big),$$

while retaining the term $\mathbf{X}^2$ produces the three-term approximation

$$\mathbf{M}^{-1}(3) = 3\mathbf{D}^{-1}\left(\mathbf{I} - \mathbf{M}\mathbf{D}^{-1} + \tfrac{1}{3}\mathbf{M}\mathbf{D}^{-1}\mathbf{M}\mathbf{D}^{-1}\right)$$

and so on.

In practice the successive approximations of such an iterative solution method can be generated very easily by means of a multi-pass algorithm. Consider the sequence of approximate solutions $\mathbf{V}_r$, $r = 0, 1, \ldots, R$, defined as follows: start from $\mathbf{V}_0 = \mathbf{0}$; then, for $r = 1, 2, \ldots, R$, determine $\mathbf{V}_r$ from $\mathbf{V}_{r-1}$ by solving the "diagonal" linear system

$$\mathbf{D}\mathbf{V}_r = \mathbf{F} - (\mathbf{M} - \mathbf{D})\mathbf{V}_{r-1}.$$

Finally assume $\mathbf{V} = \mathbf{V}_R$. It is immediately verified that the two approximations considered above are obtained by this simple procedure with $R = 2$ and $R = 3$, respectively.

The three-term approximation is particularly useful for the solution of the Taylor–Galerkin equations. In fact, for linear elements the 3-pass algorithm is found to have an amplification factor which in the limit $\xi \to 0$ coincides with that of the exact Taylor–Galerkin method (7.26). The 3-pass algorithm is explicit and thus only conditionally stable. However, its stability domain in the (c, d)-plane is found numerically to be much larger than that of the standard Euler–Galerkin method. This is confirmed by the results shown in Tables 7.1 and 7.2. The last column of the Tables shows that the 3-pass explicit TG scheme based on linear elements is practically never less accurate that its implicit counterpart, in the representation of both advection and diffusion.

Chapter 8

Incompressible Euler equations

8.1 Introduction

It may appear strange to include a chapter on the Euler equations for an incompressible ideal fluid in a work devoted to the study of the equations governing viscous flows. There are however several reasons for considering it.

- Firstly, as mentioned in the introductory chapter, the compatibility condition between boundary and initial data for the incompressible Navier–Stokes equations are due only to the incompressibility condition, irrespectively of the viscous or nonviscous character of the fluid. It is therefore interesting to study the consequences of this condition of compatibility in the particular situation of the nonviscous problem.

- Secondly, considering the formulation based on the primitive variables and using a Poisson equation for the pressure, one can discuss the fractional-step method in a nonviscous context which simplifies its presentation. In this framework, the split character of the fractional-step equations offers the opportunity of describing some new very accurate Taylor–Galerkin methods for the numerical solution of nonlinear advection problems.

- Thirdly, since the incompressible Euler equations are not limited to irrotational flows, a vortical velocity field can be specified at the initial time. Under this assumption it is convenient to express the governing equations so that the vorticity is one of the basic variables of the formulation and there are several ways of doing that. For each possible representation, the examination of the appropriate boundary conditions allows to investigate the transition from the incompressible Navier–Stokes equations to the nonviscous equations, especially as far as the boundary conditions are concerned.

- Finally, we can further explore the main theme of this study, *i.e.*, the substitution of the condition of incompressibility by a Poisson equation for a scalar or a vector variable. As it will be shown, the various different ways of formulating the Euler equations for incompressible vortical flows have a direct connection with the representations of the Navier–Stokes equations described in the previous chapters. The boundary conditions which supplement the Poisson equation replacing the condition of incompressibility are found however to be coincident only in part with those established for the viscous problem. From a numerical viewpoint, the study of the Euler equations expressed in terms of vorticity and velocity can be useful for the implementation of vortex methods in the presence of rigid (slip) boundaries.

The content of this chapter is organized as follows. We first introduce the time-dependent Euler equations governing the motion of an incompressible nonviscous fluid and describe a fractional-step method for integrating the equations in time under the constraint of mass conservation. Numerical schemes for dealing with the advection part of the momentum equation are then discussed in detail by introducing the Taylor–Galerkin method for pure advection equations. Time integration schemes of high order accuracy are described which exhibit very favourable phase-speed properties when combined with a finite-element-based spatial discretization by means of linear and multilinear basis functions. In particular, for the nonlinear vector equation of momentum in two and three dimensions a two-step fourth-order accurate Taylor–Galerkin scheme is derived which can be implemented without requiring to consider complicated extensions of the consistent mass matrix.

After discussing solution methods, alternative formulations of the Euler equations governing vortical incompressible flows are considered. The first formulation is based on the primitive variables but is characterized by the elimination of the continuity equation in favour of a Poisson equations for the pressure. The boundary condition supplementing such an equation is obtained to derive a formulation which would be uncoupled but for the presence of the nonlinear term. The weak formulation of this elliptic problem of pressure is also given for comparison with the corresponding problem in viscous flows discussed in section 3.4.7.

Formulations in which the vorticity is a primary unknowns of the problem are then addressed. The vorticity–velocity representation is discussed giving for the first time the complete boundary conditions which supplement the Poisson equation of velocity. Finally, nonprimitive variable formulations are considered including the $\boldsymbol{\zeta}$-$\boldsymbol{\phi}$ and q_S-$\boldsymbol{\zeta}$-$\boldsymbol{\psi}$ representations for three-dimensional problems, and the ζ-ψ formulation for plane flows in two dimensions. In each case the set of boundary conditions supplementing the respective velocity potential is provided.

8.2 Incompressible Euler equations

8.2.1 Basic equations

The equations governing the flow of an incompressible fluid of null viscosity, called also *ideal fluid*, are (see *e.g.*, Batchelor [3, p. 382] or Karamcheti [22, p. 178])

$$\frac{\partial \boldsymbol{u}}{\partial t} + (\boldsymbol{u} \cdot \boldsymbol{\nabla})\boldsymbol{u} = -\boldsymbol{\nabla} P, \tag{8.1}$$

$$\boldsymbol{\nabla} \cdot \boldsymbol{u} = 0, \tag{8.2}$$

where $\boldsymbol{u}(\boldsymbol{x}, t)$ is the velocity, $P(\boldsymbol{x}, t)$ is the pressure divided by the constant density of the fluid.

The statement of the problem is made complete by the specification of suitable boundary and initial conditions. A typical boundary condition consists in prescribing the value of the *normal* component of velocity b_n on the boundary,

$$\boldsymbol{n} \cdot \boldsymbol{u}|_S = b_n(\boldsymbol{x}_S, t), \tag{8.3}$$

where $\boldsymbol{n}$ denotes the outward unit normal to the boundary S and $\boldsymbol{x}_S \in S$. We use the notation b_n instead of $\boldsymbol{n} \cdot \boldsymbol{b}$ considered in the viscous problem since the tangential components of velocity on the boundary need not to be specified in the nonviscous problem.

The initial condition consists in the specification of the velocity field $\boldsymbol{u}_0$ at the initial time, $t = 0$, namely,

$$\boldsymbol{u}|_{t=0} = \boldsymbol{u}_0(\boldsymbol{x}). \tag{8.4}$$

The boundary value of the normal component of velocity b_n must satisfy, for all $t \geq 0$, the global condition

$$\oint b_n \, dS = 0, \tag{8.5}$$

which follows from integrating the continuity equation over V and using the divergence theorem. Moreover, the initial velocity field $\boldsymbol{u}_0$ is assumed to be solenoidal, *i.e.*,

$$\boldsymbol{\nabla} \cdot \boldsymbol{u}_0 = 0. \tag{8.6}$$

Finally, the boundary and initial data b_n and $\boldsymbol{u}_0$ are assumed to satisfy the following compatibility condition:

$$b_n|_{t=0} = \boldsymbol{n} \cdot \boldsymbol{u}_0|_S. \tag{8.7}$$

This compatibility condition between the boundary and initial values is necessary to prove the existence and uniqueness (up to an arbitrary function of t which may be added to P) of classical solutions of the problem in two dimensions (Kato 1967).

The boundary condition (8.3) is associated with the "elliptic character" the incompressibility. If $b_n = 0$, then no other boundary condition is requested. On the contrary, if $b_n \neq 0$ on some part of S, then the fluid enters the domain V or is flowing out of it and some other boundary condition must be imposed to comply with the "hyperbolic" character of the convective part of the problem. In the following, we will not consider these additional boundary conditions possibly requested at inflow or outflow boundaries to have a well defined problem, except when we will discuss the equations of the fractional-step method and describe a numerical scheme for the solution of the pure advection equation. As a consequence, the results of the subsequent analysis will be valid, strictly speaking, only for the particular situation of a homogeneous boundary condition, namely, $b_n = 0$.

8.2.2 Fractional-step equations

The incompressible Euler equations (8.1)–(8.2) can be solved using the fractional-step projection method developed for the incompressible Navier–Stokes equations which has been described in the previous chapter. Actually, the method lends itself to the solution of the nonviscous equations even more straightforwardly than to the equations governing viscous flows. In this section the nonviscous version of the fractional-step method is given. The variant of this method relying on a Poisson equation for the pressure is also described, considering both the differential and variational form of the problem for the incompressible half-step so as to allow a comparison with the corresponding problems for viscous flows.

Similarly to the Navier–Stokes equations, the method separates the treatment of the incompressibility from the other dynamical effects of the momentum equation. In a first half-step an intermediate velocity field $\boldsymbol{u}^{n+1/2}$ is calculated by means of a time-discretized version of the momentum equation, with the pressure term omitted. For example, considering for simplicity a fully explicit time-discretization, one has the problem

$$\begin{aligned} \frac{\boldsymbol{u}^{n+1/2} - \boldsymbol{u}^n}{\Delta t} &= -(\boldsymbol{u}^n \cdot \boldsymbol{\nabla})\boldsymbol{u}^n, \\ \boldsymbol{u}^{n+1/2}|_{S_{\text{in}}} &= \boldsymbol{b}^{n+1}|_{S_{\text{in}}}, \end{aligned} \tag{8.8}$$

where S_{in} is the portion of S on which $b_n < 0$ and $\boldsymbol{b}^{n+1}$ is the boundary value prescribed on velocity. Strictly speaking S_{in} depends in general on time since $b_n = b_n(\boldsymbol{x}_S, t)$, but the indication of this dependence is omitted here for notational simplicity. Thus, in the first half-step the value of $\boldsymbol{u}^{n+1/2}$ on the boundary is partly imposed and partly determined as a result of the solution of problem (8.8)

In the second half-step, from the intermediate velocity $\boldsymbol{u}^{n+1/2}$ the end-of-step solenoidal velocity field $\boldsymbol{u}^{n+1}$ is determined together with the unknown pressure P^{n+1} according to the following equations and boundary condition:

$$\begin{aligned} \frac{\boldsymbol{u}^{n+1}-\boldsymbol{u}^{n+1/2}}{\Delta t} &= -\boldsymbol{\nabla} P^{n+1}, \\ \boldsymbol{\nabla}\cdot \boldsymbol{u}^{n+1} &= 0, \\ \boldsymbol{n}\cdot \boldsymbol{u}^{n+1}|_S &= b_n^{n+1}. \end{aligned} \tag{8.9}$$

The problem of the second half-step can also be written in a form which involves a Poisson equation for the pressure. In fact, taking the divergence of the first equation in (8.9) and taking into account the second equation, we obtain $\nabla^2 P^{n+1} = \boldsymbol{\nabla}\cdot \boldsymbol{u}^{n+1/2}/\Delta t$. The boundary condition supplementing such an equation is obtained taking the normal component of the first equation and using the boundary condition for the normal velocity. In this manner we obtain the following Neumann problem

$$\begin{aligned} -\nabla^2 P^{n+1} &= -(\Delta t)^{-1}\boldsymbol{\nabla}\cdot \boldsymbol{u}^{n+1/2}, \\ \boldsymbol{n}\cdot \boldsymbol{\nabla} P^{n+1}|_S &= (\Delta t)^{-1}\Big(\boldsymbol{n}\cdot \boldsymbol{u}^{n+1/2}|_S - b_n^{n+1}\Big). \end{aligned}$$

We note that the solvability condition for the Neumann problem reads

$$\int \boldsymbol{\nabla}\cdot \boldsymbol{u}^{n+1/2}\, dV = \oint \Big(\boldsymbol{n}\cdot \boldsymbol{u}^{n+1/2}|_S - b_n^{n+1}\Big)\, dS,$$

which, by virtue of the global condition $\oint b_n\, dS = 0$, becomes

$$\int \boldsymbol{\nabla}\cdot \boldsymbol{u}^{n+1/2}\, dV = \oint \boldsymbol{n}\cdot \boldsymbol{u}^{n+1/2}|_S\, dS.$$

This condition is always satisfied by the divergence theorem, irrespectively of the form of the boundary condition imposed on velocity in the first half-step (8.8).

The weak form of the Neumann problem is obtained multiplying the Poisson equation by any function Q chosen in $H^1(V)$, integrating over V and applying the integration by parts. Introducing the Neumann boundary condition in the boundary integral gives the variational form of the problem: find $P^{n+1} \in H^1(V)$ such that

$$(\boldsymbol{\nabla} Q, \boldsymbol{\nabla} P^{n+1}) = (\boldsymbol{\nabla} Q, \boldsymbol{u}^{n+1/2})/\Delta t - \oint Q\, b_n^{n+1}\, dS/\Delta t,$$

for any $Q \in H^1(V)$. Of course, P^{n+1} is defined up to an arbitrary additive constant.

8.3 Taylor–Galerkin method

The most appropriate numerical technique within the finite element framework for the solution of the transient advection equation (8.8) is the Taylor–Galerkin method (Donea 1984). In this approach, the time discretization is performed before the spatial discretization of the equations by means of finite elements. In this section we will first describe the Taylor–Galerkin scheme of third-order temporal accuracy and recall its basic numerical properties as far as amplitude and phase-speed errors are concerned. After pointing out the difficulties faced by the application of this scheme to nonlinear hyperbolic equations, a two-step version of the third-order TG scheme is described which can be easily used to solve nonlinear problems without lowering the high phase accuracy of its single-step counterpart and which presents some stability advantages in multidimensional problems. This method has been introduced by Selmin (1987). The two-step strategy is then exploited a step further by introducing two new fourth-order accurate TG schemes, the second of which is found to be characterized by a domain of numerical stability which is exactly isotropic. Finally, the nonlinear vector equation governing a vector field which advects itself is examined and the second-order accurate Taylor–Galerkin scheme recently proposed for its solution by Laval (1988) (see also Laval and Quartapelle 1990) is described. This scheme is here used to construct a fourth-order accurate two-step scheme for dealing with the advection part of the momentum equation (8.8).

8.3.1 Basic third-order TG scheme

The Taylor–Galerkin scheme of third-order temporal accuracy has been introduced by Donea (1984) to solve transient advection problems by finite elements. This scheme is derived here to have at hands its amplification factor for subsequent comparison with two-step versions of TG schemes of high-order accuracy.

In contrast with the procedure commonly followed in the finite element approach for transient problems, wherein the spatial approximation precedes the time discretization, in the Taylor–Galerkin method the time discretization is taken as the first step of the discretization process and is prearranged so as to match the high spatial accuracy achieved by linear elements (Donea 1984). Consider the explicit Euler time-stepping applied to the linear advection equation

$$u_t + a u_x = 0 \tag{8.10}$$

with the spatial variable x left continuous. A Taylor series expansion in the time step provides

$$u^{n+1} = u^n + \Delta t\, u_t^u + \tfrac{1}{2}(\Delta t)^2\, u_{tt}^n + \tfrac{1}{6}(\Delta t)^3\, u_{ttt}^n + \mathrm{O}\left[(\Delta t)^4\right].$$

Now, the governing equation and its successive derivatives with respect to time imply that $u_t = -au_x$, $u_{tt} = a^2 u_{xx}$ and $u_{ttt} = a^2[u_t]_{xx}$. The derivative in the right-hand side of the third equation is written in *mixed temporal–spatial* form to avoid third-order spatial derivatives which would require C^1 continuity of the interpolation space.

Substituting these expressions into the Taylor series above and rearranging some terms provides the following *semidiscrete* equation

$$\left[1 - \tfrac{1}{6}a^2(\Delta t)^2\,\partial_x^2\right]\frac{u^{n+1} - u^n}{\Delta t} = -au_x^n + \tfrac{1}{2}a^2\Delta t\, u_{xx}^n, \tag{8.11}$$

where ∂_x denotes the differentiation with respect to the spatial variable x. The standard Galerkin formulation is now applied using a basis of linear elements $\{v_j(x)\}$ over a uniform mesh of size h. The solution $u^n(x)$ at the time level t_n is approximated by means of the expansion

$$U^n(x) = \sum_j U_j^n\, v_j(x),$$

where $U^n = \{U_j^n\}$ is the vector of the nodal values of the unknown. Substituting the expansion in the semidiscrete equation and rendering the resulting equation orthogonal to the space of the basis functions provides, after integrating by parts the terms with the second-order spatial derivative,

$$\left[1 + \tfrac{1}{6}(1 - \nu^2)\delta^2\right]\left(U^{n+1} - U^n\right) = -\nu\Delta_0 U^n + \tfrac{1}{2}\nu^2\delta^2 U^n, \tag{8.12}$$

where $\nu = a\Delta t/h$ is the Courant number and where the standard notations for the difference operators, namely, $(\Delta_0 U)_j = (U_{j+1} - U_{j-1})/2$ and $(\delta^2 U)_j = U_{j-1} - 2U_j + U_{j+1}$, have been used.

The amplification factor of this Taylor–Galerkin scheme of third-order accuracy is

$$G_{\mathrm{TG}}^3(\xi,\nu) = 1 + \frac{-i\nu\sin\xi - 2\nu^2\sin^2\frac{1}{2}\xi}{1 - \frac{2}{3}(1-\nu^2)\sin^2\frac{1}{2}\xi},$$

where $\xi = kh$ is the dimensionless wave number (k is the dimensional wave number of the Fourier mode e^{ikx}). The numerical properties of the scheme are shown in Figure 8.1 which contains the polar diagrams of the modulus of the amplification factor $|G(\xi,\nu)|$ and the phase speed ratio

$$\frac{\phi_{\mathrm{num}}}{\phi_{\mathrm{ex}}} = \frac{\arg(G(\xi,\nu))}{-\nu\xi}.$$

The condition of numerical stability is found to be $|\nu| \le 1$. It coincides, of course, with the necessary condition provided by the *modified equation* (Warming and Hyett 1974) associated with the discrete equation (8.12)

$$\begin{aligned} u_t + au_x = &-\frac{1}{24}a^2h^2\Delta t\,(1-\nu^2)u^{(\mathrm{iv})} \\ &+\frac{1}{180}ah^4(1 - 5\nu^2 + 4\nu^4)u^{(\mathrm{v})} + \cdots. \end{aligned}$$

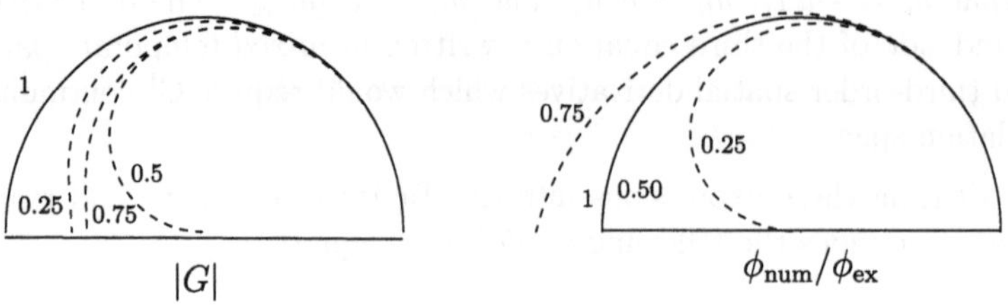

Figure 8.1: Numerical properties of the third-order Taylor–Galerkin scheme TG for some values of the Courant number ν.

The third-order TG method is referred to as TG or TG3 scheme. (It has also been called Euler/Taylor–Galerkin and Lax–Wendroff/Taylor–Galerkin scheme when its numerical properties have been compared with those of other schemes derived from the Euler and Lax–Wendroff time-stepping algorithms.) A detailed comparison of this TG method with other finite element methods for solving transient advection problems, including Semi-Lagrangian, Characteristic and Petrov–Galerkin methods, is given by Donea and Quartapelle (1992).

The basic distinguishing features of the Taylor–Galerkin method are pointed out by the following remarks.

1. The scheme has no free or adjustable parameter.

2. The scheme possesses the so-called unit CFL property, *i.e.*, the exact solution is obtained when the characteristic lines pass through the nodes.

3. The third-order term transforms the consistent mass matrix $M = 1 + \frac{1}{6}\delta^2$ into the "Taylor–Galerkin mass matrix"

$$M_{\text{TG}}(\nu) = 1 + \tfrac{1}{6}(1 - \nu^2)\delta^2.$$

   Such a generalized mass matrix depends on the Courant number but is still symmetric. For one-dimensional equations the TG mass matrix is tridiagonal and in multidimensional problems it has the same profile of the stiffness matrix. Generally, the matrix is characterized by a diagonal dominance that allows an approximate but accurate solution of the corresponding linear system with a few Jacobi iterations, as described in section 7.5.5. Other iterative methods could also be considered.

4. Despite the appearances, the last term in equation (8.12) is *not* a (second-order) numerical diffusion inherent to the scheme. In fact, in the solution

of transient problems the second-order term is only an element of the improved approximation to u_t with respect to the simple explicit Euler algorithm. This does not mean that the scheme is free from numerical diffusion but only that it has a fourth-order numerical diffusion, as shown explicitly by the modified equation above.

5. The weak variational formulation of the semidiscrete equation (8.11) reads

$$\left\langle v, \frac{u^{n+1}-u^n}{\Delta t}\right\rangle + \tfrac{1}{6}\Delta t \left\langle a v_x, (u^{n+1}-u^n)_x\right\rangle = -\left\langle v + \tfrac{1}{2}a\Delta t\, v_x, a u_x^n\right\rangle,$$

where v denotes the weighting function. Therefore, the TG scheme can also be interpreted as a kind of Petrov–Galerkin method in which the spatial derivative term of the equation is weighted by means of the modified test function

$$\widehat{v} = v + \tfrac{1}{2}\Delta t\, a v_x.$$

It must be noted, however, that in the TG method the time derivative term is weighted in a different manner, since the weighting of the various terms of the equation results from, and is dictated by, the analysis of the truncation error of the time-stepping algorithm. This is in contrast with most Petrov–Galerkin methods where the same modified test function is used consistently for all terms of the equation.

6. The Taylor–Galerkin method is applicable with very simple modifications also in the presence of a source term s, as in the equation $u_t + a u_x = s$.

Other Taylor–Galerkin schemes for linear advection equations can be obtained starting from the leap-frog and Crank–Nicolson time-stepping algorithms (see Donea *et al.* 1987).

8.3.2 Two-step third-order TG scheme

Let us now examine the computational difficulties which are encountered by the application of the Taylor–Galerkin method (8.12) to nonlinear problems. Consider, for instance, the following scalar conservation-law equation in one dimension:

$$u_t + f(u)_x = 0, \tag{8.13}$$

where the flux $f(u)$ is a known nonlinear function of u (hyperbolic equation written in conservation form). Introducing the advection velocity $a(u) = df(u)/du$, the second and third time derivative of the unknown u can be expressed as follows

$$u_{tt} = \partial_t[-f(u)_x] = -[f(u)_t]_x = -[a(u)\, u_t]_x = [a(u)\, f(u)_x]_x,$$

$$u_{ttt} = (\Delta t)^{-1}\left\{\left[a(u^{n+1})\, f(u^{n+1})_x\right]_x - \left[a(u^n)\, f(u^n)_x\right]_x\right\}.$$

Substituting them in the Taylor series yields the following third-order accurate time-integration scheme for the nonlinear equation (8.13)

$$\frac{u^{n+1}-u^n}{\Delta t} - \tfrac{1}{6}\Delta t\left[a(u^{n+1})\,f(u^{n+1})_x\right]_x$$
$$= -f(u^n)_x + \tfrac{1}{3}\Delta t\left[a(u^n)\,f(u^n)_x\right]_x .$$

The scheme would require to solve a nonlinear equation at each time step. However, this is not the only possibility for assuring third-order time accuracy in the integration of the conservation-law equation (cf. Selmin *et al.* 1985). In fact, one could also express the third-order time derivative according to the following substitutions

$$\begin{aligned}
u_{ttt} &= -\left\{[a(u)\,u_t]_x\right\}_t = -\left\{[a(u)\,u_t]_t\right\}_x \\
&= -\left\{a(u)_t\,u_t + a(u)\,u_{tt}\right\}_x \\
&= \left\{-a(u)_t\,u_t + a(u)[a(u)\,u_t]_x\right\}_x \\
&= \left\{-a(u)_t\,u_t + a(u)[a(u)_x\,u_t + a(u)\,u_{tx}]\right\}_x \\
&= \left\{-a(u)_t\,u_t + a(u)a(u)_x\,u_t + a(u)a(u)\,u_{tx}\right\}_x \\
&= \left\{a(u)_u f(u)_x\,u_t + a(u)a(u)_x\,u_t + a(u)a(u)\,u_{tx}\right\}_x .
\end{aligned}$$

Therefore, factoring out the quantity involving the time derivative gives

$$u_{ttt} = \Big\{[a(u)_u f(u)_x + a(u)a(u)_x + (a(u))^2\partial_x]u_t\Big\}_x .$$

The approximation $u_t = (u^{n+1} - u^n)/\Delta t$ can be used to express the third-order term so as to introduce it into the Taylor series and to obtain an alternative scheme for the integration of the nonlinear equation. The final equation obtained after the spatial approximation can be written as a linear system of equations for the incremental unknowns $\Delta U_j^{n+1} = U_j^{n+1} - U_j^n$. The three correction terms assuring the third-order accuracy depend on U_j^n, so that the matrix of the linear system changes at each time level making the computational cost of the scheme very high.

This drawback is still present if the third-order correction is written in another different form, using the governing equation in the first two terms (which do not involve the double space differentiation) appearing in the previous expression, to give

$$u_{ttt} = \Big\{-a(u)_u[f(u)_x]^2 - a(u)a(u)_x f(u)_x \;+ [a(u)]^2\,\partial_x u_t\Big\}_x .$$

Here, only one term is to be taken into account implicitly. In both cases, however, the various implicit or explicit third-order correction terms are rather complicated to evaluate, especially when dealing with system of conservation-law equations, not to speak of multidimensional problems. In conclusion, we see that, in order to achieve a third-order time accuracy in nonlinear problems with smooth solutions, it is necessary to include complicated generalizations of the consistent mass matrix which contain several nonsymmetric terms dependent on the solution at the previous time level. For these reasons it would be attractive to have an alternative approach to assure the high-order temporal accuracy of the Taylor–Galerkin method in these situations. As a matter of fact, these difficulties can be circumvented by formulating the Taylor–Galerkin scheme according to two-step strategy which can preserve the excellent phase-accuracy properties of the basic TG scheme (Selmin 1987).

A third-order TG scheme suitable for nonlinear advection problems can be obtained by considering the following *two-step* procedure

$$\widetilde{u}^n = u^n + \tfrac{1}{3}\Delta t\, u_t^n + \alpha(\Delta t)^2 u_{tt}^n, \tag{8.14a}$$

$$u^{n+1} = u^n + \Delta t\, u_t^n + \tfrac{1}{2}(\Delta t)^2 \widetilde{u}_{tt}^n, \tag{8.14b}$$

where the value of the parameter α is left unspecified for the time being. In fact, while the other coefficients in the equations (8.14) assume the value necessary for a third-order accuracy of the two discretizations combined together, the parameter α enters only the coefficient of the fourth-order term in the overall series. As a consequence, its value can affect only the modulus of the amplification factor of the resulting scheme but not its phase. Now, a convenient way of taking full advantage of the available degree of freedom consists in imposing that the phase speed of the two-step scheme be coincident with that of the basic Taylor–Galerkin scheme (8.12) in the linear case (Selmin 1987). The fully discrete version of equations (8.14) in the case of the linear advection equation in one dimension is

$$\left[1 + \tfrac{1}{6}\delta^2\right]\left(\widetilde{U}^n - U^n\right) = -\tfrac{1}{3}\nu\Delta_0 U^n + \alpha\nu^2\delta^2 U^n, \tag{8.15a}$$

$$\left[1 + \tfrac{1}{6}\delta^2\right]\left(U^{n+1} - U^n\right) = -\nu\Delta_0 U^n + \tfrac{1}{2}\nu^2\delta^2 \widetilde{U}^n. \tag{8.15b}$$

The amplification factor of this two-step linear scheme containing the parameter α is

$$G_{\mathrm{TTG}}^3(\xi,\nu;\alpha) = 1 + \frac{-\mathrm{i}\nu\sin\xi - 2\nu^2\sin^2\frac{1}{2}\xi \times \widetilde{G}(\xi,\nu;\alpha)}{1 - \frac{2}{3}\sin^2\frac{1}{2}\xi},$$

where

$$\widetilde{G}(\xi,\nu;\alpha) = 1 + \frac{-\frac{1}{3}\mathrm{i}\nu\sin\xi - 4\alpha\nu^2\sin^2\frac{1}{2}\xi}{1 - \frac{2}{3}\sin^2\frac{1}{2}\xi}.$$

It is immediate to verify that

$$G_{\mathrm{TTG}}^3\left(\xi,\nu;\tfrac{1}{9}\right) = G_{\mathrm{TG}}^3(\xi,\nu)\, R(\xi,\nu),$$

where $R(\xi,\nu)$ is the *real* function given by

$$R(\xi,\nu) = \frac{\left[1 - \frac{2}{3}(1-\nu^2)\sin^2\frac{1}{2}\xi\right]\left[1 - \frac{2}{3}(1+\nu^2)\sin^2\frac{1}{2}\xi\right]}{\left[1 - \frac{2}{3}\sin^2\frac{1}{2}\xi\right]^2}.$$

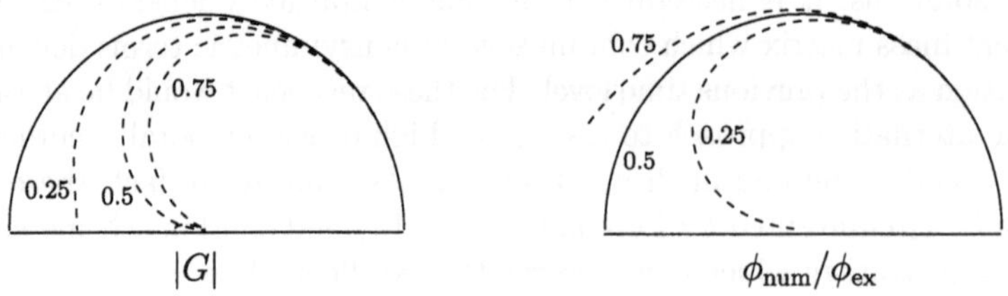

Figure 8.2: Numerical properties of the two-step third-order Taylor–Galerkin scheme TTG for some values of the Courant number ν.

Therefore, for $\alpha = \frac{1}{9}$ the two-step procedure (8.15) reproduces exactly the excellent phase-speed characteristics of the single-step third-order TG scheme. The condition of numerical stability is easily found to be $|\nu| \leq \sqrt{3}/2$. Figure 8.2 contains the diagrams of the phase and amplitude errors of the two-step scheme TTG (the interruption of the curve for $\nu = 0.75$ is due to a jump in the evaluation of an inverse trigonometric function by the computer). The new scheme is more dissipative than its one-step counterpart but this difference is negligible for all wavelengths except when ν is very close to the stability limit.

The modified equation associated with the two-step TG scheme is easily found to be

$$\begin{aligned} u_t + au_x = & -\frac{1}{72}a^2h^2\Delta t\,(3-\nu^2)u^{(\mathrm{iv})} \\ & +\frac{1}{360}ah^4(2-15\nu^2+8\nu^4)u^{(\mathrm{v})} + \cdots . \end{aligned}$$

The form assumed by the two-step scheme in the case of the nonlinear conservation-law equation $u_t + f(u)_x = 0$ is easily found to be

$$\frac{\widetilde{u}^n - u^n}{\Delta t} = -\tfrac{1}{3}\left[f(u^n) - \tfrac{1}{3}\Delta t\, a(u^n)\, f(u^n)_x\right]_x, \tag{8.16a}$$

$$\frac{u^{n+1} - u^n}{\Delta t} = -\left[f(u^n) - \tfrac{1}{2}\Delta t\, a(\widetilde{u}^n)\, f(\widetilde{u}^n)_x\right]_x. \tag{8.16b}$$

We notice the simplicity of the two equations as compared to any of the possible equations provided by the single-step TG scheme. In particular, the favourable effect of the complicated, solution-dependent, augmented mass matrix is achieved simply through a double application of the usual consistent mass matrix.

The important point now is that the phase properties of the one-step third-order TG scheme can be reproduced exactly by the two-step procedure also in two and three dimensions. To investigate these favourable numerical properties of the two-step TG scheme applied to multidimensional equations, we consider the linear advection equation

$$u_t + (\boldsymbol{a} \cdot \boldsymbol{\nabla})u = 0. \tag{8.17}$$

The single-step third-order accurate TG scheme for this equation is obtained without difficulty in the form

$$\left[1 - \tfrac{1}{6}(\Delta t)^2(\boldsymbol{a} \cdot \boldsymbol{\nabla})^2\right] \frac{u^{n+1} - u^n}{\Delta t} = -(\boldsymbol{a} \cdot \boldsymbol{\nabla})u^n + \tfrac{1}{2}\Delta t\,(\boldsymbol{a} \cdot \boldsymbol{\nabla})^2 u^n. \tag{8.18}$$

Let us assume that this equation is solved by means of finite elements, using a bilinear approximation of the unknown u^{n+1} over a uniform mesh. The amplification factor of the fully discrete equations of the single-step third-order scheme is

$$G^3_{\mathrm{TG}}(\boldsymbol{\xi}, \boldsymbol{\nu}) = 1 + \frac{-A(\boldsymbol{\xi}, \boldsymbol{\nu}) + \frac{1}{2}K(\boldsymbol{\xi}, \boldsymbol{\nu})}{M_{\mathrm{TG}}(\boldsymbol{\xi}, \boldsymbol{\nu})},$$

$$M_{\mathrm{TG}}(\boldsymbol{\xi}, \boldsymbol{\nu}) = M(\boldsymbol{\xi}) - \tfrac{1}{6}K(\boldsymbol{\xi}, \boldsymbol{\nu}),$$

where $\boldsymbol{\xi}$ is the dimensionless wave vector number and $\boldsymbol{\nu} = (\Delta t/h)\boldsymbol{a}$ is the vector counterpart of the Courant number ν. The quantities $A(\boldsymbol{\xi}, \boldsymbol{\nu})$ and $K(\boldsymbol{\xi}, \boldsymbol{\nu})$ are the Fourier transform of the operators $\Delta t(\boldsymbol{a} \cdot \boldsymbol{\nabla})$ and $\Delta t^2(\boldsymbol{a} \cdot \boldsymbol{\nabla})^2$, respectively, whereas $M(\boldsymbol{\xi})$ is the Fourier transform of the consistent mass matrix. The explicit expressions of $A(\boldsymbol{\xi}, \boldsymbol{\nu})$, $K(\boldsymbol{\xi}, \boldsymbol{\nu})$ and $M(\boldsymbol{\xi})$ for uniform meshes of multilinear elements are given in the Appendix C.

The application of the two-step procedure to the multidimensional advection equation gives immediately the equations

$$\tilde{u}^n - u^n = -\tfrac{1}{3}\Delta t\,(\boldsymbol{a} \cdot \boldsymbol{\nabla})u^n + \tfrac{1}{9}(\Delta t)^2(\boldsymbol{a} \cdot \boldsymbol{\nabla})^2 u^n, \tag{8.19a}$$

$$u^{n+1} - u^n = -\Delta t\,(\boldsymbol{a} \cdot \boldsymbol{\nabla})u^n + \tfrac{1}{2}(\Delta t)^2(\boldsymbol{a} \cdot \boldsymbol{\nabla})^2 \tilde{u}^n. \tag{8.19b}$$

The amplification factor of the two-step TG scheme is easily found to be

$$G^3_{\mathrm{TTG}}(\boldsymbol{\xi}, \boldsymbol{\nu}) = 1 + \frac{-A(\boldsymbol{\xi}, \boldsymbol{\nu}) + \frac{1}{2}K(\boldsymbol{\xi}, \boldsymbol{\nu}) \times \widetilde{G}(\boldsymbol{\xi}, \boldsymbol{\nu})}{M(\boldsymbol{\xi})},$$

where

$$\widetilde{G}(\boldsymbol{\xi}, \boldsymbol{\nu}) = 1 + \frac{-\frac{1}{3}A(\boldsymbol{\xi}, \boldsymbol{\nu}) + \frac{1}{9}K(\boldsymbol{\xi}, \boldsymbol{\nu})}{M(\boldsymbol{\xi})}.$$

It is not difficult to verify that the amplification factors of the two-step and single-step third-order accurate TG schemes are connected by the following relationship

$$G^3_{\mathrm{TTG}}(\boldsymbol{\xi}, \boldsymbol{\nu}) = G^3_{\mathrm{TG}}(\boldsymbol{\xi}, \boldsymbol{\nu})\, R(\boldsymbol{\xi}, \boldsymbol{\nu}),$$

where the factor $R(\boldsymbol{\xi},\boldsymbol{\nu})$ is the *real* quantity

$$R(\boldsymbol{\xi},\boldsymbol{\nu}) = 1 - \left[\frac{\frac{1}{6}K(\boldsymbol{\xi},\boldsymbol{\nu})}{M(\boldsymbol{\xi})}\right]^2 .$$

Therefore, also in two and three dimensions the two-step procedure reproduces exactly the phase-speed properties of the single-step third-order accurate TG scheme.

Moreover, the multidimensional two-step TG scheme presents the advantage that its stability domain is bounded by essentially the same value as the scheme for the equation in one dimension. For instance, the domain of numerical stability of the two-step scheme for the equation in two dimensions is found to be very nearly circular: the bound of the domain of numerical stability is found to be almost coincident with a circle of radius $|\boldsymbol{\nu}| = \sqrt{3}/2 = 0.866$, but for a slight flattening in the diagonal direction $\nu_x = \nu_y$ on which the stability limit reads $|\boldsymbol{\nu}| < 0.854$ (Selmin 1987).

This result is in contrast with what happens to the other single-step finite element or differene schemes, such as the (single-step) Lax–Wendroff finite difference method, the second-order Taylor–Galerkin scheme TG2 (that is the scheme with M_{TG} replaced by M) and also the third-order TG scheme: all these schemes suffer from the same drastic reduction of the stability domain in multidimensions. For instance, due to its dependence on the direction of $\boldsymbol{\nu}$, the condition of numerical stability of the two second-order schemes is $|\boldsymbol{\nu}| < 1/(2\sqrt{2})$ in 2D and $|\boldsymbol{\nu}| < 1/(3\sqrt{3})$ in 3D. These values are collected in the Table together with those of other two-step TG schemes of fourth-order accuracy to be introduced in the next section.

Bounds of the Courant number for numerical stability			
	1D	2D	3D
LW/FD	1	$\frac{1}{2\sqrt{2}} = 0.353$	$\frac{1}{3\sqrt{3}} = 0.192$
TG2, $O[(\Delta t)^2]$	$\frac{1}{\sqrt{3}} = 0.577$	$\frac{1}{2\sqrt{6}} = 0.204$	$\frac{1}{9} = 0.111$
TG, $O[(\Delta t)^3]$	1	$\frac{1}{2\sqrt{2}} = 0.353$	$\frac{1}{3\sqrt{3}} = 0.192$
TTG, $O[(\Delta t)^3]$	$0.854 \cong \frac{\sqrt{3}}{2} = 0.866$		
TTG-4A, $O[(\Delta t)^4]$	1		
TTG-4B, $O[(\Delta t)^4]$	0.84738		

To illustrate the rôle played by the surface integrals in the weak formulation of the two-step scheme, we report here the weak equations obtained in the case of the nonlinear conservation-law equation $u_t + \boldsymbol{\nabla}\cdot \boldsymbol{f}(u) = 0$ in two or three dimensions:

$$\left\langle v, \frac{\widetilde{u}^n - u^n}{\Delta t} \right\rangle = \tfrac{1}{3}\left\langle \boldsymbol{\nabla} v, \boldsymbol{f}(u^n) - \tfrac{1}{3}\Delta t\, \boldsymbol{a}(u^n)\, \boldsymbol{\nabla}\cdot \boldsymbol{f}(u^n) \right\rangle$$
$$-\tfrac{1}{3}\oint_{S_{\text{out}}} v\, \boldsymbol{n}\cdot \left[\boldsymbol{f}(u^n) - \tfrac{1}{3}\Delta t\, \boldsymbol{a}(u^n)\, \boldsymbol{\nabla}\cdot \boldsymbol{f}(u^n)\right] dS,$$

$$\left\langle v, \frac{u^{n+1} - u^n}{\Delta t} \right\rangle = \left\langle \boldsymbol{\nabla} v, \boldsymbol{f}(u^n) - \tfrac{1}{2}\Delta t\, \boldsymbol{a}(\widetilde{u}^n)\, \boldsymbol{\nabla}\cdot \boldsymbol{f}(\widetilde{u}^n) \right\rangle$$
$$-\oint_{S_{\text{out}}} v\, \boldsymbol{n}\cdot \left[\boldsymbol{f}(u^n) - \tfrac{1}{2}\Delta t\, \boldsymbol{a}(\widetilde{u}^n)\, \boldsymbol{\nabla}\cdot \boldsymbol{f}(\widetilde{u}^n)\right] dS.$$

In conclusion, the two-step formulation of the Taylor–Galerkin method, besides making high-order accuracy accessible for truly nonlinear problems, offers the additional advantage of giving an almost *isotropic* stability domain in multidimensional problems.

8.3.3 Two-step fourth-order TG schemes

The two-step strategy can be pursued further to obtain fourth-order accurate Taylor–Galerkin schemes suitable for nonlinear problems. Consider the Taylor series including terms up to the fourth-order derivative, namely,

$$u^{n+1} = u^n + \Delta t\, u_t^n + \tfrac{1}{2}(\Delta t)^2\, u_{tt}^n + \tfrac{1}{6}(\Delta t)^3\, u_{ttt}^n + \tfrac{1}{24}(\Delta t)^4\, u_{tttt}^n + O\left[(\Delta t)^5\right].$$

A first method achieving the fourth-order temporal accuracy of this expansion consists in advancing the solution from u^n to u^{n+1} by means of the following two-step procedure

$$\widetilde{u}^n = u^n + \tfrac{1}{3}\Delta t\, u_t^n + \tfrac{1}{12}(\Delta t)^2 u_{tt}^n, \tag{8.20a}$$

$$u^{n+1} = u^n + \Delta t\, u_t^n + \tfrac{1}{2}(\Delta t)^2 \widetilde{u}_{tt}^n. \tag{8.20b}$$

The application of the Galerkin method to the two equations gives, in the particular case of the linear advection equation in one dimension solved by means of linear elements,

$$\left[1 + \tfrac{1}{6}\delta^2\right]\left(\widetilde{U}^n - U^n\right) = -\tfrac{1}{3}\nu\Delta_0 U^n + \tfrac{1}{12}\nu^2\delta^2 U^n, \tag{8.21a}$$

$$\left[1 + \tfrac{1}{6}\delta^2\right]\left(U^{n+1} - U^n\right) = -\nu\Delta_0 U^n + \tfrac{1}{2}\nu^2\delta^2 \widetilde{U}^n. \tag{8.21b}$$

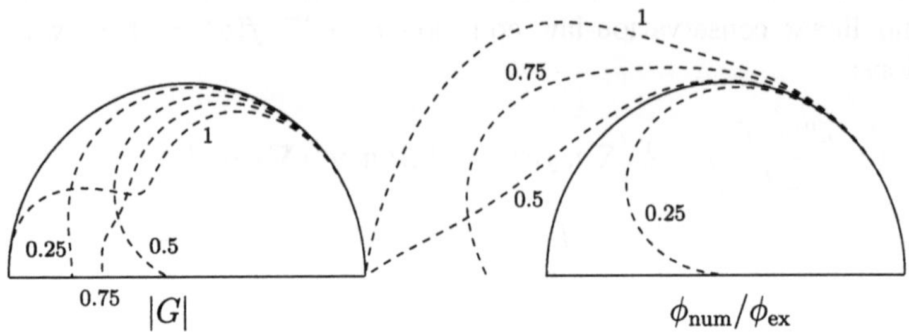

Figure 8.3: Numerical properties of the two-step fourth-order Taylor–Galerkin scheme TTG-4A for some values of the Courant number ν.

The amplification factor of this two-step fourth-order Taylor–Galerkin scheme, which will be denoted as TTG-4A, is found to be

$$G^{4A}_{\rm TTG}(\xi,\nu) = 1 + \frac{-\mathrm{i}\nu\sin\xi - 2\nu^2\sin^2\frac{1}{2}\xi \times \widetilde{G}^{\rm A}(\xi,\nu)}{1-\frac{2}{3}\sin^2\frac{1}{2}\xi},$$

where

$$\widetilde{G}^{\rm A}(\xi,\nu) = 1 + \frac{-\frac{1}{3}\mathrm{i}\nu\sin\xi - \frac{1}{3}\nu^2\sin^2\frac{1}{2}\xi}{1-\frac{2}{3}\sin^2\frac{1}{2}\xi}.$$

The scheme has a stability condition $|\nu| \le 1$ and its numerical properties can be deduced from the polar plot of $|G|$ and of the ratio $\phi_{\rm num}/\phi_{\rm ex}$ given in Figure 8.3. In comparison with the TTG scheme, the fourth-order scheme appears to reach the stability limit of 1 at the expense of lowering appreciably the phase accuracy of the propagation of the intermediate and small wavelengths as $|\nu| \to 1$.

There is still another possibility of reaching the fourth-order temporal accuracy within the two-step strategy. In fact, one can evaluate the first-order term in the second step in terms of the solution predicted in the first step. In this case to assure the fourth-order accuracy it is necessary to introduce some parameters in the equations of the two steps, as follows:

$$\begin{aligned}\widetilde{u}^n &= u^n + \alpha\Delta t\, u^n_t + \beta(\Delta t)^2 u^n_{tt},\\ u^{n+1} &= u^n + \Delta t\, \widetilde{u}^n_t + \gamma(\Delta t)^2 \widetilde{u}^n_{tt}.\end{aligned}$$

The values of the parameters α, β and γ are determined after substituting $\widetilde{u}^n$ from the first equation into the second one, which gives

$$\begin{aligned}u^{n+1} = u^n &+ \Delta t\left[u^n_t + \alpha\Delta t\, u^n_{tt} + \beta(\Delta t)^2 u^n_{ttt}\right]\\ &+ \gamma(\Delta t)^2\left[u^n_{tt} + \alpha\Delta t\, u^n_{ttt} + \beta(\Delta t)^2 u^n_{tttt}\right],\end{aligned}$$

that is

$$u^{n+1} = u^n + \Delta t\, u_t^n + (\alpha+\gamma)(\Delta t)^2 u_{tt}^n + (\beta+\gamma\alpha)(\Delta t)^3 u_{ttt}^n + \gamma\beta(\Delta t)^4 u_{tttt}^n.$$

In order for this expansion to coincide with the Taylor series truncated after the fourth-order term, the parameters in it must satisfy the following system of three equations

$$\alpha+\gamma = \tfrac{1}{2}, \qquad \beta+\gamma\alpha = \tfrac{1}{6}, \qquad \gamma\beta = \tfrac{1}{24},$$

which gives the third-order algebraic equation for γ

$$\gamma^3 - \tfrac{1}{2}\gamma^2 + \tfrac{1}{6}\gamma - \tfrac{1}{24} = 0.$$

This equation has two complex solutions and one real solution which is given by $\gamma = (k+1)/6$ with

$$k = \left[\tfrac{1}{2}\left(5+\sqrt{29}\right)\right]^{\frac{1}{3}} + \left[\tfrac{1}{2}\left(5-\sqrt{29}\right)\right]^{\frac{1}{3}} = 1.1541715.$$

Thus the three parameters are determined as a function of k from the relations

$$\alpha_k = \tfrac{1}{6}(2-k), \qquad \beta_k = \frac{1}{4(k+1)}, \qquad \gamma_k = \tfrac{1}{6}(k+1),$$

and their numerical values are found to be

$$\alpha_k = 0.1409714, \qquad \beta_k = 0.1160538, \qquad \gamma_k = 0.3590284.$$

For a linear advection equation in one dimension the resulting fourth-order accurate Taylor–Galerkin scheme (TTG-4B) assumes the form

$$\tilde{u}^n - u^n = -\alpha_k \Delta t\, a u_x^n + \beta_k (\Delta t)^2 a^2 u_{xx}^n, \tag{8.22a}$$

$$u^{n+1} - u^n = -\Delta t\, a\tilde{u}_x^n + \gamma_k (\Delta t)^2 a^2 \tilde{u}_{xx}^n. \tag{8.22b}$$

Its amplification factor for a Galerkin approximation over a uniform mesh of linear elements is given by

$$G_{\mathrm{TTG}}^{4\mathrm{B}}(\xi,\nu) = 1 + \frac{-\mathrm{i}\nu\sin\xi - 4\gamma_k\nu^2\sin^2\frac{1}{2}\xi}{1-\frac{2}{3}\sin^2\frac{1}{2}\xi}\,\tilde{G}^{\mathrm{B}}(\xi,\nu),$$

where

$$\tilde{G}^{\mathrm{B}}(\xi,\nu) = 1 + \frac{-\mathrm{i}\alpha_k\nu\sin\xi - 4\beta_k\nu^2\sin^2\frac{1}{2}\xi}{1-\frac{2}{3}\sin^2\frac{1}{2}\xi}.$$

The stability limit of the TTG-4B scheme is found to be $|\nu| \leq 0.84738$. The numerical properties of the scheme can be inferred from the polar plots of $|G|$ and $\phi_{\mathrm{num}}/\phi_{\mathrm{ex}}$ given in Figure 8.4. We observe that the (limited) reduction of the

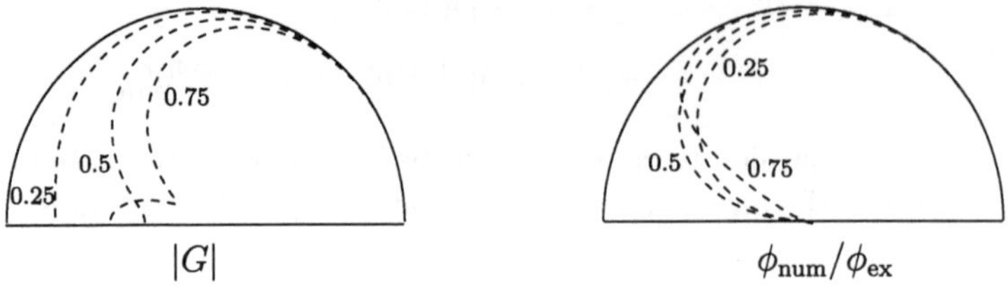

Figure 8.4: Numerical properties of the two-step fourth-order Taylor–Galerkin scheme TTG-4B for some values of the Courant number ν.

stability limit has permitted to recover very good phase response properties in the full range of the Courant number.

The scheme can be applied to multidimensional nonlinear problems with no difficulty. Considering, for an illustrative purpose, the linear advection equation

$$u_t + (\boldsymbol{a} \cdot \boldsymbol{\nabla})u = 0, \tag{8.23}$$

it can be integrated in time according to the two-step scheme

$$\widetilde{u}^n - u^n = -\alpha_k \Delta t \, (\boldsymbol{a} \cdot \boldsymbol{\nabla})u^n + \beta_k (\Delta t)^2 (\boldsymbol{a} \cdot \boldsymbol{\nabla})^2 u^n, \tag{8.24a}$$

$$u^{n+1} - u^n = -\Delta t \, (\boldsymbol{a} \cdot \boldsymbol{\nabla})\widetilde{u}^n + \gamma_k (\Delta t)^2 (\boldsymbol{a} \cdot \boldsymbol{\nabla})^2 \widetilde{u}^n. \tag{8.24b}$$

It can be noted that such a scheme is simpler to code than the previous two-step TG schemes since the equation of the second step is a replica of that of the first step, but for the change of the constants which multiply the terms of the equations.

The amplification factor of the Galerkin approximation to this multidimensional equation can be shown to be

$$G^{4\text{B}}_{\text{TTG}}(\boldsymbol{\xi}, \boldsymbol{\nu}) = 1 + \frac{-A(\boldsymbol{\xi}, \boldsymbol{\nu}) + \gamma_k K(\boldsymbol{\xi}, \boldsymbol{\nu})}{M(\boldsymbol{\xi})} \, \widetilde{G}^{\text{B}}(\boldsymbol{\xi}, \boldsymbol{\nu}),$$

where

$$\widetilde{G}^{\text{B}}(\boldsymbol{\xi}, \boldsymbol{\nu}) = 1 + \frac{-\alpha_k A(\boldsymbol{\xi}, \boldsymbol{\nu}) + \beta_k K(\boldsymbol{\xi}, \boldsymbol{\nu})}{M(\boldsymbol{\xi})}.$$

The quantities $M(\boldsymbol{\xi})$, $A(\boldsymbol{\xi}, \boldsymbol{\nu})$ and $K(\boldsymbol{\xi}, \boldsymbol{\nu})$ are given in Appendix C. The stablity study of the TTG-4B scheme shows that the domain of numerical stability is *exactly* isotropic.

The latter scheme represents a rather important achievement in the field of numerical methods for advection problems in that it is fourth-order accurate both in space and in time, is (slightly) dissipative, has a domain of numerical stability exactly isotropic and is simpler to implement than the other two-step schemes.

8.3.4 Vector advection equation

We can now examine more closely the problem of solving the momentum equation (8.8) numerically. The difficulties faced by nonlinear advection problems which have been mentioned in section 8.3.2 are here enhanced by the vector character of the advection equation, which reads

$$\boldsymbol{u}_t + (\boldsymbol{u}\cdot\boldsymbol{\nabla})\boldsymbol{u} = 0. \tag{8.25}$$

In order to apply high-order two-step TG schemes to such an equation, a second-order Taylor–Galerkin method for its advancement in time is necessary. Such a scheme has been introduced by Laval (1988 and 1990) and is described here briefly. Consider the Taylor series expansion in the time step Δt, up to second order:

$$\boldsymbol{u}^{n+1} = \boldsymbol{u}^n + \Delta t\,\boldsymbol{u}_t^n + \tfrac{1}{2}(\Delta t)^2\boldsymbol{u}_{tt}^n + \mathrm{O}[(\Delta t)^3].$$

(For notational simplicity, the intermediate velocity field $\boldsymbol{u}^{n+1/2}$ is denoted here by $\boldsymbol{u}^{n+1}$.) The first-order and second-order time derivatives in this series are then expressed from the governing equation in the form $\boldsymbol{u}_t = -(\boldsymbol{u}\cdot\boldsymbol{\nabla})\boldsymbol{u}$ and

$$\begin{aligned}\boldsymbol{u}_{tt} &= -\,[(\boldsymbol{u}\cdot\boldsymbol{\nabla})\boldsymbol{u}]_t\\ &= -\,(\boldsymbol{u}_t\cdot\boldsymbol{\nabla})\boldsymbol{u} - (\boldsymbol{u}\cdot\boldsymbol{\nabla})\boldsymbol{u}_t\\ &= ([(\boldsymbol{u}\cdot\boldsymbol{\nabla})\boldsymbol{u}]\cdot\boldsymbol{\nabla})\boldsymbol{u} + (\boldsymbol{u}\cdot\boldsymbol{\nabla})[(\boldsymbol{u}\cdot\boldsymbol{\nabla})\boldsymbol{u}],\end{aligned}$$

respectively. Substituting the two expressions into the previous Taylor series gives

$$\begin{aligned}\frac{\boldsymbol{u}^{n+1}-\boldsymbol{u}^n}{\Delta t} &= -\,(\boldsymbol{u}^n\cdot\boldsymbol{\nabla})\boldsymbol{u}^n\\ &\quad+\tfrac{1}{2}(\Delta t)^2\Big\{([(\boldsymbol{u}^n\cdot\boldsymbol{\nabla})\boldsymbol{u}^n]\cdot\boldsymbol{\nabla})\boldsymbol{u}^n + (\boldsymbol{u}^n\cdot\boldsymbol{\nabla})[(\boldsymbol{u}^n\cdot\boldsymbol{\nabla})\boldsymbol{u}^n]\Big\}.\end{aligned} \tag{8.26}$$

Equation (8.26) represents a time discretization of Lax–Wendroff type for the vector advection equation, with second-order accuracy. As for the scalar linear equation, the second-order spatial derivative terms are not a (second-order) numerical diffusion or viscosity inherent to the scheme. These terms represent instead an element of the improved finite difference approximation to the time derivative with respect to that given by the explicit Euler algorithm.

In order to employ finite elements with only C^0 continuity (linear and multilinear elements), the second-order spatial derivatives appearing in the time-discretized equation (8.26) are now transformed into first derivatives using the variational formulation. The equation is multiplied by the vector-valued weighting function $\boldsymbol{v}(\boldsymbol{x})$ belonging to a suitable space, and the terms containing second-order derivatives are integrated by parts. To simplify the calculation the weak expression of the terms within the braces is written as

$$\int \boldsymbol{v}\cdot\boldsymbol{u}_{tt}\,dV = \int \boldsymbol{v}\cdot[(\boldsymbol{a}\cdot\boldsymbol{\nabla})\boldsymbol{u} + (\boldsymbol{u}\cdot\boldsymbol{\nabla})\boldsymbol{a}]\,dV,$$

where, by definition, $\boldsymbol{a} = (\boldsymbol{u}\cdot\boldsymbol{\nabla})\boldsymbol{u}$. Now, the two terms under the integral on the right-hand side are transformd by means of the vector identity

$$\boldsymbol{\nabla}(\boldsymbol{a}\cdot\boldsymbol{b}) = (\boldsymbol{a}\cdot\boldsymbol{\nabla})\boldsymbol{b} + (\boldsymbol{b}\cdot\boldsymbol{\nabla})\boldsymbol{a} + \boldsymbol{a}\times\boldsymbol{\nabla}\times\boldsymbol{b} + \boldsymbol{b}\times\boldsymbol{\nabla}\times\boldsymbol{a},$$

which holds for arbitrarily differentiable vector fields $\boldsymbol{a}$ and $\boldsymbol{b}$, with $\boldsymbol{b} = \boldsymbol{u}$, to give

$$\int \boldsymbol{v}\cdot\boldsymbol{u}_{tt}\,dV = \int \boldsymbol{v}\cdot[-\boldsymbol{a}\times\boldsymbol{\nabla}\times\boldsymbol{u} - \boldsymbol{u}\times\boldsymbol{\nabla}\times\boldsymbol{a} + \boldsymbol{\nabla}(\boldsymbol{a}\cdot\boldsymbol{u})]\,dV.$$

Integrating by parts the last two terms on the right-hand side provides

$$\begin{aligned}\int \boldsymbol{v}\cdot\boldsymbol{u}_{tt}\,dV &= -\int \boldsymbol{v}\cdot\boldsymbol{a}\times\boldsymbol{\nabla}\times\boldsymbol{u}\,dV + \int \boldsymbol{a}\cdot\boldsymbol{\nabla}\times(\boldsymbol{u}\times\boldsymbol{v})\,dV \\ &\quad + \oint \boldsymbol{n}\cdot\boldsymbol{a}\times(\boldsymbol{u}\times\boldsymbol{v})\,dS - \int \boldsymbol{a}\cdot\boldsymbol{u}\,\boldsymbol{\nabla}\cdot\boldsymbol{v}\,dV + \oint \boldsymbol{n}\cdot\boldsymbol{v}\;\boldsymbol{a}\cdot\boldsymbol{u}\,dS.\end{aligned}$$

The term containing $\boldsymbol{\nabla}\times(\boldsymbol{u}\times\boldsymbol{v})$ is then transformed by means of the vector identity

$$\boldsymbol{\nabla}\times(\boldsymbol{a}\times\boldsymbol{b}) = \boldsymbol{a}\,\boldsymbol{\nabla}\cdot\boldsymbol{b} - \boldsymbol{b}\,\boldsymbol{\nabla}\cdot\boldsymbol{a} + (\boldsymbol{b}\cdot\boldsymbol{\nabla})\boldsymbol{a} - (\boldsymbol{a}\cdot\boldsymbol{\nabla})\boldsymbol{b}$$

with $\boldsymbol{a} = \boldsymbol{u}$ and $\boldsymbol{b} = \boldsymbol{v}$, and the double cross product $\boldsymbol{a}\times(\boldsymbol{u}\times\boldsymbol{v})$ in the surface integral is expressed through the elementary identity $\boldsymbol{a}\times(\boldsymbol{b}\times\boldsymbol{c}) = (\boldsymbol{a}\cdot\boldsymbol{c})\boldsymbol{b} - (\boldsymbol{a}\cdot\boldsymbol{b})\boldsymbol{c}$. The second-order term becomes finally

$$\begin{aligned}\int \boldsymbol{v}\cdot\boldsymbol{u}_{tt}\,dV &= -\int \boldsymbol{v}\cdot\boldsymbol{a}\times\boldsymbol{\nabla}\times\boldsymbol{u}\,dV \\ &\quad - \int \boldsymbol{a}\cdot[\boldsymbol{v}\,\boldsymbol{\nabla}\cdot\boldsymbol{u} + (\boldsymbol{u}\cdot\boldsymbol{\nabla})\boldsymbol{v} - (\boldsymbol{v}\cdot\boldsymbol{\nabla})\boldsymbol{u}]\,dV + \oint \boldsymbol{a}\cdot\boldsymbol{v}\;\boldsymbol{n}\cdot\boldsymbol{u}\,dS.\end{aligned}$$

According to this result, the weak variational formulation of the complete advection equation (8.26) yields, after some rearrangements,

$$\begin{aligned}\left\langle \boldsymbol{v}, \boldsymbol{u}^{n+1} - \boldsymbol{u}^n\right\rangle/\Delta t &= -\left\langle \boldsymbol{v}, (\boldsymbol{u}^n\cdot\boldsymbol{\nabla})\boldsymbol{u}^n\right\rangle \\ &\quad - \tfrac{1}{2}\Delta t\Big[\Big\langle(\boldsymbol{u}^n\cdot\boldsymbol{\nabla})\boldsymbol{v}, (\boldsymbol{u}^n\cdot\boldsymbol{\nabla})\boldsymbol{u}^n\Big\rangle - \oint \boldsymbol{v}\cdot(\boldsymbol{u}^n\cdot\boldsymbol{\nabla})\boldsymbol{u}^n\;\boldsymbol{n}\cdot\boldsymbol{u}^n\,dS\Big] \qquad (8.27) \\ &\quad + \tfrac{1}{2}\Delta t\Big\langle[\boldsymbol{v}\times\boldsymbol{\nabla}\times\boldsymbol{u}^n + (\boldsymbol{v}\cdot\boldsymbol{\nabla})\boldsymbol{u}^n - \boldsymbol{v}\,\boldsymbol{\nabla}\cdot\boldsymbol{u}^n], (\boldsymbol{u}^n\cdot\boldsymbol{\nabla})\boldsymbol{u}^n\Big\rangle.\end{aligned}$$

This expression represents the variational statement of the Lax–Wendroff scheme for the nonlinear vector advection equation (8.25).

The various terms ensuring the second-order temporal accuracy admit the following interpretation. The most important term is the first one, namely,

$$\Big\langle(\boldsymbol{u}^n\cdot\boldsymbol{\nabla})\boldsymbol{v},(\boldsymbol{u}^n\cdot\boldsymbol{\nabla})\boldsymbol{u}^n\Big\rangle,$$

which is present even with the linearized version of the equation, with a uniform advection velocity $\boldsymbol{u}^n=$ constant. This term is block diagonal, *i.e.*, the Cartesian components of the vector equation are left uncoupled by the term. In the spatially discrete case, each block gives rise to a symmetric matrix if $\boldsymbol{u}$ and $\boldsymbol{v}$ are chosen to belong to the same space (Galerkin method). Furthermore, each scalar block is of the form

$$\Big\langle(\boldsymbol{u}^n\cdot\boldsymbol{\nabla})v,(\boldsymbol{u}^n\cdot\boldsymbol{\nabla})u^n\Big\rangle$$

and is the weak form of the second directional derivative of u^n in the direction of $\boldsymbol{u}^n$ (times $|\boldsymbol{u}^n|^2$): therefore this term introduces a correction only when u^n has a nonzero curvature in the direction of the advection field $\boldsymbol{u}^n$. By this property, the scheme (8.26), with the third term on the right-hand side omitted, can be interpreted as a second-order characteristic method for linear advection in multidimensions, for details see Laval (1990) and also Donea and Quartapelle (1992). The surface integral

$$\oint[\boldsymbol{v}\cdot(\boldsymbol{u}^n\cdot\boldsymbol{\nabla})\boldsymbol{u}^n]\,\boldsymbol{n}\cdot\boldsymbol{u}^n\,dS$$

resulting from the integration by parts must be evaluated only on the part S_{out} of the boundary where $\boldsymbol{n}\cdot\boldsymbol{u}^n>0$ (outflow boundary). In fact, the boundary condition associated with (8.25) is $\boldsymbol{u}^{n+1}|_{S_{\text{in}}}=\boldsymbol{b}^{n+1}$, where $S_{\text{in}}\cup S_{\text{out}}=S$, so that the weighting functions $\boldsymbol{v}$ satisfy the boundary condition $\boldsymbol{v}|_{S_{\text{in}}}=0$. By its very structure, such boundary term takes into account the outflow of fluid momentum along the streamlines of the advection field and guarantees good absorbing properties at outflow boundaries in transient calculations.

Coming now to the last volume term in (8.27), namely,

$$\Big\langle[\boldsymbol{v}\times\boldsymbol{\nabla}\times\boldsymbol{u}^n+(\boldsymbol{v}\cdot\boldsymbol{\nabla})\boldsymbol{u}^n-\boldsymbol{v}\,\boldsymbol{\nabla}\cdot\boldsymbol{u}^n],(\boldsymbol{u}^n\cdot\boldsymbol{\nabla})\boldsymbol{u}^n\Big\rangle,$$

it vanishes when the velocity field is uniform, and thus it represents a correction caused by the spatial variations of the advection field. The correction depends on the nonuniformity of $\boldsymbol{u}^n$ measured by its three vector derivatives, namely, $\boldsymbol{\nabla}\times\boldsymbol{u}^n$ (vorticity), $(\boldsymbol{v}\cdot\boldsymbol{\nabla})\boldsymbol{u}^n$ (variation along the direction of $\boldsymbol{v}$) and $\boldsymbol{\nabla}\cdot\boldsymbol{u}^n$ (dilatation).

The second-order Taylor–Galerkin scheme (8.27) can be used to construct a fourth-order accurate Taylor–Galerkin method for the nonlinear advection equation (8.25) by means of a two-step procedure described in the previous section, namely,

$$\widetilde{\boldsymbol{u}}^n = \boldsymbol{u}^n + \alpha_k \Delta t\, \boldsymbol{u}_t^n + \beta_k (\Delta t)^2 \boldsymbol{u}_{tt}^n, \tag{8.28a}$$

$$\boldsymbol{u}^{n+1} = \boldsymbol{u}^n + \Delta t\, \widetilde{\boldsymbol{u}}^n + \gamma_k (\Delta t)^2 \widetilde{\boldsymbol{u}}_{tt}^n. \tag{8.28b}$$

The two-step scheme applied to the vector advection problem of interest here requires to determine a first-step solution $\widetilde{\boldsymbol{u}}^n$ from the equation

$$\begin{aligned}\Big\langle \boldsymbol{v}, \widetilde{\boldsymbol{u}}^n - \boldsymbol{u}^n \Big\rangle / \Delta t &= -\alpha_k \Big\langle \boldsymbol{v}, (\boldsymbol{u}^n \cdot \boldsymbol{\nabla}) \boldsymbol{u}^n \Big\rangle \\ &- \beta_k \Delta t \Big[\Big\langle (\boldsymbol{u}^n \cdot \boldsymbol{\nabla}) \boldsymbol{v}, (\boldsymbol{u}^n \cdot \boldsymbol{\nabla}) \boldsymbol{u}^n \Big\rangle - \oint_{S_{\text{out}}} \boldsymbol{v} \cdot (\boldsymbol{u}^n \cdot \boldsymbol{\nabla}) \boldsymbol{u}^n \; \boldsymbol{n} \cdot \boldsymbol{u}^n \, dS \Big] \\ &+ \beta_k \Delta t \Big\langle [\boldsymbol{v} \times \boldsymbol{\nabla} \times \boldsymbol{u}^n + (\boldsymbol{v} \cdot \boldsymbol{\nabla}) \boldsymbol{u}^n - \boldsymbol{v}\, \boldsymbol{\nabla} \cdot \boldsymbol{u}^n], (\boldsymbol{u}^n \cdot \boldsymbol{\nabla}) \boldsymbol{u}^n \Big\rangle, \end{aligned} \tag{8.29}$$

and then to evaluate the unknown $\boldsymbol{u}^{n+1}$ from the second-step equation

$$\begin{aligned}\Big\langle \boldsymbol{v}, \boldsymbol{u}^{n+1} - \boldsymbol{u}^n \Big\rangle / \Delta t &= - \Big\langle \boldsymbol{v}, (\widetilde{\boldsymbol{u}}^n \cdot \boldsymbol{\nabla}) \widetilde{\boldsymbol{u}}^n \Big\rangle \\ &- \gamma_k \Delta t \Big[\Big\langle (\widetilde{\boldsymbol{u}}^n \cdot \boldsymbol{\nabla}) \boldsymbol{v}, (\widetilde{\boldsymbol{u}}^n \cdot \boldsymbol{\nabla}) \widetilde{\boldsymbol{u}}^n \Big\rangle - \oint_{S_{\text{out}}} \boldsymbol{v} \cdot (\widetilde{\boldsymbol{u}}^n \cdot \boldsymbol{\nabla}) \widetilde{\boldsymbol{u}}^n \; \boldsymbol{n} \cdot \widetilde{\boldsymbol{u}}^n \, dS \Big] \\ &+ \gamma_k \Delta t \Big\langle [\boldsymbol{v} \times \boldsymbol{\nabla} \times \widetilde{\boldsymbol{u}}^n + (\boldsymbol{v} \cdot \boldsymbol{\nabla}) \widetilde{\boldsymbol{u}}^n - \boldsymbol{v}\, \boldsymbol{\nabla} \cdot \widetilde{\boldsymbol{u}}^n], (\widetilde{\boldsymbol{u}}^n \cdot \boldsymbol{\nabla}) \widetilde{\boldsymbol{u}}^n \Big\rangle. \end{aligned} \tag{8.30}$$

In both steps, the boundary integral has to be evaluated only on S_{out} and the weighting functions $\boldsymbol{v}$ are such that $\boldsymbol{v}|_{S_{\text{in}}} = 0$. This method has been implemented and used successfully to solve the incompressible Euler equations by means of the fractional-step method and using bilinear/constant finite elements to approximate velocity and pressure. The numerical results will be reported elsewhere.

8.4 Euler equations for vortical flows

Now we come back to the study of the Euler equations with a continuous dependence on time. It is well known that, if the velocity field is *irrotational*, namely, $\boldsymbol{\nabla} \times \boldsymbol{u} = 0$, and the domain V is simply connected then a scalar potential $\Phi(\boldsymbol{x}, t)$ may be introduced to represent the velocity field

$$\boldsymbol{u} = \boldsymbol{\nabla} \Phi.$$

Since $\boldsymbol{u}$ is solenoidal, the potential must be solution of the Laplace equation. The latter is supplemented by the Neumann condition obtained from the velocity boundary contition (8.3), namely,

$$-\nabla^2 \Phi = 0, \qquad \boldsymbol{n} \cdot \boldsymbol{\nabla} \Phi|_S = b_n(\boldsymbol{x}_S, t),$$

at any time $t > 0$. After this problem has been solved, the velocity field is known and the pressure field can be determined by integrating the equation

$$\boldsymbol{\nabla}\left(P + \tfrac{1}{2}|\boldsymbol{\nabla}\Phi|^2 + \frac{\partial \Phi}{\partial t}\right) = 0,$$

which is obtained directly from the momentum equation (8.1) using the well known vector identity

$$(\boldsymbol{u}\cdot\boldsymbol{\nabla})\boldsymbol{u} = (\boldsymbol{\nabla}\times\boldsymbol{u})\times\boldsymbol{u} + \boldsymbol{\nabla}(\tfrac{1}{2}u^2),$$

and the assumption $\boldsymbol{\nabla}\times\boldsymbol{u} = 0$.

If $\boldsymbol{\nabla}\times\boldsymbol{u} \neq 0$ the incompressible Euler equations cannot be reduced to the simple form above. Although the mechanism of viscous diffusion is absent from equations (8.1)–(8.2), they however contain the basic mechanisms of transport and stretching of a vorticity distribution $\boldsymbol{\zeta} = \boldsymbol{\nabla}\times\boldsymbol{u}$ whenever it is already present in the fluid. The dynamics of the vorticity is described mathematically by Kelvin circulation theorem and Helmholtz vorticity theorems (cf. *e.g.*, Acheson [1, p. 157–164]). These theorems are a direct consequence of two of the fundamental differential identities introduced in Appendix D.

Let us now assume that the initial velocity field $\boldsymbol{u}_0$ is *rotational*, so that the fluid is provided with a nonzero vorticity $\boldsymbol{\zeta}_0 = \boldsymbol{\nabla}\times\boldsymbol{u}_0 \neq 0$ at the initial time. In the subsequent instants the vorticity is redistributed within the fluid and possibly convected into or out the fluid domain. Therefore, it is convenient to express the equations with the nonlinear term written in the so-called Lamb form, which makes the quantity $\boldsymbol{\nabla}\times\boldsymbol{u}$ to appear explicitly in the mathematical formulation of the problem. Using the previous vector identity in the momentum equation, the Euler equations can be written in the form

$$\frac{\partial \boldsymbol{u}}{\partial t} + (\boldsymbol{\nabla}\times\boldsymbol{u})\times\boldsymbol{u} = -\boldsymbol{\nabla}P_{\text{tot}}, \tag{8.31}$$

$$\boldsymbol{\nabla}\cdot\boldsymbol{u} = 0, \tag{8.32}$$

where $P_{\text{tot}} = P + \frac{1}{2}u^2$. To simplify the notation of this section, the quantity $\boldsymbol{\nabla}\times\boldsymbol{u}$ is indicated by the symbol $\boldsymbol{\zeta}$, without implying the introduction of a new unknown $\boldsymbol{\zeta}$. The boundary and initial conditions for equations (8.31)–(8.32) are as in the original problem and the data are assumed to satisfy the same conditions (8.5)–(8.7).

We want now to derive an equivalent form of the equations (8.31)–(8.32) for vortical flows in which the incompressibility condition is replaced by an elliptic equation for P_{tot}.

Taking the divergence of the momentum equation (8.31), we obtain

$$-\nabla^2 P_{\text{tot}} = \boldsymbol{\nabla}\cdot(\boldsymbol{\zeta}\times\boldsymbol{u}).$$

If $P_{\rm tot}$ is calculated as solution of this Poisson equation, the incompressibility condition *is* automatically satisfied. In fact, assume $\boldsymbol{v}$ is solution of (8.31), with $P_{\rm tot}$ solution of the Poisson equation above, and satisfying the initial condition $\boldsymbol{v}|_{t=0} = \boldsymbol{u}_0$ with $\nabla\cdot\boldsymbol{u}_0 = 0$. Then, taking the divergence of the momentum equation gives

$$\frac{\partial\nabla\cdot\boldsymbol{v}}{\partial t} + \nabla\cdot(\boldsymbol{\zeta}\times\boldsymbol{v}) = -\nabla^2 P_{\rm tot}.$$

Since $P_{\rm tot}$ is solution of the Poisson equation, the previous equation implies

$$\frac{\partial\nabla\cdot\boldsymbol{v}}{\partial t} = 0.$$

Now $\nabla\cdot\boldsymbol{v}|_{t=0} = \nabla\cdot\boldsymbol{u}_0 = 0$, by the second condition on the data. Therefore $\nabla\cdot\boldsymbol{v} = 0$ for any $t > 0$.

It follows that the problem for incompressible nonviscous flows with vorticity can be written in the following equivalent form

$$\begin{aligned}-\nabla^2 P_{\rm tot} &= \nabla\cdot(\boldsymbol{\zeta}\times\boldsymbol{u}),\\ \frac{\partial\boldsymbol{u}}{\partial t} + \nabla P_{\rm tot} &= -\boldsymbol{\zeta}\times\boldsymbol{u},\\ \boldsymbol{n}\cdot\boldsymbol{u}|_S &= b_n,\\ \boldsymbol{u}|_{t=0} &= \boldsymbol{u}_0.\end{aligned}$$

It should be noted that, even assuming to neglect the nonlinear term $\boldsymbol{\zeta}\times\boldsymbol{u}$, the Poisson equation alone cannot determine the pressure field $P_{\rm tot}$ without taking into account the momentum equation, because of the lack of a boundary condition for $P_{\rm tot}$. It is however possible to obtain the appropriate boundary condition for $P_{\rm tot}$ by taking the normal component of the momentum equation on S and using the available boundary condition for the velocity, to give

$$\boldsymbol{n}\cdot\nabla P_{\rm tot}|_S = -\,\boldsymbol{n}\cdot\boldsymbol{\zeta}\times\boldsymbol{u}|_S - \frac{\partial b_n}{\partial t}.$$

Note that here it is legitimate to take the normal component of the momentum equation with a continuous time dependence since the momentum equation of the incompressible Euler problem does not contain the Lapalace operator of the viscous diffusion term. The same evaluation of the momentum equation on the boundary is not allowed in the case of the Navier–Stokes equations without increasing the regularity of the solution.

Using the Neumann boundary condition just obtained, we can write the problem for $P_{\rm tot}$ and $\boldsymbol{u}$ in a form in which each equation appears supplemented with its own boundary condition, namely:

$$\begin{aligned}-\nabla^2 P_{\rm tot} &= \nabla\cdot(\boldsymbol{\zeta}\times\boldsymbol{u}), & \boldsymbol{n}\cdot\nabla P_{\rm tot}|_S &= -\,\boldsymbol{n}\cdot\boldsymbol{\zeta}\times\boldsymbol{u}|_S - \frac{\partial b_n}{\partial t},\\ \frac{\partial\boldsymbol{u}}{\partial t} + \nabla P_{\rm tot} &= -\boldsymbol{\zeta}\times\boldsymbol{u}, & \boldsymbol{n}\cdot\boldsymbol{u}|_S &= b_n,\end{aligned} \qquad (8.33)$$

where $\boldsymbol{\zeta} = \boldsymbol{\nabla}\times\boldsymbol{u}$. The solvability condition of the Neumann problem for P_{tot} is always satisfied, as it can be easily demonstrated using the global condition $\oint b_n \, dS = 0$.

The equations for P_{tot} and $\boldsymbol{u}$ are coupled together due to the presence of the nonlinear term $\boldsymbol{\zeta}\times\boldsymbol{u}$ in the "source" terms of the two equations as well as in the boundary condition for P_{tot}. Therefore, the equations cannot be solved separately. However, in time-discretized versions of the problem, if the nonlinear term is taken into account explicitly, the two equations can be solved in sequence.

Let us now recast problem (8.33) in a weak variational form. A possible variational formulation is:

find $P_{\text{tot}} \in H^1(V)$ and $\boldsymbol{u} \in \boldsymbol{H}^1(V)$ with $\boldsymbol{n}\cdot\boldsymbol{u}|_S = b_n$, $\boldsymbol{u}|_{t=0} = \boldsymbol{u}_0$, such that

$$(\boldsymbol{\nabla}Q, \boldsymbol{\nabla}P_{\text{tot}}) = -\,(\boldsymbol{\nabla}Q, \boldsymbol{\zeta}\times\boldsymbol{u}) - \frac{d}{dt}\oint Q\, b_n \, dS,$$

for any $Q \in H^1(V)$ and

$$\left(\boldsymbol{v}, \tfrac{\partial \boldsymbol{u}}{\partial t}\right) + (\boldsymbol{v}, \boldsymbol{\nabla}P_{\text{tot}}) = -(\boldsymbol{v}, \boldsymbol{\zeta}\times\boldsymbol{u}),$$

for any $\boldsymbol{v} \in \boldsymbol{H}^1(V)$ with $\boldsymbol{n}\cdot\boldsymbol{v}|_S = 0$. The variational equation for the pressure can be compared with the analogous equation for viscous incompressible flows discussed in section 3.4.7. The viscous problem contains a boundary contribution to the weak equation which depends on the quantity $\boldsymbol{n}\cdot\boldsymbol{\nabla}\times\boldsymbol{\zeta}|_S$ and is absent in the nonviscous case.

8.5 Vorticity–velocity formulation

For vortical flows, the vorticity $\boldsymbol{\zeta}$ is a physically relevant variable of the flow field. Thus, it seems desirable to have formulations of the Euler equations for rotational incompressible flows which employ the vorticity as one of the unknowns. In this section we describe the representation of the equations for nonviscous and incompressible flows in terms of the variables vorticity and velocity. We first demonstrate the equivalence of the new formulation with the original Euler equations (8.1)–(8.4). Then we derive another equivalent formulation still based on the vorticity and velocity but such that the velocity field is solution of a Poisson equation. The boundary conditions supplementing this equation are obtained and compared with those established for the formulation of the incompressible Navier–Stokes equations expressed with the same variables considered in chapter 4.

8.5.1 Basic equations

Let us introduce the vorticity $\boldsymbol{\zeta}$ as a new unknown defined by the equation

$$\boldsymbol{\zeta} = \boldsymbol{\nabla}\times\boldsymbol{u}. \tag{8.34}$$

Taking the curl of the momentum equation (8.1), we obtain the following vorticity equation for a nonviscous fluid

$$\frac{\partial\boldsymbol{\zeta}}{\partial t} + \boldsymbol{\nabla}\times(\boldsymbol{\zeta}\times\boldsymbol{u}) = 0. \tag{8.35}$$

This equations is often written in a different form expanding the nonlinear term by means of the vector identity

$$\boldsymbol{\nabla}\times(\boldsymbol{\zeta}\times\boldsymbol{u}) = (\boldsymbol{u}\cdot\boldsymbol{\nabla})\boldsymbol{\zeta} - (\boldsymbol{\zeta}\cdot\boldsymbol{\nabla})\boldsymbol{u} + \boldsymbol{\zeta}\,\boldsymbol{\nabla}\cdot\boldsymbol{u} - \boldsymbol{u}\,\boldsymbol{\nabla}\cdot\boldsymbol{\zeta}.$$

But $\boldsymbol{\nabla}\cdot\boldsymbol{u} = 0$ and (8.34) implies $\boldsymbol{\nabla}\cdot\boldsymbol{\zeta} = 0$, hence the vorticity equation (8.35) can be written in the alternative form

$$\frac{\partial\boldsymbol{\zeta}}{\partial t} + (\boldsymbol{u}\cdot\boldsymbol{\nabla})\boldsymbol{\zeta} = (\boldsymbol{\zeta}\cdot\boldsymbol{\nabla})\boldsymbol{u}$$

which displays the terms describing the advection and the stretching of the vorticity $\boldsymbol{\zeta}$ by the velocity field $\boldsymbol{u}$.

From the initial condition for the velocity it is immediate to derive the initial condition for the vorticity field:

$$\boldsymbol{\zeta}|_{t=0} = \boldsymbol{\nabla}\times\boldsymbol{u}_0. \tag{8.36}$$

Combining equations (8.34)–(8.35) and the initial condition (8.36) with the incompressibility condition (8.2) and the velocity boundary condition (8.3), we obtain an equivalent formulation of the Euler equations for incompressible flows, as stated by the following theorem. We remind again that additional boundary conditions for the velocity or the vorticity, possibly required to cope with the hyperbolic character of the dynamical equation, are not considered here.

Theorem 8.1. *The primitive variable incompressible Euler problem (8.1)–(8.4) is equivalent to the problem for the variables $\boldsymbol{\zeta}$ and $\boldsymbol{u}$, defined by the set of equations and conditions*

$$\begin{aligned} &\frac{\partial\boldsymbol{\zeta}}{\partial t} + \boldsymbol{\nabla}\times(\boldsymbol{\zeta}\times\boldsymbol{u}) = 0, && \boldsymbol{\zeta}|_{t=0} = \boldsymbol{\nabla}\times\boldsymbol{u}_0,\\ &\boldsymbol{\nabla}\times\boldsymbol{u} = \boldsymbol{\zeta}, \qquad \boldsymbol{\nabla}\cdot\boldsymbol{u} = 0, && \boldsymbol{n}\cdot\boldsymbol{u}|_S = b_n, \end{aligned} \tag{8.37}$$

assuming the following conditions on the data

$$\oint b_n\,dS = 0, \qquad \boldsymbol{\nabla}\cdot\boldsymbol{u}_0 = 0, \qquad b_n|_{t=0} = \boldsymbol{n}\cdot\boldsymbol{u}_0|_S. \tag{8.38}$$

Proof. The implication has just been shown. Conversely, let $(\boldsymbol{\zeta}, \boldsymbol{v})$ be a solution to the set (8.37) with the data b_n and $\boldsymbol{u}_0$ satisfying the conditions (8.38).

First, let us note that $\boldsymbol{v}$ is uniquely defined for all $t > 0$ because the problem

$$\nabla\times\boldsymbol{v} = \boldsymbol{\zeta}, \qquad \nabla\cdot\boldsymbol{v} = 0, \qquad \boldsymbol{n}\cdot\boldsymbol{v}|_S = b_n,$$

admits a unique solution provided that $\nabla\cdot\boldsymbol{\zeta} = 0$ and $\oint b_n\,dS = 0$, for all $t > 0$. Both conditions are satisfied here by the assumptions: the latter coincides with the first condition in (8.38), whereas the former, *i.e.*, $\nabla\cdot\boldsymbol{\zeta} = 0$, is satisfied for all $t > 0$ by virtue of the first equation satisfied by $\boldsymbol{v}$, namely, $\nabla\times\boldsymbol{v} = \boldsymbol{\zeta}$. Then, the vorticity equation means

$$\frac{\partial\nabla\times\boldsymbol{v}}{\partial t} + \nabla\times[(\nabla\times\boldsymbol{v})\times\boldsymbol{v}] = 0,$$

that is

$$\nabla\times\left[\frac{\partial\boldsymbol{v}}{\partial t} + (\nabla\times\boldsymbol{v})\times\boldsymbol{v}\right] = 0,$$

which means

$$\frac{\partial\boldsymbol{v}}{\partial t} + (\nabla\times\boldsymbol{v})\times\boldsymbol{v} = \nabla Q,$$

where Q is some function. To identify the field $\boldsymbol{v}$ with the velocity field $\boldsymbol{u}$ solution of the primitive variable incompressible Euler equations (and Q with $-P - \frac{1}{2}u^2$), it is necessary to show that the initial value of $\boldsymbol{v}$ implied by the set (8.37) coincides with $\boldsymbol{u}_0$. The well-posedness of the problem $\nabla\times\boldsymbol{v} = \boldsymbol{\zeta}$, $\nabla\cdot\boldsymbol{v} = 0$, $\boldsymbol{n}\cdot\boldsymbol{v}|_S = b_n$ under the conditions $\nabla\cdot\boldsymbol{\zeta} = 0$ and $\oint b_n\,dS = 0$, together with the initial condition $\boldsymbol{\zeta}|_{t=0} = \nabla\times\boldsymbol{u}_0$ and the assumed continuity of b_n as a function of time as $t \to 0^+$, implies that $\boldsymbol{v}|_{t=0} = \boldsymbol{v}_0$, where $\boldsymbol{v}_0$ is the solution of the problem

$$\nabla\times\boldsymbol{v}_0 = \boldsymbol{\zeta}_0 = \nabla\times\boldsymbol{u}_0, \qquad \nabla\cdot\boldsymbol{v}_0 = 0, \qquad \boldsymbol{n}\cdot\boldsymbol{v}_0|_S = b_n|_{t=0}.$$

The first equation means that $\nabla\times(\boldsymbol{v}_0 - \boldsymbol{u}_0) = 0$, namely, $\boldsymbol{v}_0 - \boldsymbol{u}_0 = \nabla\alpha$, for some function α, which is harmonic since $\nabla\cdot\boldsymbol{v}_0 = 0$ and also $\nabla\cdot\boldsymbol{u}_0 = 0$ by the second of the conditions (8.38). Taking the normal component of the equation $\nabla\times(\boldsymbol{v}_0 - \boldsymbol{u}_0) = 0$ on S gives:

$$\boldsymbol{n}\cdot(\boldsymbol{v}_0 - \boldsymbol{u}_0)|_S = \boldsymbol{n}\cdot\boldsymbol{v}_0|_S - \boldsymbol{n}\cdot\boldsymbol{u}_0|_S = b_n|_{t=0} - \boldsymbol{n}\cdot\boldsymbol{u}_0|_S = \boldsymbol{n}\cdot\nabla\alpha|_S.$$

Then, by the third of the conditions (8.17), $\boldsymbol{n}\cdot\nabla\alpha|_S = 0$, so that $\alpha =$ constant. It follows that $\boldsymbol{v}_0 = \boldsymbol{u}_0$ and hence $\boldsymbol{v} = \boldsymbol{u}$, which completes the proof. □

8.5.2 An equivalent formulation

The equation relating $\boldsymbol{u}$ to $\boldsymbol{\zeta}$ in the formulation above is of first order and its solution cannot be attempted by means of standard numerical techniques. It is however possible to make use of the second first-order equation for $\boldsymbol{u}$, which imposes that the velocity is solenoidal, to derive a second-order elliptic equation which can be solved by standard techniques. By means of such a transformation, one obtains a vorticity–velocity formulation which enforces the condition of incompressibility through the Poisson equation for the velocity, as stated by the following theorem.

Theorem 8.2. *The set of equations (8.37) is equivalent to the following system of equations and boundary conditions:*

$$\begin{aligned} &\frac{\partial \boldsymbol{\zeta}}{\partial t} + \boldsymbol{\nabla}\times(\boldsymbol{\zeta}\times\boldsymbol{u}) = 0, && \boldsymbol{\zeta}|_{t=0} = \boldsymbol{\nabla}\times\boldsymbol{u}_0, \\ &-\nabla^2\boldsymbol{u} = \boldsymbol{\nabla}\times\boldsymbol{\zeta}, \qquad \boldsymbol{n}\cdot\boldsymbol{u}|_S = b_n, && \boldsymbol{n}\times\boldsymbol{\nabla}\times\boldsymbol{u}|_S = \boldsymbol{n}\times\boldsymbol{\zeta}|_S. \end{aligned} \tag{8.39}$$

Proof. The implication is a consequence of the following two facts:

i) the elliptic equation $-\nabla^2\boldsymbol{u} = \boldsymbol{\zeta}$ follows from taking the curl of the equation $\boldsymbol{\zeta} = \boldsymbol{\nabla}\times\boldsymbol{u}$, using the vector identity $\boldsymbol{\nabla}\times\boldsymbol{\nabla}\times\boldsymbol{\alpha} = -\nabla^2\boldsymbol{\alpha} + \boldsymbol{\nabla}(\boldsymbol{\nabla}\cdot\boldsymbol{\alpha})$ and the incompressibility condition;

ii) the derivative boundary condition $\boldsymbol{n}\times\boldsymbol{\nabla}\times\boldsymbol{u}|_S = \boldsymbol{n}\times\boldsymbol{\zeta}|_S$ is the trace of the tangential components of the equation $\boldsymbol{\nabla}\times\boldsymbol{u} = \boldsymbol{\zeta}$ on the boundary.

Conversely, let $(\boldsymbol{\omega}, \boldsymbol{v})$ be a solution of the set of equations and conditions (8.39). Then, $\boldsymbol{v}$ is solenoidal. In fact, using the identity $-\nabla^2\boldsymbol{\alpha} = \boldsymbol{\nabla}\times\boldsymbol{\nabla}\times\boldsymbol{\alpha} - \boldsymbol{\nabla}(\boldsymbol{\nabla}\cdot\boldsymbol{\alpha})$, the equation $-\nabla^2\boldsymbol{v} = \boldsymbol{\nabla}\times\boldsymbol{\omega}$ can be written as $\boldsymbol{\nabla}\times\boldsymbol{\nabla}\times\boldsymbol{v} - \boldsymbol{\nabla}(\boldsymbol{\nabla}\cdot\boldsymbol{v}) = \boldsymbol{\nabla}\times\boldsymbol{\omega}$ or, equivalently,

$$\boldsymbol{\nabla}\times(\boldsymbol{\nabla}\times\boldsymbol{v} - \boldsymbol{\omega}) = \boldsymbol{\nabla}(\boldsymbol{\nabla}\cdot\boldsymbol{v}).$$

We can now take the component normal to S of this equation since the tangential components of $\boldsymbol{\nabla}\times\boldsymbol{v}-\boldsymbol{\omega}$ are prescribed (equal to zero) by the boundary condition $\boldsymbol{n}\times\boldsymbol{\nabla}\times\boldsymbol{v}|_S = \boldsymbol{n}\times\boldsymbol{\omega}|_S$. Thus the normal component of the equation gives

$$\boldsymbol{n}\cdot\boldsymbol{\nabla}(\boldsymbol{\nabla}\cdot\boldsymbol{v})|_S = 0.$$

On the other hand, the divergence of the equation $-\nabla^2\boldsymbol{v} = \boldsymbol{\nabla}\times\boldsymbol{\omega}$ gives

$$-\nabla^2(\boldsymbol{\nabla}\cdot\boldsymbol{v}) = \boldsymbol{\nabla}\cdot\boldsymbol{\nabla}\times\boldsymbol{\omega} = 0,$$

which means that $\boldsymbol{\nabla}\cdot\boldsymbol{v}$ is harmonic in V. It follows that $\boldsymbol{\nabla}\cdot\boldsymbol{v} = C(t)$, where $C(t)$ is an arbitrary function of time. By the divergence theorem and the boundary contion $\boldsymbol{n}\cdot\boldsymbol{v}|_S = b_n$ we have

$$\int C\,dV = \int \boldsymbol{\nabla}\cdot\boldsymbol{v}\,dV = \oint \boldsymbol{n}\cdot\boldsymbol{v}\,dS = \oint b_n\,dS.$$

Then, the global condition $\oint b_n\,dS = 0$ implies $C = 0$ and therefore $\boldsymbol{\nabla}\cdot\boldsymbol{v} = 0$.

Since $\boldsymbol{\nabla}\boldsymbol{\cdot}\boldsymbol{v} = 0$, the previous vector identity allows to write the equation $-\nabla^2\boldsymbol{v} = \boldsymbol{\nabla}\times\boldsymbol{\omega}$ in the form $\boldsymbol{\nabla}\times\boldsymbol{\nabla}\times\boldsymbol{v} = \boldsymbol{\nabla}\times\boldsymbol{\omega}$ which means

$$\boldsymbol{\nabla}\times(\boldsymbol{\nabla}\times\boldsymbol{v} - \boldsymbol{\omega}) = 0,$$

so that $\boldsymbol{\nabla}\times\boldsymbol{v} - \boldsymbol{\omega} = \boldsymbol{\nabla}\alpha$, for some scalar function α. The boundary condition $\boldsymbol{n}\times\boldsymbol{\nabla}\times\boldsymbol{v}|_S = \boldsymbol{n}\times\boldsymbol{\omega}|_S$ implies $\boldsymbol{n}\times\boldsymbol{\nabla}\alpha|_S = 0$ and thus $\alpha|_S = 0$. Moreover, α must be harmonic since $\boldsymbol{\nabla}\boldsymbol{\cdot}\boldsymbol{\omega} = 0$ by virtue of the form of the dynamical equation of $\boldsymbol{\zeta}$ and of the initial condition $\boldsymbol{\omega}|_{t=0} = \boldsymbol{\nabla}\times\boldsymbol{u}_0$. It follows that $\alpha =$ constant in all V so that $\boldsymbol{\nabla}\times\boldsymbol{v} = \boldsymbol{\omega}$. Thus $\boldsymbol{\omega}$ and $\boldsymbol{v}$ satisfy the same equations and conditions satisfied by $\boldsymbol{\zeta}$ and $\boldsymbol{u}$, so that $\boldsymbol{\omega} = \boldsymbol{\zeta}$ and $\boldsymbol{v} = \boldsymbol{u}$. □

We note that the boundary conditions for the velocity in (8.39) are different from those prescribed for viscous flows in section 4.2.1 in two respects: firstly, in the nonviscous problem the tangential components of velocity are not imposed on the boundary; secondly, here the boundary condition which assures the fulfillment of the incompressibility condition is $\boldsymbol{n}\times\boldsymbol{\nabla}\times\boldsymbol{u}|_S = \boldsymbol{n}\times\boldsymbol{\zeta}|_S$ instead of the scalar condition $\boldsymbol{\nabla}\boldsymbol{\cdot}\boldsymbol{u}|_S = 0$.

The velocity boundary condition

$$\boldsymbol{n}\times\boldsymbol{\nabla}\times\boldsymbol{u}|_S = \boldsymbol{n}\times\boldsymbol{\zeta}|_S$$

is important because it allows to replace the elliptic equation $\boldsymbol{\nabla}\times\boldsymbol{\nabla}\times\boldsymbol{u} = \boldsymbol{\nabla}\times\boldsymbol{\zeta}$ for nonviscous flows by the Poisson equation $-\nabla^2\boldsymbol{u} = \boldsymbol{\nabla}\times\boldsymbol{\zeta}$, guaranteeing at the same time that incompressibility is automatically satisfied. Considering that such Poisson equation is one of the governing equations of the vortex methods for the solution of the Euler equations for vortical incompressible flows (see, *e.g.*, [9, p. 61]), it is worthwhile to know the complete set of velocity boundary conditions in this $\boldsymbol{\zeta}$-$\boldsymbol{u}$ formulation. It is also interesting to see how this velocity boundary condition fits into a variational formulation of the problem (8.39). In this case, the derivative condition disappears from the weak equations when the two terms of the Poisson equation for velocity are integrated by parts. By standard calculation it is immediate to show that problem (8.39) can be expressed in the following variational form:

find $\boldsymbol{\zeta} \in \boldsymbol{Z}(V)$ with $\boldsymbol{\zeta}|_{t=0} = \boldsymbol{\nabla}\times\boldsymbol{u}_0$ and $\boldsymbol{u} \in \boldsymbol{H}^1(V)$ with $\boldsymbol{n}\boldsymbol{\cdot}\boldsymbol{u}|_S = b_n$ such that

$$\begin{aligned} &\left(\boldsymbol{\xi}, \tfrac{\partial}{\partial t}\boldsymbol{\zeta}\right) + (\boldsymbol{\nabla}\times\boldsymbol{\xi}, \boldsymbol{\zeta}\times\boldsymbol{u}) = \oint \boldsymbol{n}\times\boldsymbol{\xi}\boldsymbol{\cdot}\boldsymbol{\zeta}\times\boldsymbol{u}\, dS, \quad \forall\boldsymbol{\xi} \in \boldsymbol{Z}(V), \\ &(\boldsymbol{\nabla}\times\boldsymbol{v}, \boldsymbol{\nabla}\times\boldsymbol{u}) + (\boldsymbol{\nabla}\boldsymbol{\cdot}\boldsymbol{v}, \boldsymbol{\nabla}\boldsymbol{\cdot}\boldsymbol{u}) = (\boldsymbol{\nabla}\times\boldsymbol{v}, \boldsymbol{\zeta}), \qquad \forall\boldsymbol{v} \in \boldsymbol{H}^1(V),\ \boldsymbol{n}\boldsymbol{\cdot}\boldsymbol{v}|_S = 0. \end{aligned} \tag{8.40}$$

A definite characterization of the space $\boldsymbol{Z}(V)$, which the vorticity field belongs to, requires to specify the complete set of boundary conditions for $\boldsymbol{\zeta}$—a target beyond the purposes of the present study.

8.6 Nonprimitive variable formulations

The vorticity-velocity formulation is not the only representation of the incompressible Euler equation in which the vorticity can play the rôle of a dependent variable. It is also possible to formulate the equations replacing the velocity unknown by means of velocity potentials, very similarly to the incompressible Navier–Stokes equations. In this section we describe the nonprimitive variable formulations of the incompressible Euler equations giving the complete set of boundary conditions which supplement the velocity potentials. These conditions are not easily available in textbooks on fluid dynamics, see, *e.g.*, [9, p. 30]. However, they can be derived without encountering major difficulties once the corresponding viscous problem has been investigated, as in chapter 3 of the present study.

For three-dimensional flows we will consider the two formulations which have been denoted by $\boldsymbol{\zeta}$-$\boldsymbol{\psi}$ and q_S-$\boldsymbol{\zeta}$-$\boldsymbol{\phi}$ for the incompressible Navier–Stokes equations: both are based on a representation of the velocity field in terms of a stream vector. For two-dimensional problems we will recall the governing equations of the classical formulation based on the scalar unknowns vorticity and stream function. The presentation will be rather sketchy and is aimed only at providing the basic elements that allow to compare the equations and conditions governing the incompressible flow of a viscous fluid with those of its nonviscous counterpart.

8.6.1 ζ-ϕ formulation

Let us start from the incompressible Euler equations (8.37) which govern the vorticity and velocity variables and contain the two first-order velocity equations $\nabla\times\boldsymbol{u} = \boldsymbol{\zeta}$ and $\nabla\cdot\boldsymbol{u} = 0$. Introducing a stream vector $\boldsymbol{\phi}$ to represent the solenoidal velocity field by means of the relation

$$\boldsymbol{u} = \nabla\times\boldsymbol{\phi},$$

it is immediate to deduce that $\boldsymbol{\phi}$ will satisfy the second-order equation

$$\nabla\times\nabla\times\boldsymbol{\phi} = \boldsymbol{\zeta}.$$

A single boundary condition supplementing this stream vector is obtained from the velocity condition in (8.37), in the form

$$\boldsymbol{n}\cdot\nabla\times\boldsymbol{\phi}|_S = b_n.$$

Of course $\boldsymbol{\phi}$ is indeterminate unless supplementary conditions are introduced to fix the the arbitrariness allowed by the well known gauge transformation

$$\boldsymbol{\phi} \longrightarrow \boldsymbol{\phi} + \nabla\Theta,$$

where Θ is an arbitrary function. As for the viscous equations discussed in section 3.6, one can impose on $\boldsymbol{\phi}$ the boundary condition

$$\boldsymbol{n} \cdot \boldsymbol{\phi}|_S = a,$$

where a is a function defined on S, $a = a(\boldsymbol{x}_S)$, chosen arbitrarily but for the single global condition $\oint a\, dS = 0$. This kind of boundary condition is suitable for dealing with multiply connected domains, although we do not intend to discuss the real situation of domains of this kind. In consequence of the aforementioned gauge transformation, after imposing this boundary condition, the field $\boldsymbol{\phi}$ remains still indeterminate up to the gradient of a function with zero normal derivative on the boundary. To eliminate this residual indeterminacy, the usual Euclid gauge condition

$$\boldsymbol{\nabla} \cdot \boldsymbol{\phi} = 0$$

can be imposed on $\boldsymbol{\phi}$, so that the equation for $\boldsymbol{\phi}$ simplifies to

$$-\nabla^2 \boldsymbol{\phi} = \boldsymbol{\zeta}.$$

On the other hand, the gauge condition $\boldsymbol{\nabla} \cdot \boldsymbol{\phi} = 0$ can be satisfied by imposing only the boundary condition

$$\boldsymbol{\nabla} \cdot \boldsymbol{\phi}|_S = 0.$$

In fact, taking the divergence of the equation $-\nabla^2 \boldsymbol{\phi} = \boldsymbol{\zeta}$ gives $-\nabla^2 \boldsymbol{\nabla} \cdot \boldsymbol{\phi} = 0$ since $\boldsymbol{\nabla} \cdot \boldsymbol{\zeta} = 0$ by virtue of the equation and initial condition for the vorticity. Thus $\boldsymbol{\nabla} \cdot \boldsymbol{\phi}$ is harmonic and identically zero thanks to the boundary condition $\boldsymbol{\nabla} \cdot \boldsymbol{\phi}|_S = 0$.

In conclusion, an equivalent formulation of the incompressible Euler equations for vortical flows is provided by the following set of equations and conditions

$$\begin{aligned} &\frac{\partial \boldsymbol{\zeta}}{\partial t} + \boldsymbol{\nabla} \times (\boldsymbol{\zeta} \times \boldsymbol{\nabla} \times \boldsymbol{\phi}) = 0, \qquad \boldsymbol{\zeta}|_{t=0} = \boldsymbol{\nabla} \times \boldsymbol{u}_0, \\ &-\nabla^2 \boldsymbol{\phi} = \boldsymbol{\zeta}, \qquad \boldsymbol{n} \cdot \boldsymbol{\phi}|_S = a, \qquad \boldsymbol{\nabla} \cdot \boldsymbol{\phi}|_S = 0, \qquad \boldsymbol{n} \cdot \boldsymbol{\nabla} \times \boldsymbol{\phi}|_S = b_n. \end{aligned} \tag{8.41}$$

In this formulation the data comprise the arbitrary function a defined on S together with the boundary and initial values b_n and $\boldsymbol{u}_0$ specified in the original Euler problem. These data must satisfy the two global conditions

$$\oint a\, dS = 0, \qquad \oint b_n\, dS = 0,$$

and the conditions

$$\boldsymbol{\nabla} \cdot \boldsymbol{u}_0 = 0, \qquad b_n|_{t=0} = \boldsymbol{n} \cdot \boldsymbol{u}_0|_S.$$

It is interesting to compare the boundary conditions for $\boldsymbol{\phi}$ of the present nonviscous problem with the conditions of the viscous case, see problem (3.85) in section 3.6.1. First, only three (scalar) conditions are present here whereas five boundary conditions have been found for the stream vector in the Navier–Stokes equations. The two additional boundary conditions of the viscous case are evidently those imposing the tangential components of the velocity on the boundary, namely, $\boldsymbol{n}\times\nabla\times\boldsymbol{\phi}|_S = \boldsymbol{n}\times\boldsymbol{b}$, where $\boldsymbol{b}$ denotes the velocity prescribed on the boundary (of course $\boldsymbol{n}\cdot\boldsymbol{b} = b_n$). The remaining three conditions are instead identical in the viscous and nonviscous situations. It is however to be noted that for nonviscous flows the boundary condition $\nabla\cdot\boldsymbol{\phi}|_S = 0$ has to be imposed in the actual solution of the equation for the stream vector, while for viscous flows the explicit imposition of this condition is avoided when the Navier–Stokes problem is formulated in the equivalent form (3.87). This was possible by virtue of theorem 3.4 of section 3.6.2, which is not applicable in the nonviscous case because the tangential components of velocity on the boundary cannot be prescribed.

It is also to be noted that the first two boundary conditions for $\boldsymbol{\phi}$ are faced with the same difficulty which has been encountered with the solution of the vorticity equation when the latter was supplemented by the irreducible integral conditions (cf. the discussion in section 3.5.2). Therefore, this difficulty "moves" from the vorticity variable $\boldsymbol{\zeta}$ to the stream vector $\boldsymbol{\phi}$ as $\nu \to 0$.

8.6.2 q_S-$\boldsymbol{\zeta}$-$\boldsymbol{\psi}$ formulation

These difficulties are absent in the other nonprimitive variable formulation of the incompressible equations which uses a stream vector $\boldsymbol{\psi}$ subject to a different choice of boundary conditions, as it has been done with the Navier–Stokes equations in section 3.4.

Let the solenoidal velocity field $\boldsymbol{u}$ be represented by the stream vector $\boldsymbol{\psi}$, $\boldsymbol{u} = \nabla\times\boldsymbol{\psi}$, so that the velocity boundary condition for the nonviscous problem can be imposed by means of the condition

$$\boldsymbol{n}\cdot\nabla\times\boldsymbol{\psi}|_S = b_n.$$

Similarly to the viscous problem, this boundary condition can be satisfied by imposing the tangential components of $\boldsymbol{\psi}$. The boundary values of the latter are obtained from the datum b_n by means of a surface scalar potential q_S which is obtained by solving the elliptic equation

$$-\nabla_S^2 q_S = b_n$$

over the surface S, which is assumed to be a closed surface. It has been shown in section 3.4.1 that the solution q_S determines a vector field $\boldsymbol{a}$ tangential to S by means of the relation

$$\boldsymbol{a} = -\boldsymbol{n}\times\nabla_S q_S$$

such that $\boldsymbol{a}$ gives the tangential components of $\boldsymbol{\psi}$ on S. Thus, once the surface elliptic problem for q_S has been solved, the stream vector $\boldsymbol{\psi}$ can be supplemented by the boundary condition

$$\boldsymbol{n}\times\boldsymbol{\psi}|_S = \boldsymbol{n}\times\boldsymbol{a}.$$

Under this boundary condition, the vector field $\boldsymbol{\psi}$ remains still indeterminate by the gradient of an arbitrary function which vanishes on the boundary. To eliminate this residual arbitrariness, the stream vector is required to satisfy the usual Euclid gauge condition $\nabla\cdot\boldsymbol{\psi} = 0$, and the latter can be satisfied simply by imposing the boundary condition

$$\nabla\cdot\boldsymbol{\psi}|_S = 0,$$

for the same reasons we have described in connection with the $\boldsymbol{\zeta}$-$\boldsymbol{\phi}$ formulation.

Therefore, we arrive at the following set of equations and conditions which constitute an equivalent formulation of the problem governing the motion of a nonviscous incompressible fluid with nonzero vorticity:

$$\begin{aligned}
&-\nabla_S^2 q_S = b_n, && [\Longrightarrow \boldsymbol{a} = -\boldsymbol{n}\times\nabla_S q_S\ ] \\
&\frac{\partial\boldsymbol{\zeta}}{\partial t} + \nabla\times(\boldsymbol{\zeta}\times\nabla\times\boldsymbol{\psi}) = 0, && \boldsymbol{\zeta}|_{t=0} = \nabla\times\boldsymbol{u}_0, \\
&-\nabla^2\boldsymbol{\psi} = \boldsymbol{\zeta}, && \boldsymbol{n}\times\boldsymbol{\psi}|_S = \boldsymbol{n}\times\boldsymbol{a}, \qquad \nabla\cdot\boldsymbol{\psi}|_S = 0.
\end{aligned} \tag{8.42}$$

In this case, the boundary conditions supplementing the stream vector $\boldsymbol{\psi}$ in nonviscous flows are coincident with the conditions which are to be imposed in the correspondent viscous problem after the vorticity field has been determined subject to the integral conditions (cf. the last line of the uncoupled formulation (3.38) of the q_S-$\boldsymbol{\zeta}$-$\boldsymbol{\psi}$ equations in section 3.4.3).

To complete our analysis of the nonprimitive variable formulations of the incompressible Euler equations for three-dimensional flows, we can note that the representation in terms of the scalar and vector potentials ϕ and $\boldsymbol{A}$ can be also adopted and the respective equations and conditions can be established similarly to the previously discussed representation relying on the stream vector $\boldsymbol{\psi}$.

8.6.3 Vorticity–stream function equations

It is fair to conclude this section on nonprimitive formulations of the incompressible Euler equations and also the entire chapter on the equations governing nonviscous flows by considering the equations for two-dimensional problems expressed in terms of the vorticity and stream function, which were the starting point of our analysis of the incompressible Navier–Stokes equations. The ζ-ψ formulation can be very easily derived as a particular case of the q_S-$\boldsymbol{\zeta}$-$\boldsymbol{\psi}$ formulation in which the surface elliptic problem for q_S (to determine the tangential boundary value of ψ) reduces itself to a simple quadrature along the boundary of the domain V of the plane. Without going into the details, we note that the global condition $\oint b_n \, dS = 0$ implies that the integration of the datum $b_n(s,t)$ along the boundary defines a single-valued function of s, $a(s,t)$, up to an arbitrary additive function of time, namely,

$$a(s,t) = \int_{s_1}^{s} b_n(s',t)\, ds' + A(t),$$

where s_1 is the coordinate of any fixed point of S and s' is the integration variable. Since the relationship between the stream function ψ and the velocity $\boldsymbol{u}$ is $\boldsymbol{\nabla}\psi \times \boldsymbol{k} = \boldsymbol{u}$, the velocity boundary condition for the normal component of the nonviscous problem becomes a Dirichlet condition for ψ, namely, $\psi|_S = a(s,t)$. Therefore, the nonprimitive variable formulation of the Euler equations for incompressible flows in two dimensions reads

$$\begin{aligned} &\frac{\partial \zeta}{\partial t} + J(\zeta,\psi) = 0, \qquad && \zeta|_{t=0} = \boldsymbol{\nabla}\times \boldsymbol{u}_0 \cdot \boldsymbol{k}, \\ &-\nabla^2 \psi = \zeta, && \psi|_S = a(s,t), \end{aligned} \tag{8.43}$$

where $J(\zeta,\psi)$ denotes the Jacobian determinant. Of course the data of this problems are subject to the two remaining conditions of the incompressible problem, that is, the incompressibility of the initial velocity field, $\boldsymbol{\nabla}\cdot \boldsymbol{u}_0 = 0$, and the compatibility of the boundary and initial data, $b_n|_{t=0} = \boldsymbol{n}\cdot \boldsymbol{u}_0|_S$, the latter assuming here the form

$$\frac{\partial a(s,0)}{\partial s} = \boldsymbol{n}(s)\cdot \boldsymbol{u}_0(\boldsymbol{x}_s).$$

Appendix A

Vector differential operators

In this Appendix we introduce orthogonal curvilinear coordinates and derive the general expressions of the vector differential operators in this kind of coordinates. Moreover, we give the expressions of the differential operators for the particular cases of cylindrical and spherical coordinates. The presentation follows closely the treatment given in *Calculus of Several Variables* (Adams [2, p. 336]) from which the figures have been borrowed, by gentle permission of the author.

A.1 Orthogonal curvilinear coordinates

We assume that (u, v, w) are a set of *orthogonal* curvilinear coordinates in xyz-space defined via the transformation

$$x = x(u, v, w), \qquad y = y(u, v, w), \qquad z = z(u, v, w).$$

We also assume that the coordinate surfaces are smooth at any nonsingular point and that the local basis vectors $\widehat{\boldsymbol{u}}$, $\widehat{\boldsymbol{v}}$ and $\widehat{\boldsymbol{w}}$ at any such point form a right-handed triad.

The position vector $\boldsymbol{r}$ of a point P in xyz-space can be expressed in terms of the curvilinear coordinates:

$$\boldsymbol{r} = x(u, v, w)\,\boldsymbol{i} + y(u, v, w)\,\boldsymbol{j} + z(u, v, w)\,\boldsymbol{k}.$$

If we hold $v = v_0$ and $w = w_0$ fixed and let u vary, then $\boldsymbol{r} = \boldsymbol{r}(u, v_0, w_0)$ defines a u-curve in xyz-space. At any point P on this curve, the vector

$$\frac{\partial \boldsymbol{r}}{\partial u} = \frac{\partial x}{\partial u}\boldsymbol{i} + \frac{\partial y}{\partial u}\boldsymbol{j} + \frac{\partial z}{\partial u}\boldsymbol{k}$$

is tangent to the u-curve at P. In general, the three vectors

$$\frac{\partial \boldsymbol{r}}{\partial u}, \qquad \frac{\partial \boldsymbol{r}}{\partial v}, \qquad \frac{\partial \boldsymbol{r}}{\partial w}$$

are tangent, respectively, to the the u-curve, the v-curve and the w-curve through P. They are also normal, respectively, to the u-surface, the v-surface and the w-surface through P, so they are mutually perpendicular. (See Figure A.1) The lengths of these tangent vectors are called the *scale factors* of the coordinate system and are therefore defined by

$$h_u = \left|\frac{\partial \boldsymbol{r}}{\partial u}\right|, \qquad h_v = \left|\frac{\partial \boldsymbol{r}}{\partial v}\right|, \qquad h_w = \left|\frac{\partial \boldsymbol{r}}{\partial w}\right|.$$

The scale factors are nonzero at a nonsingular point P of the coordinate system, so a local basis for the coordinate system at P can be obtained by dividing the tangent vectors to the coordinate curves by their lengths. Denoting the local basis vectors by $\widehat{\boldsymbol{u}}$, $\widehat{\boldsymbol{v}}$ and $\widehat{\boldsymbol{w}}$, we have

$$\frac{\partial \boldsymbol{r}}{\partial u} = h_u \widehat{\boldsymbol{u}}, \qquad \frac{\partial \boldsymbol{r}}{\partial v} = h_v \widehat{\boldsymbol{v}}, \qquad \frac{\partial \boldsymbol{r}}{\partial w} = h_w \widehat{\boldsymbol{w}}.$$

The basis vectors $\widehat{\boldsymbol{u}}$, $\widehat{\boldsymbol{v}}$ and $\widehat{\boldsymbol{w}}$ will form a right-handed triad provided we have chosen a suitable order for the coordinates u, v and w.

The volume element in an orthogonal curvilinear coordinate system is the volume of an infinitesimal *coordinate box* bounded by pairs of u-, v- and w-surfaces corresponding to values u and $u + du$, v and $v + dv$ and w and $w + dw$, respectively. See Figure A.1. Since these coordinate surfaces are assumed smooth, and since they intersect at right angles, the coordinate box is rectangular and is spanned by the vectors

$$\frac{\partial \boldsymbol{r}}{\partial u}\, du = h_u\, du\, \widehat{\boldsymbol{u}}, \qquad \frac{\partial \boldsymbol{r}}{\partial v}\, dv = h_v\, dv\, \widehat{\boldsymbol{v}}, \qquad \frac{\partial \boldsymbol{r}}{\partial w}\, dw = h_w\, dw\, \widehat{\boldsymbol{w}}.$$

Therefore, the volume element is given by

$$dV = h_u h_v h_w\, du\, dv\, dw.$$

Furthermore, the surface area elements on the u-, v- and w-surfaces are the areas of the appropriate faces of the coordinate box:

$$dS_u = h_v h_w\, dv\, dw, \qquad dS_v = h_u h_w\, du\, dw, \qquad dS_w = h_u h_v\, du\, dv.$$

The arc length elements along the u-, v- and w-coordinate curves are the edges of the coordinate box:

$$ds_u = h_u\, du, \qquad ds_v = h_v\, dv, \qquad ds_w = h_w\, dw.$$

A.2 Differential operators

The gradient $\boldsymbol{\nabla} f$ of a scalar field f can be expressed in terms of the local basis at any point P with curvilinear coordinates (u, v, w) in the form

$$\boldsymbol{\nabla} f = F_u\, \widehat{\boldsymbol{u}} + F_v\, \widehat{\boldsymbol{v}} + F_w\, \widehat{\boldsymbol{w}}.$$

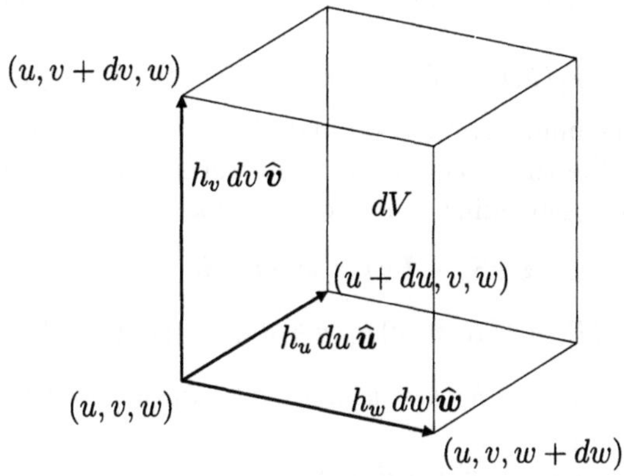

Figure A.1: Infinitesimal coordinate box of an orthogonal coordinate system.

In order to determine the coefficients F_u, F_v and F_w in this formula, we will compare two expressions for the directional derivative of f along an arbitrary curve in xyz-space.

If the curve $\mathcal{C}$ has parametrization $\boldsymbol{r} = \boldsymbol{r}(s)$ in terms of arc length s, then the directional derivative of f along $\mathcal{C}$ is given by

$$\frac{df}{ds} = \frac{\partial f}{\partial u}\frac{du}{ds} + \frac{\partial f}{\partial v}\frac{dv}{ds} + \frac{\partial f}{\partial w}\frac{dw}{ds}.$$

On the other hand, this directional derivative is also given by $\frac{df}{ds} = \boldsymbol{\nabla} f \cdot \widehat{\boldsymbol{\gamma}}$, where $\widehat{\boldsymbol{\gamma}}$ is the unit tangent vector to $\mathcal{C}$. We have

$$\begin{aligned}\widehat{\boldsymbol{\gamma}} = \frac{d\boldsymbol{r}}{ds} &= \frac{\partial \boldsymbol{r}}{\partial u}\frac{du}{ds} + \frac{\partial \boldsymbol{r}}{\partial v}\frac{dv}{ds} + \frac{\partial \boldsymbol{r}}{\partial w}\frac{dw}{ds} \\ &= h_u \frac{du}{ds}\,\widehat{\boldsymbol{u}} + h_v \frac{dv}{ds}\,\widehat{\boldsymbol{v}} + h_w \frac{dw}{ds}\,\widehat{\boldsymbol{w}}.\end{aligned}$$

Thus

$$\frac{df}{ds} = \boldsymbol{\nabla} f \cdot \widehat{\boldsymbol{\gamma}} = F_u h_u \frac{du}{ds} + F_v h_v \frac{dv}{ds} + F_w h_w \frac{dw}{ds}.$$

Comparing these two expressions for df/ds along $\mathcal{C}$, we see that

$$F_u h_u = \frac{\partial f}{\partial u}, \qquad F_v h_v = \frac{\partial f}{\partial v}, \qquad F_w h_w = \frac{\partial f}{\partial w}.$$

Therefore, we have shown that

$$\boldsymbol{\nabla} f(u, v, w) = \frac{1}{h_u}\frac{\partial f}{\partial u}\,\widehat{\boldsymbol{u}} + \frac{1}{h_v}\frac{\partial f}{\partial v}\,\widehat{\boldsymbol{v}} + \frac{1}{h_w}\frac{\partial f}{\partial w}\,\widehat{\boldsymbol{w}}.$$

Now consider a vector field $\boldsymbol{F}$ expressed in terms of the curvilinear coordinates:

$$\boldsymbol{F}(u,v,w) = F_u(u,v,w)\,\widehat{\boldsymbol{u}} + F_v(u,v,w)\,\widehat{\boldsymbol{v}} + F_w(u,v,w)\,\widehat{\boldsymbol{w}}.$$

The flux of $\boldsymbol{F}$ out of the infinitesimal coordinate box of Figure A.1 is the sum of the fluxes of $\boldsymbol{F}$ out of the three pairs of opposite surfaces of the box. The flux out of the u-surfaces corresponding to u and $u+du$ is

$$\begin{aligned}
\boldsymbol{F}(u+du,v,w)\cdot\widehat{\boldsymbol{u}}\,dS_u &- \boldsymbol{F}(u,v,w)\cdot\widehat{\boldsymbol{u}}\,dS_u \\
&= (F_u(u+du,v,w)\,h_v(u+du,v,w)\,h_w(u+du,v,w) \\
&\qquad -F_u(u,v,w)\,h_v(u,v,w)\,h_w(u,v,w))\,dv\,dw \\
&= \frac{\partial}{\partial u}(h_v h_w F_u)\,du\,dv\,dw.
\end{aligned}$$

Similar expressions hold for the fluxes out of the other pairs of coordinate surfaces.

The divergence at P of $\boldsymbol{F}$ is the flux *per unit volume* out of the infinitesimal coordinate box at P. Thus it is given by

$$\begin{aligned}
\boldsymbol{\nabla}\cdot\boldsymbol{F}(u,v,w) = \frac{1}{h_u h_v h_w}\Big[&\frac{\partial}{\partial u}(h_v h_w F_u(u,v,w)) + \frac{\partial}{\partial v}(h_u h_w F_v(u,v,w)) \\
&+\frac{\partial}{\partial w}(h_u h_v F_w(u,v,w))\Big].
\end{aligned}$$

To calculate the curl of a vector field expressed in terms of orthogonal curvilinear coordinates we can make use of some previously obtained vector identities. First, observe that the gradient of the scalar field $f(u,v,w)=u$ is $\widehat{\boldsymbol{u}}/h_u$, so that $\widehat{\boldsymbol{u}} = h_u\boldsymbol{\nabla}u$. Similarly, $\widehat{\boldsymbol{v}} = h_v\boldsymbol{\nabla}v$ and $\widehat{\boldsymbol{w}} = h_w\boldsymbol{\nabla}w$. Therefore, the vector field

$$\boldsymbol{F} = F_u\,\widehat{\boldsymbol{u}} + F_v\,\widehat{\boldsymbol{v}} + F_w\,\widehat{\boldsymbol{w}}$$

can be written in the form

$$\boldsymbol{F} = F_u h_u\boldsymbol{\nabla}u + F_v h_v\boldsymbol{\nabla}v + F_w h_w\boldsymbol{\nabla}w.$$

Using the identity $\boldsymbol{\nabla}\times(f\boldsymbol{\nabla}g) = \boldsymbol{\nabla}f\times\boldsymbol{\nabla}g$, we can calculate the curl of each term in the expression above. We have

$$\begin{aligned}
\boldsymbol{\nabla}\times(F_u h_u\boldsymbol{\nabla}u) &= \boldsymbol{\nabla}(F_u h_u)\times\boldsymbol{\nabla}u \\
&= \left[\frac{1}{h_u}\frac{\partial}{\partial u}(F_u h_u)\,\widehat{\boldsymbol{u}} + \frac{1}{h_v}\frac{\partial}{\partial v}(F_u h_u)\,\widehat{\boldsymbol{v}} + \frac{1}{h_w}\frac{\partial}{\partial w}(F_u h_u)\,\widehat{\boldsymbol{w}}\right]\times\frac{\widehat{\boldsymbol{u}}}{h_u} \\
&= \frac{1}{h_u h_w}\frac{\partial}{\partial w}(F_u h_u)\,\widehat{\boldsymbol{v}} - \frac{1}{h_u h_v}\frac{\partial}{\partial v}(F_u h_u)\,\widehat{\boldsymbol{w}} \\
&= \frac{1}{h_u h_v h_w}\left[\frac{\partial}{\partial w}(F_u h_u)\,(h_v\widehat{\boldsymbol{v}}) - \frac{\partial}{\partial v}(F_u h_u)\,(h_w\widehat{\boldsymbol{w}})\right].
\end{aligned}$$

We have used the facts that $\widehat{\boldsymbol{u}}\times\widehat{\boldsymbol{u}} = \boldsymbol{0}$, $\widehat{\boldsymbol{v}}\times\widehat{\boldsymbol{u}} = -\widehat{\boldsymbol{w}}$ and $\widehat{\boldsymbol{w}}\times\widehat{\boldsymbol{u}} = \widehat{\boldsymbol{v}}$ to obtain the result above. This is why we assumed that the curvilinear coordinate system was right-handed. Corresponding expressions can be calculated for the other two terms in the formula for $\boldsymbol{\nabla}\times\boldsymbol{F}$.

Combining the three terms, we conclude that the curl of

$$\boldsymbol{F} = F_u\,\widehat{\boldsymbol{u}} + F_v\,\widehat{\boldsymbol{v}} + F_w\,\widehat{\boldsymbol{w}}$$

is given by

$$\boldsymbol{\nabla}\times\boldsymbol{F}(u,v,w) = \frac{1}{h_u h_v h_w}\begin{vmatrix} h_u\widehat{\boldsymbol{u}} & h_v\widehat{\boldsymbol{v}} & h_w\widehat{\boldsymbol{w}} \\ \dfrac{\partial}{\partial u} & \dfrac{\partial}{\partial v} & \dfrac{\partial}{\partial w} \\ F_u h_u & F_v h_v & F_w h_w \end{vmatrix}.$$

A.3 Cylindrical coordinates

A.3.1 Definition

For cylindrical coordinates we have $\boldsymbol{r} = r\cos\phi\,\boldsymbol{i} + r\sin\phi\,\boldsymbol{j} + z\,\boldsymbol{k}$, so

$$\frac{\partial\boldsymbol{r}}{\partial r} = \cos\phi\,\boldsymbol{i} + \sin\phi\,\boldsymbol{j}, \qquad \frac{\partial\boldsymbol{r}}{\partial\phi} = -r\sin\phi\,\boldsymbol{i} + r\cos\phi\,\boldsymbol{j}, \qquad \frac{\partial\boldsymbol{r}}{\partial z} = \boldsymbol{k}.$$

It should be noted that here r does not denote the magnitude of the position vector $\boldsymbol{r}$, but the distance of $\boldsymbol{r}$ from the z-axis. Similarly, the vector $\widehat{\boldsymbol{r}}(\phi)$ does not represent the unit vector in direction of $\boldsymbol{r}$, but the directions orthogonal to the z-axis.

The scale factors for the cylindrical coordinate system are given by

$$h_r = \left|\frac{\partial\boldsymbol{r}}{\partial r}\right| = 1, \qquad h_\phi = \left|\frac{\partial\boldsymbol{r}}{\partial\phi}\right| = r, \qquad h_z = \left|\frac{\partial\boldsymbol{r}}{\partial z}\right| = 1,$$

and the local basis consists of the vectors

$$\widehat{\boldsymbol{r}} = \cos\phi\,\boldsymbol{i} + \sin\phi\,\boldsymbol{j}, \qquad \widehat{\boldsymbol{\phi}} = -\sin\phi\,\boldsymbol{i} + \cos\phi\,\boldsymbol{j}, \qquad \widehat{\boldsymbol{z}} = \boldsymbol{k}.$$

See Figure A.2. The local basis is right-handed. The unit vectors do not change with r or z, but

$$\frac{\partial\widehat{\boldsymbol{r}}}{\partial\phi} = \widehat{\boldsymbol{\phi}}, \qquad \frac{\partial\widehat{\boldsymbol{\phi}}}{\partial\phi} = -\widehat{\boldsymbol{r}}, \qquad \frac{\partial\widehat{\boldsymbol{z}}}{\partial\phi} = 0.$$

The volume element is given by

$$dV = h_r h_\phi h_z\,dr\,d\phi\,dz = r\,dr\,d\phi\,dz.$$

Its boundary consists of surface elements on the cylinder r = constant, the half-plane ϕ = constant and the plane z = constant. These elements are given, respectively, by

$$dS_r = r\,d\phi\,dz, \qquad dS_\phi = dr\,dz, \qquad dS_z = r\,dr\,d\phi.$$

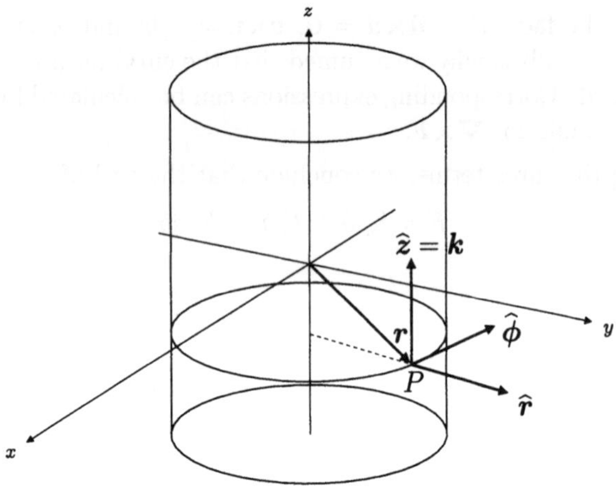

Figure A.2: Cylindrical coordinate system.

A.3.2 Gradient, divergence and curl

In terms of cylindrical coordinates, the gradient of the scalar field $f(r, \phi, z)$ is given by

$$\nabla f(r, \phi, z) = \frac{\partial f}{\partial r}\widehat{\boldsymbol{r}} + \frac{1}{r}\frac{\partial f}{\partial \phi}\widehat{\boldsymbol{\phi}} + \frac{\partial f}{\partial z}\widehat{\boldsymbol{z}}.$$

Since for cylindrical coordinates $h_r = h_z = 1$ and $h_\theta = r$, the divergence of $\boldsymbol{F} = F_r\widehat{\boldsymbol{r}} + F_\phi\widehat{\boldsymbol{\phi}} + F_z\widehat{\boldsymbol{z}}$ is

$$\begin{aligned}\nabla \cdot \boldsymbol{F} &= \frac{1}{r}\left[\frac{\partial}{\partial r}(rF_r) + \frac{\partial}{\partial \phi}F_\phi + \frac{\partial}{\partial z}(rF_z)\right] \\ &= \frac{\partial F_r}{\partial r} + \frac{1}{r}F_r + \frac{1}{r}\frac{\partial F_\phi}{\partial \phi} + \frac{\partial F_z}{\partial z}.\end{aligned}$$

The curl of $\boldsymbol{F}$ is given by

$$\begin{aligned}\nabla \times \boldsymbol{F} &= \frac{1}{r}\begin{vmatrix} \widehat{\boldsymbol{r}} & r\widehat{\boldsymbol{\phi}} & \widehat{\boldsymbol{z}} \\ \dfrac{\partial}{\partial r} & \dfrac{\partial}{\partial \phi} & \dfrac{\partial}{\partial z} \\ F_r & rF_\phi & F_z \end{vmatrix} \\ &= \left(\frac{1}{r}\frac{\partial F_z}{\partial \phi} - \frac{\partial F_\phi}{\partial z}\right)\widehat{\boldsymbol{r}} + \left(\frac{\partial F_r}{\partial z} - \frac{\partial F_z}{\partial r}\right)\widehat{\boldsymbol{\phi}} + \left(\frac{\partial F_\phi}{\partial r} + \frac{F_\phi}{r} - \frac{1}{r}\frac{\partial F_r}{\partial \phi}\right)\widehat{\boldsymbol{z}}.\end{aligned}$$

A.3.3 Laplace and advection operators

Remembering that the Laplacian of a function f is defined by the relationship

$$\nabla^2 f = \boldsymbol{\nabla} \cdot \boldsymbol{\nabla} f,$$

the results just obtained fòr the gradient and the divergence lead to the following expression for the Laplacian in cylindrical coordinates

$$\begin{aligned}\nabla^2 f &= \frac{1}{r}\frac{\partial}{\partial r}\left(r\frac{\partial f}{\partial r}\right) + \frac{1}{r^2}\frac{\partial^2 f}{\partial \phi^2} + \frac{\partial^2 f}{\partial z^2} \\ &= \frac{\partial^2 f}{\partial r^2} + \frac{1}{r}\frac{\partial f}{\partial r} + \frac{1}{r^2}\frac{\partial^2 f}{\partial \phi^2} + \frac{\partial^2 f}{\partial z^2}. + \frac{\partial^2 f}{\partial z^2}.\end{aligned}$$

As far as the Laplacian of a vector field $\boldsymbol{F}$ is concerned, its explicit expression can be derived from the vector identity

$$\nabla^2 \boldsymbol{F} = -\boldsymbol{\nabla} \times (\boldsymbol{\nabla} \times \boldsymbol{F}) + \boldsymbol{\nabla}(\boldsymbol{\nabla} \cdot \boldsymbol{F}),$$

and using some previously obtained results. We obtain

$$\begin{aligned}\nabla^2 \boldsymbol{F} = &\left(\nabla^2 F_r - \frac{F_r}{r^2} - \frac{2}{r^2}\frac{\partial F_\phi}{\partial \phi}\right)\widehat{\boldsymbol{r}} \\ &+ \left(\nabla^2 F_\phi - \frac{F_\phi}{r^2} + \frac{2}{r^2}\frac{\partial F_r}{\partial \phi}\right)\widehat{\boldsymbol{\phi}} \\ &+ \left(\nabla^2 F_z\right)\widehat{\boldsymbol{z}}.\end{aligned}$$

Finally, we give the expression in cylindrical coordinates of the advection operators for both a scalar and a vector field:

$$\boldsymbol{a} \cdot \boldsymbol{\nabla} u = a_r \frac{\partial u}{\partial r} + \frac{a_\phi}{r}\frac{\partial u}{\partial \phi} + a_z \frac{\partial u}{\partial z},$$

$$\begin{aligned}(\boldsymbol{a} \cdot \boldsymbol{\nabla})\boldsymbol{u} = &\left[(\boldsymbol{a} \cdot \boldsymbol{\nabla})u_r - \frac{a_\phi u_\phi}{r}\right]\widehat{\boldsymbol{r}} \\ &+ \left[(\boldsymbol{a} \cdot \boldsymbol{\nabla})u_\phi + \frac{a_\phi u_r}{r}\right]\widehat{\boldsymbol{\phi}} \\ &+ \left[(\boldsymbol{a} \cdot \boldsymbol{\nabla})u_z\right]\widehat{\boldsymbol{z}}.\end{aligned}$$

A.4 Spherical coordinates

A.4.1 Definition

For spherical coordinates we have

$$\boldsymbol{r} = r\sin\theta\cos\phi\,\boldsymbol{i} + r\sin\theta\sin\phi\,\boldsymbol{j} + r\cos\theta\,\widehat{\boldsymbol{z}}.$$

Thus the tangent vectors to the coordinate curves are

$$\begin{aligned}
\frac{\partial \boldsymbol{r}}{\partial r} &= \sin\theta\cos\phi\,\boldsymbol{i} + \sin\theta\sin\phi\,\boldsymbol{j} + \cos\theta\,\widehat{\boldsymbol{z}} \\
\frac{\partial \boldsymbol{r}}{\partial \theta} &= r\cos\theta\cos\phi\,\boldsymbol{i} + r\cos\theta\sin\phi\,\boldsymbol{j} - r\sin\theta\,\widehat{\boldsymbol{z}} \\
\frac{\partial \boldsymbol{r}}{\partial \phi} &= -r\sin\theta\sin\phi\,\boldsymbol{i} + r\sin\theta\cos\phi\,\boldsymbol{j},
\end{aligned}$$

and the scale factors are given by

$$h_r = \left|\frac{\partial \boldsymbol{r}}{\partial r}\right| = 1, \qquad h_\theta = \left|\frac{\partial \boldsymbol{r}}{\partial \theta}\right| = r, \qquad h_\phi = \left|\frac{\partial \boldsymbol{r}}{\partial \phi}\right| = r\sin\theta.$$

The local basis consists of the vectors

$$\begin{aligned}
\widehat{\boldsymbol{r}} &= \sin\theta\cos\phi\,\boldsymbol{i} + \sin\theta\sin\phi\,\boldsymbol{j} + \cos\theta\,\widehat{\boldsymbol{z}} \\
\widehat{\boldsymbol{\theta}} &= \cos\theta\cos\phi\,\boldsymbol{i} + \cos\theta\sin\phi\,\boldsymbol{j} - \sin\theta\,\widehat{\boldsymbol{z}} \\
\widehat{\boldsymbol{\phi}} &= -\sin\phi\,\boldsymbol{i} + \cos\phi\,\boldsymbol{j}.
\end{aligned}$$

See Figure A.3. The local basis is right-handed. The unit vectors do not change with r, but

$$\frac{\partial \widehat{\boldsymbol{r}}}{\partial \theta} = \widehat{\boldsymbol{\theta}}, \qquad \frac{\partial \widehat{\boldsymbol{\theta}}}{\partial \theta} = -\widehat{\boldsymbol{r}}, \qquad \frac{\partial \widehat{\boldsymbol{\phi}}}{\partial \theta} = 0,$$

$$\frac{\partial \widehat{\boldsymbol{r}}}{\partial \phi} = \sin\theta\,\widehat{\boldsymbol{\phi}}, \qquad \frac{\partial \widehat{\boldsymbol{\theta}}}{\partial \phi} = \cos\theta\,\widehat{\boldsymbol{\phi}}, \qquad \frac{\partial \widehat{\boldsymbol{\phi}}}{\partial \phi} = -\sin\theta\,\widehat{\boldsymbol{r}} - \cos\theta\,\widehat{\boldsymbol{\theta}}.$$

The volume element in spherical coordinates is

$$dV = h_r h_\theta h_\phi\, dr\, d\theta\, d\phi = r^2 \sin\theta\, dr\, d\theta\, d\phi.$$

The area element on the sphere $r =$ constant is

$$dS_r = h_\theta h_\phi\, d\theta\, d\phi = r^2 \sin\theta\, d\theta\, d\phi.$$

The area element on the cone $\theta =$ constant is

$$dS_\theta = h_r h_\phi\, dr\, d\phi = r\sin\theta\, dr\, d\phi.$$

The area element on the half-plane $\phi =$ constant is

$$dS_\phi = h_r h_\theta\, dr\, d\theta = r\, dr\, d\theta.$$

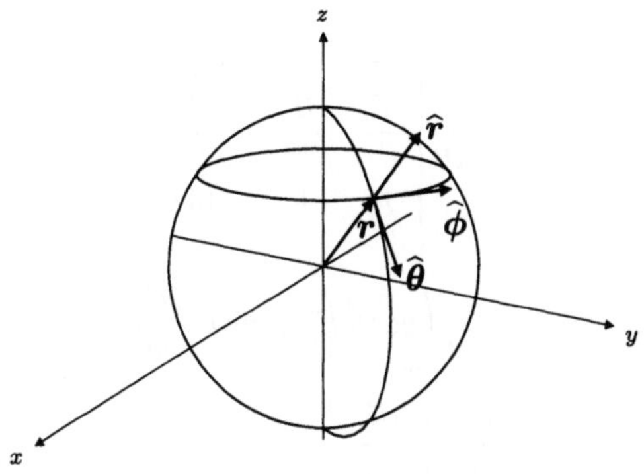

Figure A.3: Spherical coordinate system.

A.4.2 Gradient, divergence and curl

In terms of spherical coordinates, the gradient of the scalar field $f(r,\theta,\phi)$ is expressed by

$$\nabla f(r,\theta,\phi) = \frac{\partial f}{\partial r}\,\widehat{\boldsymbol{r}} + \frac{1}{r}\frac{\partial f}{\partial \theta}\,\widehat{\boldsymbol{\theta}} + \frac{1}{r\sin\theta}\frac{\partial f}{\partial \phi}\,\widehat{\boldsymbol{\phi}}.$$

For spherical coordinates, $h_r = 1$, $h_\theta = r$ and $h_\phi = r\sin\theta$. The divergence of the vector field $\boldsymbol{F} = F_r\,\widehat{\boldsymbol{r}} + F_\theta\,\widehat{\boldsymbol{\theta}} + F_\phi\,\widehat{\boldsymbol{\phi}}$ is

$$\begin{aligned}\nabla\cdot\boldsymbol{F} &= \frac{1}{r^2\sin\theta}\left[\frac{\partial}{\partial r}(r^2\sin\theta\,F_r) + \frac{\partial}{\partial\theta}(r\sin\theta\,F_\theta) + \frac{\partial}{\partial\phi}(r\,F_\phi)\right]\\ &= \frac{\partial}{\partial r}(r^2\,F_r) + \frac{1}{r\sin\theta}\frac{\partial}{\partial\theta}(\sin\theta\,F_\theta) + \frac{1}{r\sin\theta}\frac{\partial F_\phi}{\partial\phi}\\ &= \frac{\partial F_r}{\partial r} + \frac{2}{r}F_r + \frac{1}{r}\frac{\partial F_\theta}{\partial\theta} + \frac{\cot\theta}{r}F_\theta + \frac{1}{r\sin\theta}\frac{\partial F_\phi}{\partial\phi}.\end{aligned}$$

The curl of $\boldsymbol{F}$ is given by

$$\nabla\times\boldsymbol{F}=\frac{1}{r^2\sin\theta}\begin{vmatrix}\hat{\boldsymbol{r}} & r\hat{\boldsymbol{\theta}} & r\sin\theta\,\hat{\boldsymbol{\phi}}\\ \dfrac{\partial}{\partial r} & \dfrac{\partial}{\partial\theta} & \dfrac{\partial}{\partial\phi}\\ F_r & rF_\theta & r\sin\theta F_\phi\end{vmatrix}$$

$$=\frac{1}{r\sin\theta}\left(\cos\theta\,F_\phi+\sin\theta\,\frac{\partial F_\phi}{\partial\theta}-\frac{\partial F_\theta}{\partial\phi}\right)\hat{\boldsymbol{r}}$$

$$+\frac{1}{r\sin\theta}\left(\frac{\partial F_r}{\partial\phi}-\sin\theta\,F_\phi-r\sin\theta\,\frac{\partial F_\phi}{\partial r}\right)\hat{\boldsymbol{\theta}}$$

$$+\frac{1}{r}\left(F_\theta+r\,\frac{\partial F_\theta}{\partial r}-\frac{\partial F_r}{\partial\theta}\right)\hat{\boldsymbol{\phi}}.$$

A.4.3 Laplace and advection operators

Using the results above, the expressions of the Laplacian operators in spherical coordinates are easily found. The Laplacian of the scalar field $f(r,\theta,\phi)$ assumes the form

$$\nabla^2 f=\frac{1}{r^2}\frac{\partial}{\partial r}\left(r^2\frac{\partial f}{\partial r}\right)+\frac{1}{r^2\sin\theta}\frac{\partial}{\partial\theta}\left(\sin\theta\,\frac{\partial f}{\partial\theta}\right)+\frac{1}{r^2\sin^2\theta}\frac{\partial^2 f}{\partial\phi^2}$$

$$=\frac{\partial^2 f}{\partial r^2}+\frac{2}{r}\frac{\partial f}{\partial r}+\frac{\cot\theta}{r^2}\frac{\partial f}{\partial\theta}+\frac{1}{r^2}\frac{\partial^2 f}{\partial\theta^2}+\frac{1}{r^2\sin^2\theta}\frac{\partial^2 f}{\partial\phi^2},$$

whereas the Laplacian of the vector field $\boldsymbol{F}=F_r\hat{\boldsymbol{r}}+F_\theta\hat{\boldsymbol{\theta}}+F_\phi\hat{\boldsymbol{\phi}}$ is

$$\nabla^2\boldsymbol{F}=\left(\nabla^2F_r-\frac{2F_r}{r^2}-\frac{2}{r^2\sin\theta}\frac{\partial(\sin\theta\,F_\theta)}{\partial\theta}-\frac{2}{r^2\sin\theta}\frac{\partial F_\phi}{\partial\phi}\right)\hat{\boldsymbol{r}}$$

$$+\left(\nabla^2F_\theta-\frac{F_\theta}{r^2\sin^2\theta}-\frac{2\cos\theta}{r^2\sin^2\theta}\frac{\partial F_\phi}{\partial\phi}+\frac{2}{r^2}\frac{\partial F_r}{\partial\theta}\right)\hat{\boldsymbol{\theta}}$$

$$+\left(\nabla^2F_\phi-\frac{F_\phi}{r^2\sin^2\theta}+\frac{2\cos\theta}{r^2\sin^2\theta}\frac{\partial F_\theta}{\partial\phi}+\frac{2}{r^2\sin\theta}\frac{\partial F_r}{\partial\phi}\right)\hat{\boldsymbol{\phi}}.$$

Finally, the advection operators in spherical coordinates assume the form

$$\boldsymbol{a}\cdot\nabla u=a_r\frac{\partial u}{\partial r}+\frac{a_\theta}{r}\frac{\partial u}{\partial\theta}+\frac{a_\phi}{r\sin\theta}\frac{\partial u}{\partial\phi},$$

$$
\begin{aligned}
(\boldsymbol{a}\cdot\boldsymbol{\nabla})\boldsymbol{u} = {} & \left[(\boldsymbol{a}\cdot\boldsymbol{\nabla})u_r - \frac{a_\theta u_\theta + a_\phi u_\phi}{r}\right]\widehat{\boldsymbol{r}} \\
& + \left[(\boldsymbol{a}\cdot\boldsymbol{\nabla})u_\theta + \frac{a_\theta u_r - \cot\theta\, a_\phi u_\phi}{r}\right]\widehat{\boldsymbol{\theta}} \\
& + \left[(\boldsymbol{a}\cdot\boldsymbol{\nabla})u_\phi + \frac{a_\phi u_r + \cot\theta\, a_\phi u_\theta}{r}\right]\widehat{\boldsymbol{\phi}}.
\end{aligned}
$$

Appendix B

Separation of vector elliptic equations

B.1 Introduction

Many of the methods discussed in this study for the approximate solution of the incompressible Navier–Stokes require to solve Poisson and Helmholtz equations of vector type, supplemented by Dirichlet boundary conditions. As well known, this kind of boundary value problems for a vector field separates into uncoupled Dirichlet problems for the scalar components of the unknown only in Cartesian coordinates. On the contrary, when the vector elliptic equation is expressed in polar, cylindrical or spherical coordinates, the Laplace operator acting on a vector field produces a system of coupled elliptic equations for the orthogonal components of the unknown. The aim of the present Appendix is to describe the similarity transformations which reduce the vector Poisson equation expressed in these orthogonal coordinate systems into a set of uncoupled elliptic equations for scalar unknowns.

Four different geometrical situations will be analyzed. Firstly, the vector two-dimensional Poisson equation in polar coordinates is examined (section B.2). Secondly, the case of elliptic equations for a vector field defined on, and tangential to, a spherical surface is considered, using the latitude and the longitude as orthogonal coordinates (section B.3). For both cases, the same similarity transformation is found to be effective.

Three-dimensional equations are then addressed, starting with the examination of cylindrical coordinates (section B.4). Here, the solution domain of the Poisson equation is assumed to consist of an entire annular region of the three-dimensional space, so that the dependence on the angular variable can be represented by Fourier analysis. A convenient complex representation of the vector field is introduced to obtain a real (*i.e.*, noncomplex) characterization of the operator occurring in the Fourier-transformed problem. In this way, the three-dimensional vector problem is reduced to a set of two-dimensional problems, but always of vector type. Two of the three cylindrical components of each Fourier mode, with the exception of the first mode, are found to be coupled together by the elliptic operator, like the cylindrical components of the original three-dimensional vector problem. This coupling is eliminated by a similarity transformation which reduces the operator to a diagonal form.

A similar analysis is finally conducted for the 3D vector Poisson equation in annular regions radially bounded by two concentric spherical surfaces, using spherical coordinates (section B.5). In this case, after the Fourier analysis of the longitudinal dependence, the three spherical components of each Fourier mode of the vector unknown are found to be solution of a system of three two-dimensional elliptic equations which are coupled together. The coupling is eliminated by an appropriate similarity transformation and the solution of the original problem is thus reduced to a sequence of purely scalar elliptic equations in two dimensions, plus the operations for performing the transformation, the Fouries analysis of the data as well as the synthesis of the solution.

As it will be shown, the similarity transformations which allow the uncoupling (in 3D after Fourier analysis) of the orthogonal components of the vector unknown are so simple that some of them, if not all, may have been already considered and employed in other studies. However, since the complete set of these transformations is not easily available in the computational literature, it is collected here, also with the purpose of showing how the use of solution algorithms for scalar two-dimensional elliptic equations can be extended to the solution of Dirichlet problems for vector fields in two and three dimensions.

B.2 Polar coordinates

Let us consider the vector Poisson equation $\nabla^2 \boldsymbol{u} = \boldsymbol{f}$ in two dimensions expressed in polar coordinates (r, ϕ). The Laplace operator in this coordinate system is

$$\nabla^2 = \frac{1}{r}\frac{\partial}{\partial r}\left(r\frac{\partial}{\partial r}\right) + \frac{1}{r^2}\frac{\partial^2}{\partial \phi^2},$$

the vector field $\boldsymbol{u}$ being expressed in its polar components (u_r, u_ϕ) according to

$$\boldsymbol{u}(r, \phi) = u_r(r, \phi)\,\widehat{\boldsymbol{r}}(\phi) + u_\phi(r, \phi)\,\widehat{\boldsymbol{\phi}}(\phi),$$

where $\widehat{\boldsymbol{r}}(\phi)$ and $\widehat{\boldsymbol{\phi}}(\phi)$ are the unit vectors of the polar coordinate system. Due to the dependence of the unit vectors on the angle ϕ, the action of the Laplace operator on a vector field is such that the vector Poisson equation becomes, in polar components, (see, e.g., Acheson [1])

$$\left(\nabla^2 - \frac{1}{r^2}\right) u_r - \frac{2}{r^2}\frac{\partial u_\phi}{\partial \phi} = f_r,$$

$$\left(\nabla^2 - \frac{1}{r^2}\right) u_\phi + \frac{2}{r^2}\frac{\partial u_r}{\partial \phi} = f_\phi.$$

Thus, a system of two coupled scalar elliptic equations for the polar components of the unknown is obtained. Introducing the matrix differential operator

$$\boldsymbol{\nabla}^2 = \begin{pmatrix} \nabla^2 - \frac{1}{r^2} & -\frac{2}{r^2}\frac{\partial}{\partial \phi} \\ \frac{2}{r^2}\frac{\partial}{\partial \phi} & \nabla^2 - \frac{1}{r^2} \end{pmatrix},$$

the vector Poisson equation $\nabla^2 \boldsymbol{u} = \boldsymbol{f}$ can be written in the matrix form:

$$\boldsymbol{\nabla}^2 \begin{pmatrix} u_r \\ u_\phi \end{pmatrix} = \begin{pmatrix} f_r \\ f_\phi \end{pmatrix}.$$

Let us now consider the change of variables defined by the following linear trasformation

$$\boldsymbol{t}(\phi) = \begin{pmatrix} \sin\phi & \cos\phi \\ \cos\phi & -\sin\phi \end{pmatrix}, \qquad \begin{pmatrix} u_1 \\ u_2 \end{pmatrix} = \boldsymbol{t}(\phi) \begin{pmatrix} u_r \\ u_\phi \end{pmatrix}.$$

The matrix $\boldsymbol{t}(\phi)$ is such that $\boldsymbol{t}(\phi) = \boldsymbol{t}(\phi)^{\text{tr}} = \boldsymbol{t}(\phi)^{-1}$, and thus is both symmetric and orthogonal. As usual, the superscript "tr" denotes transposition. By simple calculations one can show that the similarity transformation provided by matrix $\boldsymbol{t}(\phi)$ is such that

$$\boldsymbol{t}(\phi)\, \boldsymbol{\nabla}^2\, \boldsymbol{t}(\phi) = \begin{pmatrix} \nabla^2 & 0 \\ 0 & \nabla^2 \end{pmatrix},$$

i.e., it diagonalizes the matrix operator of the vector Poisson equation in polar coordinates. As a consequence, the solution of such an equation can be calculated by first applying the transformation $\boldsymbol{t}(\phi)$ to the source field $\boldsymbol{f}$, then solving the two independent equations

$$\nabla^2 u_1 = f_1,$$

$$\nabla^2 u_2 = f_2,$$

and finally backtrasforming the two computed solutions u_1 and u_2 according to

$$\begin{pmatrix} u_r \\ u_\phi \end{pmatrix} = \boldsymbol{t}(\phi) \begin{pmatrix} u_1 \\ u_2 \end{pmatrix}.$$

Of course, the transformation $\boldsymbol{t}(\phi)$ must be applied also to the prescribed data of the boundary conditions associated with the original vector elliptic problem.

The same trasformation can be obviously applied also to the Dirichlet vector problem for the Helmholtz operator $(\nabla^2 - \gamma)$, with $\gamma > 0$. More generally, the similarity trasformation is effective in separating the polar components in any Dirichlet problem for the operator $(\nabla^2 - \Gamma(r))$, where $\Gamma(r)$ is a given function, possibly satisfying some suitable conditions.

In the particular case of an annular domain, $[r_1 \le r \le r_2, 0 \le \phi < 2\pi]$, the two-dimensional scalar Poisson equation $\nabla^2 u = f$ could be solved by introducing a Fourier representation of the dependence on ϕ in f and u by means of complex Fourier series, namely,

$$f(r,\phi) = \sum_{m=-\infty}^{\infty} f_m(r)\, e^{im\phi}, \qquad u(r,\phi) = \sum_{m=-\infty}^{\infty} u_m(r)\, e^{im\phi}.$$

This decomposition reduces the two-dimensional Poisson equation to a set of ordinary differential equations for the second-order operator

$$\mathcal{D}_m^2 = \frac{1}{r}\frac{d}{dr}\left(r\frac{d}{dr}\right) - \frac{m^2}{r^2} = \frac{d^2}{dr^2} + \frac{1}{r}\frac{d}{dr} - \frac{m^2}{r^2}.$$

The equation governing the Fourier coefficient $u_m(r)$, namely,

$$\mathcal{D}_m^2 u_m = f_m,$$

is an Euler (equidimensional) equation whose solution can be written as

$$u_m(r) = A_m r^m + B_m r^{-m} + F_m(r),$$

where A_m and B_m are constants determined by the boundary conditions, and $F_m(r)$ is any particular solution to the equation (14).

It is interesting to note that, always with reference to the annular domain, the solution method above is not the only method allowing the reduction of the vector elliptic equation to a set of uncloupled ordinary differential equations for scalar unknowns. Another method consists in introducing the Fourier decomposition *before* uncoupling the vector components of the unknown by means of the similarity trasformation. In fact, let the vector field $\boldsymbol{u}(r,\phi)$, $0 \le \phi < 2\pi$, be represented by means of the following complex Fourier series:

$$\boldsymbol{u}(r,\phi) = \boldsymbol{u}_0(r;\phi) + \sum_{\substack{m=-\infty \\ m\neq 0}}^{\infty} \boldsymbol{u}_m(r;\phi)\, e^{im\phi},$$

where

$$\boldsymbol{u}_0(r;\phi) = u_{r,0}(r)\,\widehat{\boldsymbol{r}}(\phi) + u_{\phi,0}(r)\,\widehat{\boldsymbol{\phi}}(\phi)$$

and

$$\boldsymbol{u}_m(r;\phi) = u_{r,m}(r)\,\widehat{\boldsymbol{r}}(\phi) + u_{\phi,m}(r)\,\widehat{\boldsymbol{\phi}}(\phi)\,\mathrm{i}.$$

In the expansion above, the coefficient of the angular component of the Fourier mode, with the exclusion of the first mode, has been multiplied by the imaginary unit in order to obtain a purely real representation of the second-order matrix operator acting on the Fourier coefficients of the vector mode. The condition that the vector function $\boldsymbol{u}$ be real-valued implies that the complex coefficients

$$u_{r,m}(r) = v_{r,m}(r) + \mathrm{i}\, w_{r,m}(r),$$

$$u_{\phi,m}(r) = v_{\phi,m}(r) + \mathrm{i}\, w_{\phi,m}(r),$$

for $m > 0$, satify the following conditions:

$$v_{r,-m} = v_{r,m}, \qquad w_{r,-m} = -w_{r,m};$$

$$v_{\phi,-m} = -v_{\phi,m}, \qquad w_{\phi,-m} = w_{\phi,m}.$$

According to this Fourier representation, the Laplace operator $\boldsymbol{\nabla}^2$ acting on the Fourier coefficients of the mth vector mode becomes

$$\boldsymbol{\mathcal{D}}_m^2 = \begin{pmatrix} \mathcal{D}_m^2 - \frac{1}{r^2} & \frac{2m}{r^2} \\ \frac{2m}{r^2} & \mathcal{D}_m^2 - \frac{1}{r^2} \end{pmatrix}.$$

Therefore, the conditions of reality mean that the complex system

$$\boldsymbol{\mathcal{D}}_m^2 \begin{pmatrix} u_{r,m} \\ u_{\phi,m} \end{pmatrix} = \begin{pmatrix} f_{r,m} \\ f_{\phi,m} \end{pmatrix}$$

has to be solved only for $m > 0$. Of course, such a complex system is equivalent to the two real systems

$$\boldsymbol{\mathcal{D}}_m^2 \begin{pmatrix} v_{r,m} \\ v_{\phi,m} \end{pmatrix} = \begin{pmatrix} g_{r,m} \\ g_{\phi,m} \end{pmatrix}, \qquad \boldsymbol{\mathcal{D}}_m^2 \begin{pmatrix} w_{r,m} \\ w_{\phi,m} \end{pmatrix} = \begin{pmatrix} h_{r,m} \\ h_{\phi,m} \end{pmatrix},$$

where the right hand sides are the real and imaginary parts of the complex Fourier coefficients of $\boldsymbol{f}$, *i.e.*, $\boldsymbol{f}_m = \boldsymbol{g}_m + \mathrm{i}\, \boldsymbol{h}_m$.

Introducing the similarity trasformation

$$\boldsymbol{s} = \frac{1}{\sqrt{2}} \begin{pmatrix} 1 & 1 \\ 1 & -1 \end{pmatrix},$$

the matrix ordinary differential operator $\boldsymbol{\mathcal{D}}_m$ is diagonalized as follows

$$\boldsymbol{s}\, \boldsymbol{\mathcal{D}}_m^2\, \boldsymbol{s} = \begin{pmatrix} \mathcal{D}_{m-1}^2 & 0 \\ 0 & \mathcal{D}_{m+1}^2 \end{pmatrix}.$$

This uncoupling is exploited in section 4 to deal with the three-dimensional vector Poisson equation expressed in cylindrical coordinates.

Once the pairs $v_{r,m}, v_{\phi,m}$, and $w_{r,m}, w_{\phi,m}$, have been determined, the solution $\boldsymbol{u}$ can be expressed in real form according to

$$\boldsymbol{u}(r,\phi) = \left\{ u_{r,0}(r) + 2 \sum_{m=1}^{\infty} \left[v_{r,m}(r) \cos(m\phi) - w_{r,m}(r) \sin(m\phi) \right] \right\} \hat{\boldsymbol{r}}(\phi)$$
$$+ \left\{ u_{\phi,0}(r) - 2 \sum_{m=1}^{\infty} \left[v_{\phi,m}(r) \sin(m\phi) + w_{\phi,m}(r) \cos(m\phi) \right] \right\} \hat{\boldsymbol{\phi}}(\phi).$$

B.3 Spherical coordinates on the unit sphere

The similarity trasformation $\boldsymbol{t}(\phi)$ is effective also in diagonalizing the Poisson equation which governs a vector field defined on, and tangential to, the surface of a sphere. Consider the unit sphere and an ortogonal coordinate system on it, consisting of the angular variables latitude θ $(0 \leq \theta \leq \pi)$ and longitude ϕ $(0 \leq \phi < 2\pi)$. The Laplace operator over the unit sphere is

$$\nabla^2 = \frac{1}{\sin\theta}\frac{\partial}{\partial\theta}\left(\sin\theta\frac{\partial}{\partial\theta}\right) + \frac{1}{\sin^2\theta}\frac{\partial^2}{\partial\phi^2},$$

and a vector field $\boldsymbol{u}$ "belonging" to the spherical surface can be expressed in terms of its spherical components u_θ and u_ϕ according to

$$\boldsymbol{u}(\theta,\phi) = u_\theta(\theta,\phi)\,\widehat{\boldsymbol{\theta}}(\theta,\phi) + u_\phi(\theta,\phi)\,\widehat{\boldsymbol{\phi}}(\theta,\phi),$$

where $\widehat{\boldsymbol{\theta}}(\theta,\phi)$ and $\widehat{\boldsymbol{\phi}}(\theta,\phi)$ denote the two unit vectors of the spherical coordinate system over the spherical surface.

It is to be noted that the Laplacian and the other related second-order differential operators introduced in each section of this Appendix represent the operators in different coordinates, but are indicated by identical symbols in the various sections to avoid a cumbersome notation. In other words, the definite meaning of these operators changes in each section although they are denoted identically, since the range of validity of each definition is limited only to a specific section.

Considering now the vector Poisson equation $\nabla^2\boldsymbol{u} = \boldsymbol{f}$, in terms of its spherical components it becomes

$$\left(\nabla^2 - \frac{1}{\sin^2\theta}\right)u_\theta - \frac{2\cos\theta}{\sin^2\theta}\frac{\partial u_\phi}{\partial\phi} = f_\theta,$$

$$\left(\nabla^2 - \frac{1}{\sin^2\theta}\right)u_\phi + \frac{2\cos\theta}{\sin^2\theta}\frac{\partial u_\theta}{\partial\phi} = f_\phi.$$

Introducing the matrix differential operator

$$\boldsymbol{\nabla}^2 = \begin{pmatrix} \nabla^2 - \frac{1}{\sin^2\theta} & -\frac{2\cos\theta}{\sin^2\theta}\frac{\partial}{\partial\phi} \\ \frac{2\cos\theta}{\sin^2\theta}\frac{\partial}{\partial\phi} & \nabla^2 - \frac{1}{\sin^2\theta} \end{pmatrix},$$

the Poisson equation can be written in the matrix form

$$\boldsymbol{\nabla}^2\begin{pmatrix} u_\theta \\ u_\phi \end{pmatrix} = \begin{pmatrix} f_\theta \\ f_\phi \end{pmatrix}.$$

Now, the partial derivative $\frac{\partial}{\partial\phi}$ occurs in this matrix operator $\boldsymbol{\nabla}^2$ as well as in the Laplacian ∇^2 exactly as it occurrs in the corresponding operators in polar coordinates. Therefore, the similarity trasformation $\boldsymbol{t}(\phi)$ introduced in the previous section,

$$\boldsymbol{t}(\phi) = \begin{pmatrix} \sin\phi & \cos\phi \\ \cos\phi & -\sin\phi \end{pmatrix}, \qquad \begin{pmatrix} u_1 \\ u_2 \end{pmatrix} = \boldsymbol{t}(\phi) \begin{pmatrix} u_\theta \\ u_\phi \end{pmatrix},$$

diagonalize also the matrix operator $\boldsymbol{\nabla}^2$ for spherical coordinates, namely,

$$\boldsymbol{t}(\phi)\,\boldsymbol{\nabla}^2\,\boldsymbol{t}(\phi) = \begin{pmatrix} \nabla^2 & 0 \\ 0 & \nabla^2 \end{pmatrix}.$$

Similarly to polar coordinates, the solution of the equation $\nabla^2\boldsymbol{u} = \boldsymbol{f}$ on the unit sphere can be calculated by applying the transformation $\boldsymbol{t}(\phi)$ to the source field $\boldsymbol{f}$, solving the two independent equations

$$\nabla^2 u_1 = f_1,$$
$$\nabla^2 u_2 = f_2,$$

and backtransforming the calculated solutions through

$$\begin{pmatrix} u_\theta \\ u_\phi \end{pmatrix} = \boldsymbol{t}(\phi) \begin{pmatrix} u_1 \\ u_2 \end{pmatrix}.$$

The same trasformation can be applied to more general elliptic equations, such as, for instance, the Helmholtz equation $(\nabla^2 - \gamma)\boldsymbol{u} = \boldsymbol{f}$, with $\gamma > 0$, or the equation $(\nabla^2 - \Gamma(\theta))\boldsymbol{u} = \boldsymbol{f}$, where $\Gamma(\theta)$ is a given function possibly satisfying convenient conditions.

As for polar coordinates, if the domain is such that $0 \le \phi < 2\pi$, it would be possible to reduce each scalar Poisson equation $\nabla^2 u = f$ to a set of second-order ordinary differential equations by introducing a Fourier decomposition of the involved functions.

B.4 Cylindrical coordinates

Coming to three-dimensional problems, we examine first the case of cylindrical coordinates (r, z, ϕ). Let us consider the Poisson equation $\nabla^2\boldsymbol{u} = \boldsymbol{f}$, with the Laplace operator defined by

$$\nabla^2 = \frac{1}{r}\frac{\partial}{\partial r}\left(r\frac{\partial}{\partial r}\right) + \frac{\partial^2}{\partial z^2} + \frac{1}{r^2}\frac{\partial^2}{\partial \phi^2}.$$

If the vector field $\boldsymbol{u}$ is expressed in terms of its cylindrical components u_r, u_z and u_ϕ, the Laplacian pertaining to the vector problem is found to be given by the matrix (see section A.3.3 of Appendix A)

$$\boldsymbol{\nabla}^2 = \begin{pmatrix} \nabla^2 - \frac{1}{r^2} & 0 & -\frac{2}{r^2}\frac{\partial}{\partial\phi} \\ 0 & \nabla^2 & 0 \\ \frac{2}{r^2}\frac{\partial}{\partial\phi} & 0 & \nabla^2 - \frac{1}{r^2} \end{pmatrix}.$$

We assume now that the domain of definition of the elliptic equation goes all around the z-axis, that is, $0 \leq \phi < 2\pi$. Under these circumstances, the vector field $\boldsymbol{u}(r, z, \phi)$ can be represented by means of a complex Fourier series, which we will write in the following form

$$\boldsymbol{u}(r, z, \phi) = \boldsymbol{u}_0(r, z; \phi) + \sum_{\substack{m=-\infty \\ m\neq 0}}^{\infty} \boldsymbol{u}_m(r, z; \phi)\, e^{im\phi},$$

where

$$\boldsymbol{u}_0(r, z; \phi) = u_0^r(r, z)\, \widehat{\boldsymbol{r}}(\phi) + u_0^z(r, z)\, \widehat{\boldsymbol{z}} + u_0^\phi(r, z)\, \widehat{\boldsymbol{\phi}}(\phi)$$

and

$$\boldsymbol{u}_m(r, z; \phi) = u_m^r(r, z)\, \widehat{\boldsymbol{r}}(\phi) + u_m^z(r, z)\, \widehat{\boldsymbol{z}} + u_m^\phi(r, z)\, \widehat{\boldsymbol{\phi}}(\phi)\, \mathrm{i}.$$

In the two expressions above, $\widehat{\boldsymbol{r}}(\phi)$, $\widehat{\boldsymbol{z}}$ and $\widehat{\boldsymbol{\phi}}(\phi)$ denote the unit vectors of the cylindrical coordinate system. (Note that here $\widehat{\boldsymbol{r}}(\phi)$ represents the direction orthogonal to the z-axis and is not the radial unit vector pointing out from the origin.)

Being interested in the solution of only *real* vector fields, we have multiplied the coefficient of the angular component of the Fourier mode, with the exclusion of the first mode, by the imaginary unit in order to obtain a purely real representation of the second-order matrix operator acting on the Fourier coefficients of the vector mode (see later). Then, the condition of reality for $\boldsymbol{u}$ expressed by the Fourier series above implies that the real and imaginary parts of its Fourier coefficients $\boldsymbol{u}_m = \boldsymbol{v}_m + \mathrm{i}\,\boldsymbol{w}_m$ with $m \neq 0$ must satisfy the conditions:

$$\begin{aligned} v_{-m}^r &= v_m^r, & w_{-m}^r &= -w_m^r; \\ v_{-m}^z &= v_m^z, & w_{-m}^z &= -w_m^z; \\ v_{-m}^\phi &= -v_m^\phi, & w_{-m}^\phi &= w_m^\phi. \end{aligned}$$

It follows that the Fourier expansion of a real vector field in cylindrical coordinates can be written also in the following form

$$\begin{aligned}\boldsymbol{u}(r,z,\phi) = {} & \left\{u_0^r(r,z) + 2\sum_{m=1}^{\infty}[v_m^r(r,z)\cos(m\phi) - w_m^r(r,z)\sin(m\phi)]\right\}\hat{\boldsymbol{r}}(\phi) \\ & + \left\{u_0^z(r,z) + 2\sum_{m=1}^{\infty}[v_m^z(r,z)\cos(m\phi) - w_m^z(r,z)\sin(m\phi)]\right\}\hat{\boldsymbol{z}} \\ & + \left\{u_0^\phi(r,z) - 2\sum_{m=1}^{\infty}\left[v_m^\phi(r,z)\sin(m\phi) + w_m^\phi(r,z)\cos(m\phi)\right]\right\}\hat{\boldsymbol{\phi}}(\phi).\end{aligned}$$

According to the adopted Fourier representation, the Laplace operator for the Fourier coefficient of the mth mode assumes the form

$$\partial_m^2 = \frac{1}{r}\frac{\partial}{\partial r}\left(r\frac{\partial}{\partial r}\right) + \frac{\partial^2}{\partial z^2} - \frac{m^2}{r^2}.$$

Furthermore, the matrix Laplace operator for the Fourier coefficients of a vector field becomes

$$\boldsymbol{\partial}_m^2 = \begin{pmatrix} \partial_m^2 - \frac{1}{r^2} & 0 & \frac{2m}{r^2} \\ 0 & \partial_m^2 & 0 \\ \frac{2m}{r^2} & 0 & \partial_m^2 - \frac{1}{r^2} \end{pmatrix}.$$

Thus, for $m \neq 0$ there is a coupling between the two components of the vector mode normal to the z-axis, whereas the axial component is always uncoupled. Consider the change of variables given by the following linear transformation:

$$\boldsymbol{S} = \frac{1}{\sqrt{2}}\begin{pmatrix} 1 & 0 & 1 \\ 0 & \sqrt{2} & 0 \\ 1 & 0 & -1 \end{pmatrix}, \qquad \begin{pmatrix} u_m^1 \\ u_m^2 \\ u_m^3 \end{pmatrix} = \boldsymbol{S}\begin{pmatrix} u_m^r \\ u_m^z \\ u_m^\phi \end{pmatrix},$$

which lineary combines only the first and the third components ($u_m^2 = u_m^z$). The matrix $\boldsymbol{S}$ is such that $\boldsymbol{S} = \boldsymbol{S}^{\text{tr}} = \boldsymbol{S}^{-1}$ and that

$$\boldsymbol{S}\,\boldsymbol{\partial}_m^2\,\boldsymbol{S} = \begin{pmatrix} \partial_{m-1}^2 & 0 & 0 \\ 0 & \partial_m^2 & 0 \\ 0 & 0 & \partial_{m+1}^2 \end{pmatrix}.$$

The solution of the vector mode with $m \neq 0$ can be determined by first applying the trasformation $\boldsymbol{S}$ to the the Fourier coefficient $\boldsymbol{f}_m$ of the source term $\boldsymbol{f}$, then solving the three uncoupled equations

$$\begin{aligned} \partial_{m-1}^2 u_m^1 &= f_m^1, \\ \partial_m^2 u_m^2 &= f_m^2, \\ \partial_{m+1}^2 u_m^3 &= f_m^3, \end{aligned}$$

and finally backtrasforming the solutions according to

$$\begin{pmatrix} u_m^r \\ u_m^z \\ u_m^\phi \end{pmatrix} = \boldsymbol{S} \begin{pmatrix} u_m^1 \\ u_m^2 \\ u_m^3 \end{pmatrix}.$$

For the first Fourier mode $m = 0$, the similarity trasformation is not needed since the three equations for the cylindrical components are already uncoupled; in fact, the operator of the first mode is already in the diagonal form

$$\boldsymbol{\partial}_0^2 = \begin{pmatrix} \partial_1^2 & 0 & 0 \\ 0 & \partial_0^2 & 0 \\ 0 & 0 & \partial_1^2 \end{pmatrix}.$$

Of course, a Fourier analysis of the source term $\boldsymbol{f}$ and of the data for the boundary conditions is required before the solution process, and a final Fourier synthesis of the solutions $\boldsymbol{u}_m$, $m = 1, 2, \ldots, M$, is to be performed as the last step of the solution process.

B.5 Spherical coordinates

We are now ready for the most interesting case, *i.e.*, the three-dimensional vector Poisson equation in spherical coordinates (r, θ, ϕ), assuming that the domain goes all around the polar axis, namely, $0 \leq \phi < 2\pi$. The Laplace operator in spherical coordinates is (cf. section A.4.3 of Appendix A)

$$\nabla^2 = \frac{1}{r^2}\frac{\partial}{\partial r}\left(r^2 \frac{\partial}{\partial r}\right) + \frac{1}{r^2 \sin\theta}\frac{\partial}{\partial \theta}\left(\sin\theta \frac{\partial}{\partial \theta}\right) + \frac{1}{r^2 \sin^2\theta}\frac{\partial^2}{\partial \phi^2}.$$

When the Laplacian acts on a vector field $\boldsymbol{u}$ expressed in terms of its spherical coordinates u_r, u_θ and u_ϕ, the following matrix differential operator has to be considered

$$\boldsymbol{\nabla}^2 = \begin{pmatrix} \nabla^2 - \frac{2}{r^2} & -\frac{2}{r^2 \sin\theta}\frac{\partial}{\partial\theta}(\sin\theta \ldots) & -\frac{2}{r^2\sin\theta}\frac{\partial}{\partial\phi} \\ \frac{2}{r^2}\frac{\partial}{\partial\theta} & \nabla^2 - \frac{1}{r^2\sin^2\theta} & -\frac{2\cos\theta}{r^2\sin^2\theta}\frac{\partial}{\partial\phi} \\ \frac{2}{r^2\sin\theta}\frac{\partial}{\partial\phi} & \frac{2\cos\theta}{r^2\sin^2\theta}\frac{\partial}{\partial\phi} & \nabla^2 - \frac{1}{r^2\sin^2\theta} \end{pmatrix}.$$

Let us now represent the dependence on the cyclic coordinate ϕ by means of the Fourier series

$$\boldsymbol{u}(r, \theta, \phi) = \boldsymbol{u}_0(r, \theta; \theta, \phi) + \sum_{\substack{m=-\infty \\ m \neq 0}}^{\infty} \boldsymbol{u}_m(r, \theta; \theta, \phi)\, e^{im\phi},$$

where

$$\boldsymbol{u}_0(r,\theta;\theta,\phi) = u_0^r(r,\theta)\,\widehat{\boldsymbol{r}}(\theta,\phi) + u_0^\theta(r,\theta)\,\widehat{\boldsymbol{\theta}}(\theta,\phi) + u_0^\phi(r,\theta)\,\widehat{\boldsymbol{\phi}}(\theta,\phi)$$

and

$$\boldsymbol{u}_m(r,\theta;\theta,\phi) = u_m^r(r,\theta)\,\widehat{\boldsymbol{r}}(\theta,\phi) + u_m^\theta(r,\theta)\,\widehat{\boldsymbol{\theta}}(\theta,\phi) + u_m^\phi(r,\theta)\,\widehat{\boldsymbol{\phi}}(\theta,\phi)\,\mathrm{i}.$$

In the two expressions above, $\widehat{\boldsymbol{r}}(\theta,\phi)$, $\widehat{\boldsymbol{\theta}}(\theta,\phi)$ and $\widehat{\boldsymbol{\phi}}(\theta,\phi)$ denote the unit vectors of the spherical coordinate system. The coefficients of the ϕ component of the Fourier modes with $m \neq 0$ have been multiplied by the imaginary unit to obtain a purely real representation of the second-order matrix operator acting on the Fourier coefficients of the vector mode.

The condition of reality of the vector field $\boldsymbol{u}$ expressed in spherical coordinates implies that its complex Fourier coefficients $\boldsymbol{u}_m = \boldsymbol{v}_m + \mathrm{i}\,\boldsymbol{w}_m$ with $m \neq 0$ have real and imaginary parts satisfying the following conditions

$$\begin{aligned} v_{-m}^r &= v_m^r, & w_{-m}^r &= -w_m^r;\\ v_{-m}^\theta &= v_m^\theta, & w_{-m}^\theta &= -w_m^\theta;\\ v_{-m}^\phi &= -v_m^\phi, & w_{-m}^\phi &= w_m^\phi. \end{aligned}$$

As a consequence, the Fourier series above of a real vector field $\boldsymbol{u}$ can be written also in the form

$$\begin{aligned} &\boldsymbol{u}(r,\theta,\phi)\\ &= \left\{u_0^r(r,\theta) + 2\sum_{m=1}^{\infty}\Big[v_m^r(r,\theta)\cos(m\phi) - w_m^r(r,\theta)\sin(m\phi)\Big]\right\}\widehat{\boldsymbol{r}}(\theta,\phi)\\ &+ \left\{u_0^\theta(r,\theta) + 2\sum_{m=1}^{\infty}\Big[v_m^\theta(r,\theta)\cos(m\phi) - w_m^\theta(r,\theta)\sin(m\phi)\Big]\right\}\widehat{\boldsymbol{\theta}}(\theta,\phi)\\ &+ \left\{u_0^\phi(r,\theta) - 2\sum_{m=1}^{\infty}\Big[v_m^\phi(r,\theta)\sin(m\phi) + w_m^\phi(r,\theta)\cos(m\phi)\Big]\right\}\widehat{\boldsymbol{\phi}}(\theta,\phi) \end{aligned}$$

which may be more convenient for the computations.

By virtue of the adopted Fourier expansion, the operator representing the Laplace operator ∇^2 in the space of the Fourier coefficients of a scalar function is

$$\partial_m^2 = \frac{1}{r^2}\frac{\partial}{\partial r}\left(r^2\frac{\partial}{\partial r}\right) + \frac{1}{r^2\sin\theta}\frac{\partial}{\partial\theta}\left(\sin\theta\frac{\partial}{\partial\theta}\right) - \frac{m^2}{r^2\sin^2\theta},$$

whereas the operator representing the Laplace matrix operator $\boldsymbol{\nabla}^2$ in the space of the Fourier coefficients of a vector field is

$$\boldsymbol{\partial}_m^2 = \begin{pmatrix} \partial_m^2 - \frac{2}{r^2} & -\frac{2}{r^2\sin\theta}\frac{\partial}{\partial\theta}(\sin\theta\ldots) & \frac{2m}{r^2\sin\theta}\\ \frac{2}{r^2}\frac{\partial}{\partial\theta} & \partial_m^2 - \frac{1}{r^2\sin^2\theta} & \frac{2m\cos\theta}{r^2\sin^2\theta}\\ \frac{2m}{r^2\sin\theta} & \frac{2m\cos\theta}{r^2\sin^2\theta} & \partial_m^2 - \frac{1}{r^2\sin^2\theta} \end{pmatrix}.$$

Thus, for $m \neq 0$, the three spherical components of the vector Fourier mode are coupled together, whilst for the first mode, $m = 0$, only the first two componenst are coupled.

The similarity trasformation which uncouples the three equations is constructed in two steps: first, the coupling between the r and θ components due to the off-diagonal terms containing the first-order derivative $\frac{\partial}{\partial\theta}$ is eliminated; then, the remaining coupling between the first new variable and the third unchanged variable (the ϕ component) is eliminated by means of the same similarity transformation considered in the analysis of the cylindrical coordinates.

Let us consider the (partial) change of variables defined by the linear transformation

$$\boldsymbol{T}(\theta) = \begin{pmatrix} \sin\theta & \cos\theta & 0 \\ \cos\theta & -\sin\theta & 0 \\ 0 & 0 & 1 \end{pmatrix}, \qquad \begin{pmatrix} \overline{u}_m^1 \\ \overline{u}_m^2 \\ u_m^\phi \end{pmatrix} = \boldsymbol{T}(\theta) \begin{pmatrix} u_m^r \\ u_m^\theta \\ u_m^\phi \end{pmatrix}.$$

As for matrix $\boldsymbol{t}(\phi)$ occurring in the previous analysis of polar coordinates, matrix $\boldsymbol{T}(\theta)$ is such that $\boldsymbol{T}(\theta) = \boldsymbol{T}^{\mathrm{tr}}(\theta) = \boldsymbol{T}^{-1}(\theta)$. Furthermore, by standard calculations, it is possible to show that the similarity trasformation provided by matrix $\boldsymbol{T}(\theta)$ gives the following partial diagonalization of the matrix operator $\boldsymbol{\partial}_m^2$:

$$\boldsymbol{T}(\theta)\,\boldsymbol{\partial}_m^2\,\boldsymbol{T}(\theta) = \begin{pmatrix} \partial_m^2 - \frac{1}{r^2\sin^2\theta} & 0 & \frac{2m}{r^2\sin\theta} \\ 0 & \partial_m^2 & 0 \\ \frac{2m}{r^2\sin\theta} & 0 & \partial_m^2 - \frac{1}{r^2\sin^2\theta} \end{pmatrix}.$$

It is important to note that, even though the matrix $\boldsymbol{T}(\theta)$ is coincident (apart from its dimension) with $\boldsymbol{t}(\phi)$, this demonstration is different from, and less elementary than, that given for the polar coordinates because of the presence of the function $\sin\theta$ inside and outside the derivative $\frac{\partial}{\partial\theta}$ in the operator ∂_m^2 as well as in its matrix counterpart $\boldsymbol{\partial}_m^2$.

Very easy is instead the elimination of the coupling still remaining between the first (new) and the third (old) components of the vector mode, the structure of the coupling being identical to that encountered for the cylindrical coordinates. Thus, introducing a second linear trasformation defined by the matrix

$$\boldsymbol{S} = \frac{1}{\sqrt{2}} \begin{pmatrix} 1 & 0 & 1 \\ 0 & \sqrt{2} & 0 \\ 1 & 0 & -1 \end{pmatrix}, \qquad \begin{pmatrix} u_m^1 \\ u_m^2 \\ u_m^3 \end{pmatrix} = \boldsymbol{S} \begin{pmatrix} \overline{u}_m^1 \\ \overline{u}_m^2 \\ u_m^\phi \end{pmatrix},$$

with, as before, $\boldsymbol{S} = \boldsymbol{S}^{\mathrm{tr}} = \boldsymbol{S}^{-1}$, one obtains immediately

$$\boldsymbol{S}\boldsymbol{T}(\theta)\,\boldsymbol{\partial}_m^2\,\boldsymbol{T}(\theta)\,\boldsymbol{S} = \begin{pmatrix} \partial_{m-1}^2 & 0 & 0 \\ 0 & \partial_m^2 & 0 \\ 0 & 0 & \partial_{m+1}^2 \end{pmatrix}.$$

Therefore, defining the complete trasformation $\boldsymbol{R}(\theta) = \boldsymbol{S}\boldsymbol{T}(\theta)$, it is immediate to obtain

$$\boldsymbol{R}(\theta) = \frac{1}{\sqrt{2}} \begin{pmatrix} \sin\theta & \cos\theta & 1 \\ \sqrt{2}\cos\theta & -\sqrt{2}\sin\theta & 0 \\ \sin\theta & \cos\theta & -1 \end{pmatrix}, \qquad \begin{pmatrix} u_m^1 \\ u_m^2 \\ u_m^3 \end{pmatrix} = \boldsymbol{R}(\theta) \begin{pmatrix} u_m^r \\ u_m^\theta \\ u_m^\phi \end{pmatrix},$$

and the previous complete diagonalization of the matrix operator $\boldsymbol{\partial}_m^2$ can be written as

$$\boldsymbol{R}(\theta)\, \boldsymbol{\partial}_m^2\, \boldsymbol{R}(\theta)^{\mathrm{tr}} = \begin{pmatrix} \partial_{m-1}^2 & 0 & 0 \\ 0 & \partial_m^2 & 0 \\ 0 & 0 & \partial_{m+1}^2 \end{pmatrix},$$

since $\boldsymbol{T}(\theta)\,\boldsymbol{S} = \boldsymbol{T}(\theta)^{\mathrm{tr}}\,\boldsymbol{S}^{\mathrm{tr}} = (\boldsymbol{S}\boldsymbol{T}(\theta))^{\mathrm{tr}} = \boldsymbol{R}(\theta)^{\mathrm{tr}}$, by the properties of matrices $\boldsymbol{T}(\theta)$ and $\boldsymbol{S}$. It is interesting to note that the same properties imply $\boldsymbol{T}(\theta)\,\boldsymbol{S} = \boldsymbol{T}(\theta)^{-1}\,\boldsymbol{S}^{-1} = (\boldsymbol{S}\boldsymbol{T}(\theta))^{-1} = \boldsymbol{R}(\theta)^{-1}$, so that $\boldsymbol{R}(\theta)^{\mathrm{tr}} = \boldsymbol{R}(\theta)^{-1}$ and $\boldsymbol{R}(\theta)$ is orthogonal, but $\boldsymbol{R}(\theta) \neq \boldsymbol{R}(\theta)^{\mathrm{tr}}$ since $\boldsymbol{T}(\theta)$ and $\boldsymbol{S}$ do not commute.

The solution of the equations for the mth Fourier mode with $m \neq 0$ proceedes as follows: first, the trasformation $\boldsymbol{R}(\theta)$ is applied to the Fourier coefficient $\boldsymbol{f}_m$ of the source term $\boldsymbol{f}$ of the Poisson equation; then, the three uncoupled elliptic 2D equations

$$\begin{aligned} \partial_{m-1}^2 u_m^1 &= f_m^1, \\ \partial_m^2 u_m^2 &= f_m^2, \\ \partial_{m+1}^2 u_m^3 &= f_m^3, \end{aligned}$$

are solved; finally, the spherical components of the mth Fourier mode are obtained through the backtransformation

$$\begin{pmatrix} u_m^r \\ u_m^\theta \\ u_m^\phi \end{pmatrix} = \boldsymbol{R}(\theta)^{\mathrm{tr}} \begin{pmatrix} u_m^1 \\ u_m^2 \\ u_m^3 \end{pmatrix}.$$

The problem for the first Fourier mode $m = 0$ is simpler, since the corresponding matrix operator in the Fourier space is

$$\boldsymbol{\partial}_0^2 = \begin{pmatrix} \partial_0^2 - \frac{2}{r^2} & -\frac{2}{r^2 \sin\theta}\frac{\partial}{\partial\theta}(\sin\theta ...) & 0 \\ \frac{2}{r^2}\frac{\partial}{\partial\theta} & \partial_1^2 & 0 \\ 0 & 0 & \partial_1^2 \end{pmatrix}.$$

Thus, the similarity trasformation performing the diagonalization is easily found to be

$$\boldsymbol{T}(\theta)\, \boldsymbol{\partial}_0^2\, \boldsymbol{T}(\theta) = \begin{pmatrix} \partial_1^2 & 0 & 0 \\ 0 & \partial_0^2 & 0 \\ 0 & 0 & \partial_1^2 \end{pmatrix}.$$

The solution precedure for the first mode $m = 0$ amounts therefore to transforming the source term $\boldsymbol{f}_0$ by means of matrix $\boldsymbol{T}(\theta)$, solving the three uncoupled equations

$$\partial_1^2 u_0^1 = f_0^1,$$
$$\partial_0^2 u_0^2 = f_0^2,$$
$$\partial_1^2 u_0^3 = f_0^3,$$

and backtrasforming the computed solutions through

$$\begin{pmatrix} u_0^r \\ u_0^\theta \\ u_0^\phi \end{pmatrix} = \boldsymbol{T}(\theta) \begin{pmatrix} u_0^1 \\ u_0^2 \\ u_0^3 \end{pmatrix}.$$

Needless to say, all of the transformations above must be performed also on the data of the boundary conditions. Moreover, steps of Fourier analysis and synthesis must precede and follow the solution procedure.

The same procedure can be applied to solve Dirichlet vector problems for the Helmholtz operator $(\nabla^2 - \gamma)$, with $\gamma > 0$, as well as for more general elliptic operators of the form $(\nabla^2 - \Gamma(r, \phi))$, where the function $\Gamma(r, \phi)$ can be subject to suitable conditions.

To conclude, we recall that the similarity transformation described in this Appendix can be combined with existing algorithms for the numerical solution of scalar elliptic equations in two dimensions to obtain solution methods for vector Dirichlet problems in any plane region and in annular three-dimensional regions.

Appendix C

Spatial difference operators

C.1 Introduction

This Appendix contains the various difference operators which are involved in the spatial discretization of the transient advection equation for a scalar unknown by means of finite elements. We consider the equations both in two and three dimensions and assume a uniform mesh of square and cubic elements, respectively, with a multinear interpolation of the unknown variable. We will give also the explicit expressions of the operators produced by the application of the Taylor–Galerkin method to the advection equation. The Fourier representation of the various spatial operators is provided to allow the study of numerical stability and response properties of integration schemes for multidimensional problems.

C.2 2D equation: four-node bilinear element

Let a variable $u(x,y)$ be approximated over a uniform mesh of square elements of size h by means of bilinear functions which match with continuity at interelement boundaries. When the Galerkin method is applied to solve an advection equation of the type $u_t + (\boldsymbol{a} \cdot \boldsymbol{\nabla})u = s$, approximately, one obtains a semidiscrete equation which involves spatial difference operators of first order and also of second order due to the presence of the consistent mass matrix. Such operators are obtained from the following elementary first-order operators:

$$
\begin{aligned}
(\Delta_x U)_{j,k} &= \tfrac{1}{2}(U_{j+1,k} - U_{j-1,k}), \\
(\Delta_y U)_{j,k} &= \tfrac{1}{2}(U_{j,k+1} - U_{j,k-1}), \\
(\Delta_{xy} U)_{j,k} &= \tfrac{1}{2}(U_{j+1,k+1} - U_{j-1,k-1}), \\
(\Delta_{yx} U)_{j,k} &= \tfrac{1}{2}(U_{j+1,k-1} - U_{j-1,k+1}),
\end{aligned}
$$

$$\widehat{\Delta}_x = \tfrac{2}{3}\left(\Delta_x + \tfrac{1}{4}\Delta_{xy} + \tfrac{1}{4}\Delta_{yx}\right),$$
$$\widehat{\Delta}_y = \tfrac{2}{3}\left(\Delta_y + \tfrac{1}{4}\Delta_{xy} - \tfrac{1}{4}\Delta_{yx}\right);$$

and from the following elementary second-order operators:

$$\left(\delta_x^2 U\right)_{j,k} = U_{j+1,k} - 2U_{j,k} + U_{j-1,k},$$
$$\left(\delta_y^2 U\right)_{j,k} = U_{j,k+1} - 2U_{j,k} + U_{j,k-1},$$
$$\left(\delta_{xy}^2 U\right)_{j,k} = U_{j+1,k+1} - 2U_{j,k} + U_{j-1,k-1},$$
$$\left(\delta_{yx}^2 U\right)_{j,k} = U_{j+1,k-1} - 2U_{j,k} + U_{j-1,k+1},$$

$$\widehat{\delta_x^2} = \tfrac{2}{3}\delta_x^2 + \tfrac{1}{6}\left(\delta_{xy}^2 + \delta_{yx}^2\right) - \tfrac{1}{3}\delta_y^2,$$
$$\widehat{\delta_y^2} = \tfrac{2}{3}\delta_y^2 + \tfrac{1}{6}\left(\delta_{xy}^2 + \delta_{yx}^2\right) - \tfrac{1}{3}\delta_x^2.$$

The Galerkin method applied to the equation $u_t + (\boldsymbol{a}\cdot\boldsymbol{\nabla})u = 0$ gives the semidiscrete equation

$$M\frac{dU}{dt} = -\frac{1}{h}\boldsymbol{a}\cdot\widehat{\boldsymbol{\Delta}}\,U,$$

where

$$M = 1 + \tfrac{1}{9}\left(\delta_x^2 + \delta_y^2 + \tfrac{1}{4}\delta_{xy}^2 + \tfrac{1}{4}\delta_{yx}^2\right)$$
$$= \left(1 + \tfrac{1}{6}\delta_x^2\right)\left(1 + \tfrac{1}{6}\delta_y^2\right),$$

$$\widehat{\boldsymbol{\Delta}} = \left(\widehat{\Delta}_x, \widehat{\Delta}_y\right).$$

It is worth noting the factorization of the consistent mass matrix operator, which allowed to develop a very efficient method for the solution of the linear system of the consistent mass on Cartesian (possibly nonuniform) meshes (Staniforth and Mitchell 1978 and 1987).

The application of the Taylor–Galerkin method would lead instead to the fully discrete equation

$$M_{\mathrm{TG}}(\boldsymbol{\nu})(U^{n+1} - U^n) = -\boldsymbol{\nu}\cdot\widehat{\boldsymbol{\Delta}}\,U^n + K(\boldsymbol{\nu})U^n,$$

where $\boldsymbol{\nu} = (\Delta t/h)\boldsymbol{a}$ and

$$\widehat{\boldsymbol{\Delta}} = \left(\widehat{\Delta}_x, \widehat{\Delta}_y\right),$$
$$K(\boldsymbol{\nu}) = \nu_x^2\widehat{\delta_x^2} + 2\nu_x\nu_y\Delta_x\Delta_y + \nu_y^2\widehat{\delta_y^2},$$
$$M_{\mathrm{TG}}(\boldsymbol{\nu}) = M - \tfrac{1}{6}K(\boldsymbol{\nu}).$$

In other words, one has the following correspondences

$$\Delta t\, \boldsymbol{a}\cdot\boldsymbol{\nabla} \longrightarrow \boldsymbol{\nu}\cdot\widehat{\boldsymbol{\Delta}},$$
$$(\Delta t)^2(\boldsymbol{a}\cdot\boldsymbol{\nabla})^2 \longrightarrow K(\boldsymbol{\nu}).$$

Introducing a Fourier analysis with mode $\exp(\mathrm{i}\boldsymbol{k}\cdot\boldsymbol{x})$ and the dimensionless wave number $\boldsymbol{\xi} = h\boldsymbol{k}$, it is not difficult to find the Fourier image of the relevant difference operators:

$$\begin{aligned} M(\boldsymbol{\xi}) &= \tfrac{4}{9}\left(1+\tfrac{1}{2}\cos\xi\right)\left(1+\tfrac{1}{2}\cos\eta\right) \\ &= \left(1-\tfrac{2}{3}\sin^2\tfrac{1}{2}\xi\right)\left(1-\tfrac{2}{3}\sin^2\tfrac{1}{2}\eta\right), \end{aligned}$$

$$A(\boldsymbol{\xi},\boldsymbol{\nu}) = \tfrac{2}{3}\mathrm{i}\nu_x\sin\xi\left(1+\tfrac{1}{2}\cos\eta\right) + \tfrac{2}{3}\mathrm{i}\nu_y\sin\eta\left(1+\tfrac{1}{2}\cos\xi\right),$$

$$\begin{aligned} K(\boldsymbol{\xi},\boldsymbol{\nu}) = &-\tfrac{4}{3}\nu_x^2(1-\cos\xi)\left(1+\tfrac{1}{2}\cos\eta\right) - 2\nu_x\nu_y\sin\xi\,\sin\eta \\ &-\tfrac{4}{3}\nu_y^2(1-\cos\eta)\left(1+\tfrac{1}{2}\cos\xi\right), \end{aligned}$$

$$M_{\mathrm{TG}}(\boldsymbol{\xi},\boldsymbol{\nu}) = M(\boldsymbol{\xi}) - \tfrac{1}{6}K(\boldsymbol{\xi},\boldsymbol{\nu}).$$

Of course $A(\boldsymbol{\xi},\boldsymbol{\nu})$ denotes the Fourier transform of $\boldsymbol{\nu}\cdot\widehat{\boldsymbol{\Delta}}$.

C.3 3D equation: eight-node trilinear element

The approximate solution of the three-dimensional equation $u_t + (\boldsymbol{a}\cdot\boldsymbol{\nabla})u = 0$ on a uniform cubic mesh of trilinear elements requires to consider the following elementary spatial difference operators. The first-order operators are

$$\begin{aligned} (\Delta_x U)_{j,k,l} &= \tfrac{1}{2}(U_{j+1,k,l} - U_{j-1,k,l}), \\ (\Delta_y U)_{j,k,l} &= \tfrac{1}{2}(U_{j,k+1,l} - U_{j,k-1,l}), \\ (\Delta_z U)_{j,k,l} &= \tfrac{1}{2}(U_{j,k,l+1} - U_{j,k,l-1}), \end{aligned}$$

$$\begin{aligned} (\Delta_{xy} U)_{j,k,l} &= \tfrac{1}{2}(U_{j+1,k+1,l} - U_{j-1,k-1,l}), \\ (\Delta_{yx} U)_{j,k,l} &= \tfrac{1}{2}(U_{j+1,k-1,l} - U_{j-1,k+1,l}), \\ (\Delta_{zx} U)_{j,k,l} &= \tfrac{1}{2}(U_{j+1,k,l+1} - U_{j-1,k,l-1}), \\ (\Delta_{xz} U)_{j,k,l} &= \tfrac{1}{2}(U_{j+1,k,l-1} - U_{j-1,k,l+1}), \\ (\Delta_{yz} U)_{j,k,l} &= \tfrac{1}{2}(U_{j,k+1,l+1} - U_{j,k-1,l-1}), \\ (\Delta_{zy} U)_{j,k,l} &= \tfrac{1}{2}(U_{j,k+1,l-1} - U_{j,k-1,l+1}), \end{aligned}$$

$$(\Delta_{xyz}U)_{j,k,l} = \tfrac{1}{2}(U_{j+1,k+1,l+1} - U_{j-1,k-1,l-1}),$$
$$(\Delta_{xy\bar{z}}U)_{j,k,l} = \tfrac{1}{2}(U_{j+1,k+1,l-1} - U_{j-1,k-1,l+1}),$$
$$(\Delta_{x\bar{y}z}U)_{j,k,l} = \tfrac{1}{2}(U_{j+1,k-1,l+1} - U_{j-1,k+1,l-1}),$$
$$(\Delta_{\bar{x}yz}U)_{j,k,l} = \tfrac{1}{2}(U_{j-1,k+1,l+1} - U_{j+1,k-1,l-1}),$$

$$\widehat{\Delta}_x = \tfrac{4}{9}\left[\Delta_x + \tfrac{1}{4}(\Delta_{xy} + \Delta_{yx} + \Delta_{zx} + \Delta_{xz}) + \tfrac{1}{16}(\Delta_{xyz} + \Delta_{xy\bar{z}} + \Delta_{x\bar{y}z} - \Delta_{\bar{x}yz})\right],$$
$$\widehat{\Delta}_y = \tfrac{4}{9}\left[\Delta_y + \tfrac{1}{4}(\Delta_{xy} - \Delta_{yx} + \Delta_{yz} + \Delta_{zy}) + \tfrac{1}{16}(\Delta_{xyz} + \Delta_{xy\bar{z}} - \Delta_{x\bar{y}z} + \Delta_{\bar{x}yz})\right],$$
$$\widehat{\Delta}_z = \tfrac{4}{9}\left[\Delta_z + \tfrac{1}{4}(\Delta_{zx} - \Delta_{xz} + \Delta_{yz} - \Delta_{zy}) + \tfrac{1}{16}(\Delta_{xyz} - \Delta_{xy\bar{z}} + \Delta_{x\bar{y}z} + \Delta_{\bar{x}yz})\right].$$

The second-order operators are

$$\left(\delta_x^2 U\right)_{j,k,l} = U_{j+1,k,l} - 2U_{j,k,l} + U_{j-1,k,l},$$
$$\left(\delta_y^2 U\right)_{j,k,l} = U_{j,k+1,l} - 2U_{j,k,l} + U_{j,k-1,l},$$
$$\left(\delta_z^2 U\right)_{j,k,l} = U_{j,k,l+1} - 2U_{j,k,l} + U_{j,k,l-1},$$

$$\left(\delta_{xy}^2 U\right)_{j,k,l} = U_{j+1,k+1,l} - 2U_{j,k,l} + U_{j-1,k-1,l},$$
$$\left(\delta_{yx}^2 U\right)_{j,k,l} = U_{j+1,k-1,l} - 2U_{j,k,l} + U_{j-1,k+1,l},$$
$$\left(\delta_{zx}^2 U\right)_{j,k,l} = U_{j+1,k,l+1} - 2U_{j,k,l} + U_{j-1,k,l-1},$$
$$\left(\delta_{xz}^2 U\right)_{j,k,l} = U_{j+1,k,l-1} - 2U_{j,k,l} + U_{j-1,k,l+1},$$
$$\left(\delta_{yz}^2 U\right)_{j,k,l} = U_{j,k+1,l+1} - 2U_{j,k,l} + U_{j,k-1,l-1},$$
$$\left(\delta_{zy}^2 U\right)_{j,k,l} = U_{j,k+1,l-1} - 2U_{j,k,l} + U_{j,k-1,l+1},$$

$$\left(\delta_{xyz}^2 U\right)_{j,k,l} = U_{j+1,k+1,l+1} - 2U_{j,k,l} + U_{j-1,k-1,l-1},$$
$$\left(\delta_{xy\bar{z}}^2 U\right)_{j,k,l} = U_{j+1,k+1,l-1} - 2U_{j,k,l} + U_{j-1,k-1,l+1},$$
$$\left(\delta_{x\bar{y}z}^2 U\right)_{j,k,l} = U_{j+1,k-1,l+1} - 2U_{j,k,l} + U_{j-1,k+1,l-1},$$
$$\left(\delta_{\bar{x}yz}^2 U\right)_{j,k,l} = U_{j-1,k+1,l+1} - 2U_{j,k,l} + U_{j+1,k-1,l-1},$$

$$\widetilde{\delta^2} = \delta_{xy}^2 + \delta_{yx}^2 + \delta_{zx}^2 + \delta_{xz}^2 + \delta_{yz}^2 + \delta_{zy}^2,$$
$$\overline{\delta^2} = \delta_{xyz}^2 + \delta_{xy\bar{z}}^2 + \delta_{x\bar{y}z}^2 + \delta_{\bar{x}yz}^2,$$

$$
\begin{aligned}
\widehat{\delta_x^2} &= \tfrac{4}{9}\left[\delta_x^2 - \tfrac{1}{2}\left(\delta_y^2 + \delta_z^2\right) + \tfrac{1}{4}\widetilde{\delta^2} - \tfrac{3}{8}\left(\delta_{yz}^2 + \delta_{zy}^2\right) + \tfrac{1}{16}\overline{\delta^2}\right],\\
\widehat{\delta_y^2} &= \tfrac{4}{9}\left[\delta_y^2 - \tfrac{1}{2}\left(\delta_y^2 + \delta_z^2\right) + \tfrac{1}{4}\widetilde{\delta^2} - \tfrac{3}{8}\left(\delta_{zx}^2 + \delta_{xz}^2\right) + \tfrac{1}{16}\overline{\delta^2}\right],\\
\widehat{\delta_z^2} &= \tfrac{4}{9}\left[\delta_z^2 - \tfrac{1}{2}\left(\delta_x^2 + \delta_y^2\right) + \tfrac{1}{4}\widetilde{\delta^2} - \tfrac{3}{8}\left(\delta_{xy}^2 + \delta_{yz}^2\right) + \tfrac{1}{16}\overline{\delta^2}\right].
\end{aligned}
$$

The application of the Galerkin method to the advection equation in three dimensions gives the semidiscrete equation

$$M\frac{dU}{dt} = -\frac{1}{h}\boldsymbol{a}\cdot\widehat{\boldsymbol{\Delta}}\,U,$$

where

$$
\begin{aligned}
M &= 1 + \tfrac{2}{27}\left[\delta_x^2 + \delta_y^2 + \delta_z^2 + \tfrac{1}{4}\widetilde{\delta^2} + \tfrac{1}{16}\overline{\delta^2}\right],\\
\widehat{\boldsymbol{\Delta}} &= \left(\widehat{\Delta}_x, \widehat{\Delta}_y, \widehat{\Delta}_z\right).
\end{aligned}
$$

As in two dimensions, the consistent mass matrix allows a factorization in one-dimensional operators

$$M = \left(1 + \tfrac{1}{6}\delta_x^2\right)\left(1 + \tfrac{1}{6}\delta_y^2\right)\left(1 + \tfrac{1}{6}\delta_z^2\right)$$

which can be exploited for the efficient solution of advection problems over a Cartesian mesh of brick elements.

The application of the Taylor–Galerkin method introduces the additional operators

$$
\begin{aligned}
K(\boldsymbol{\nu}) &= \nu_x^2\widehat{\delta_x^2} + \nu_y^2\widehat{\delta_y^2} + \nu_z^2\widehat{\delta_z^2} + 2\nu_x\nu_y\Delta_x\Delta_y + 2\nu_x\nu_z\Delta_x\Delta_z + 2\nu_y\nu_z\Delta_y\Delta_z,\\
M_{\mathrm{TG}}(\boldsymbol{\nu}) &= M - \tfrac{1}{6}K(\boldsymbol{\nu}),
\end{aligned}
$$

where $\boldsymbol{\nu} = (\Delta t/h)\boldsymbol{a}$. Finally, the Fourier analysis of the numerical schemes can be done by means of the following explicit expressions

$$
\begin{aligned}
M(\boldsymbol{\xi}) &= \tfrac{8}{27}\left(1 + \tfrac{1}{2}\cos\xi\right)\left(1 + \tfrac{1}{2}\cos\eta\right)\left(1 + \tfrac{1}{2}\cos\zeta\right)\\
&= \left(1 - \tfrac{2}{3}\sin^2\tfrac{1}{2}\xi\right)\left(1 - \tfrac{2}{3}\sin^2\tfrac{1}{2}\eta\right)\left(1 - \tfrac{2}{3}\sin^2\tfrac{1}{2}\zeta\right),
\end{aligned}
$$

$$
\begin{aligned}
A(\boldsymbol{\xi},\boldsymbol{\nu}) &= \tfrac{4}{9}\mathrm{i}\nu_x\sin\xi\left(1 + \tfrac{1}{2}\cos\eta\right)\left(1 + \tfrac{1}{2}\cos\zeta\right)\\
&\quad + \tfrac{4}{9}\mathrm{i}\nu_y\sin\eta\left(1 + \tfrac{1}{2}\cos\xi\right)\left(1 + \tfrac{1}{2}\cos\zeta\right)\\
&\quad + \tfrac{4}{9}\mathrm{i}\nu_y\sin\zeta\left(1 + \tfrac{1}{2}\cos\xi\right)\left(1 + \tfrac{1}{2}\cos\eta\right),
\end{aligned}
$$

$$
\begin{aligned}
K(\boldsymbol{\xi},\boldsymbol{\nu}) &= -\tfrac{8}{9}\nu_x^2(1 - \cos\xi)\left(1 + \tfrac{1}{2}\cos\eta\right)\left(1 + \tfrac{1}{2}\cos\zeta\right)\\
&\quad - \tfrac{8}{9}\nu_y^2(1 - \cos\eta)\left(1 + \tfrac{1}{2}\cos\xi\right)\left(1 + \tfrac{1}{2}\cos\zeta\right)\\
&\quad - \tfrac{8}{9}\nu_z^2(1 - \cos\zeta)\left(1 + \tfrac{1}{2}\cos\xi\right)\left(1 + \tfrac{1}{2}\cos\eta\right)\\
&\quad - 2\nu_x\nu_y\sin\xi\,\sin\eta - 2\nu_x\nu_z\sin\xi\,\sin\zeta - 2\nu_y\nu_z\sin\eta\,\sin\eta.
\end{aligned}
$$

Appendix D

Time derivative of integrals over moving domains

In this Appendix we provide some vector differential identities expressing the time derivative of integrals over domains which move and change in shape in an arbitrary manner. We consider the integrals over curves and surfaces moving and deforming themselves in the three-dimensional space, as well as integrals over three-dimensional regions whose boundary has a velocity specified at each point.

These identities generalize to the three-dimensional space the well known theorem of calculus expressing the derivative of a definite integral with respect to a variable which the limits of integration and possibly the integrand depend on. If the variable involved in the derivation is denoted by t and referred to as "time," and the integration interval $\mathcal{I}_t = [a(t), b(t)]$ depends on time, the theorem states

$$\frac{d}{dt}\int_{a(t)}^{b(t)} f(x,t)\,dx = \int_{a(t)}^{b(t)} \frac{\partial f(x,t)}{\partial t}\,dx + f(b(t),t)\frac{db(t)}{dt} - f(a(t),t)\frac{da(t)}{dt}.$$

The time derivative of $a(t)$ and $b(t)$ represents the "velocity" of the end points of $\mathcal{I}_t$, according to the definition:

$$v_a(t) = \frac{da(t)}{dt} \qquad \text{and} \qquad v_b(t) = \frac{db(t)}{dt},$$

so that the previous identity can also be written in the form

$$\frac{d}{dt}\int_{a(t)}^{b(t)} f(x,t)\,dx = \int_{a(t)}^{b(t)} \frac{\partial f(x,t)}{\partial t}\,dx + f(b(t),t)\,v_b(t) - f(a(t),t)\,v_a(t).$$

This expression is now generalized to integrals expressing the circulation along a curve and the flux across a surface both evolving in the three-dimensional space in a known manner. The expression of the time derivative of the three-dimensional

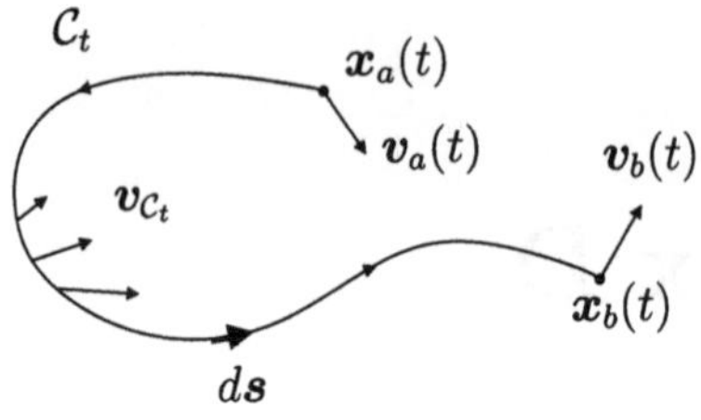

Figure D.1: Curve moving and deforming in space.

integrals of scalar and vector fields over a moving and deforming volume will be also given.

The scalar functions and vector fields to be integrated in the following are defined not only on the integration domain but in the entire spatial region where the integration domain is evolving. We assume that the various functions are smooth enough so that the standard operations of calculus may be performed on them.

D.1 Circulation along a moving curve

Let us consider a continuous curve $\mathcal{C}_t$ which moves and deforms itself with time and let $\boldsymbol{x}_{\mathcal{C}_t}(s)$ be its parametric representation at time t with s the arclength parameter. The end points of $\mathcal{C}_t$ will be (see Figure D.1)

$$\boldsymbol{x}_a(t) = \boldsymbol{x}_{\mathcal{C}_t}(s_a(t)) \qquad \text{and} \qquad \boldsymbol{x}_b(t) = \boldsymbol{x}_{\mathcal{C}_t}(s_b(t)).$$

The velocity of points of $\mathcal{C}_t$ at time t will be indicated by

$$\boldsymbol{v}_{\mathcal{C}_t} = \boldsymbol{v}_{\mathcal{C}_t}(\boldsymbol{x}_{\mathcal{C}_t}),$$

where $\boldsymbol{x}_{\mathcal{C}_t} \in \mathcal{C}_t$.

Let $\boldsymbol{F}(\boldsymbol{x}, t)$ denote a time-dependent vector field which is defined in the three-dimensional region of evolution of $\mathcal{C}_t$. The generalization of the previous differential identity to the circulation integral of $\boldsymbol{F}(\boldsymbol{x}, t)$ along the moving curve $\mathcal{C}_t$ is:

$$\frac{d}{dt}\int_{\mathcal{C}_t} \boldsymbol{F} \cdot d\boldsymbol{s} = \int_{\mathcal{C}_t} \frac{\partial \boldsymbol{F}}{\partial t} \cdot d\boldsymbol{s} + \int_{\mathcal{C}_t} (\boldsymbol{\nabla} \times \boldsymbol{F}) \times \boldsymbol{v}_{\mathcal{C}_t} \cdot d\boldsymbol{s}$$
$$+ \boldsymbol{F}(\boldsymbol{x}_b(t), t) \cdot \boldsymbol{v}_b(t) - \boldsymbol{F}(\boldsymbol{x}_a(t), t) \cdot \boldsymbol{v}_a(t),$$

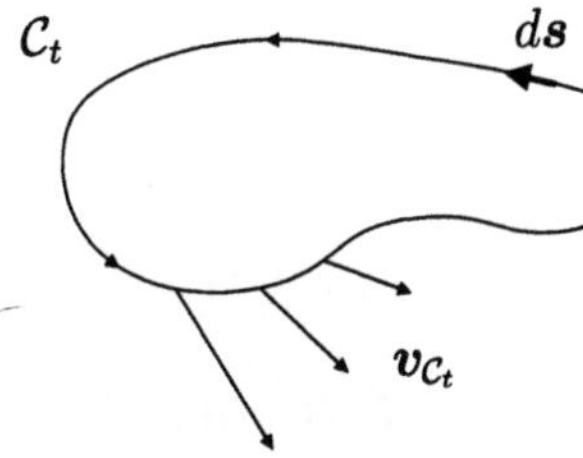

Figure D.2: Closed curve moving in space.

where, of course,

$$v_a(t) = v_{C_t}(x_a(t)) \qquad \text{and} \qquad v_b(t) = v_{C_t}(x_b(t)).$$

In particular, if the curve $\mathcal{C}_t$ is assumed to be *closed* (see Figure D.2), this result simplifies to

$$\frac{d}{dt}\oint_{\mathcal{C}_t} \boldsymbol{F}\cdot d\boldsymbol{s} = \oint_{\mathcal{C}_t} \frac{\partial \boldsymbol{F}}{\partial t}\cdot d\boldsymbol{s} + \oint_{\mathcal{C}_t} (\boldsymbol{\nabla}\times\boldsymbol{F})\times\boldsymbol{v}_{\mathcal{C}_t}\cdot d\boldsymbol{s}.$$

Notice that this identity gives a straightforward proof of Kelvin circulation theorem.

D.2 Flux across a moving surface

Let us consider the flux of a vector field across a surface $\mathcal{S}_t$ which has a boundary $\mathcal{C}_t$ and which moves with a velocity given by

$$\boldsymbol{v}_{\mathcal{S}_t} = \boldsymbol{v}_{\mathcal{S}_t}(\boldsymbol{x}_{\mathcal{S}_t}),$$

where $\boldsymbol{x}_{\mathcal{S}_t} \in \mathcal{S}_t$ (see Figure D.3).

The time derivative of the flux of $\boldsymbol{F}(\boldsymbol{x}, t)$ across $\mathcal{S}_t$ is given by the following identity

$$\frac{d}{dt}\iint_{\mathcal{S}_t} \boldsymbol{F}\cdot\boldsymbol{n}\,dS = \iint_{\mathcal{S}_t} \frac{\partial \boldsymbol{F}}{\partial t}\cdot\boldsymbol{n}\,dS + \iint_{\mathcal{S}_t} (\boldsymbol{\nabla}\cdot\boldsymbol{F})\,\boldsymbol{v}_{\mathcal{S}_t}\cdot\boldsymbol{n}\,dS + \oint_{\mathcal{C}_t} \boldsymbol{F}\times\boldsymbol{v}_{\mathcal{C}_t}\cdot d\boldsymbol{s},$$

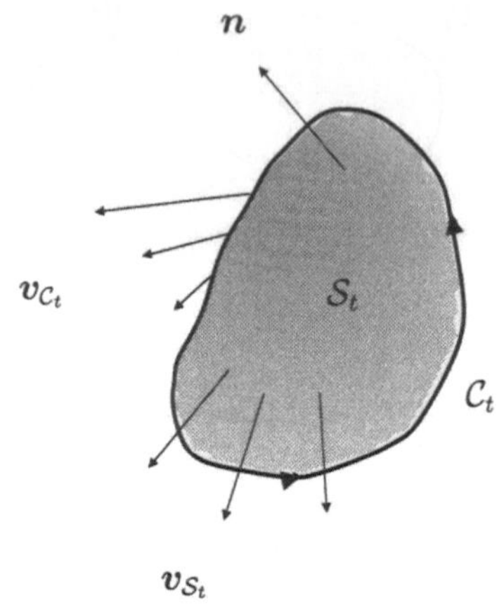

Figure D.3: Bounded surface moving and deforming in space.

where, of course, $\boldsymbol{v}_{\mathcal{C}_t} = \boldsymbol{v}_{\mathcal{S}_t}(\boldsymbol{x}_{\mathcal{C}_t})$. This identity allows a direct demonstration of Helmholtz first vorticity theorem, see, *e.g.*, Acheson [1, p. 162].

In particular, if the surface $\mathcal{S}_t$ is *closed* (see Figure D.4), the curvilinear integral along the boundary $\mathcal{C}_t$ disappears and the time derivative of the flux becomes

$$\frac{d}{dt}\oint\!\!\!\oint_{\mathcal{S}_t} \boldsymbol{F}\cdot\boldsymbol{n}\,dS = \oint\!\!\!\oint_{\mathcal{S}_t} \frac{\partial \boldsymbol{F}}{\partial t}\cdot\boldsymbol{n}\,dS + \oint\!\!\!\oint_{\mathcal{S}_t} (\boldsymbol{\nabla}\cdot\boldsymbol{F})\,\boldsymbol{v}_{\mathcal{S}_t}\cdot\boldsymbol{n}\,dS.$$

D.3 Integrals over a moving volume

Consider finally a varying three-dimensional domain $\mathcal{V}_t$ bounded by the closed surface $\mathcal{S}_t$ whose points at time t move with the velocity

$$\boldsymbol{v}_{\mathcal{S}_t} = \boldsymbol{v}_{\mathcal{S}_t}(\boldsymbol{x}_{\mathcal{S}_t}).$$

The derivative of the integral of a scalar function $f(\boldsymbol{x},t)$ over the varying domain $\mathcal{V}_t$ with respect to time is given by the well known expression:

$$\frac{d}{dt}\iiint_{\mathcal{V}_t} f(\boldsymbol{x},t)\,dV = \iiint_{\mathcal{V}_t} \frac{\partial f(\boldsymbol{x},t)}{\partial t}\,dV + \oint\!\!\!\oint_{\mathcal{S}_t} f\,\boldsymbol{v}_{\mathcal{S}_t}\cdot\boldsymbol{n}\,dS.$$

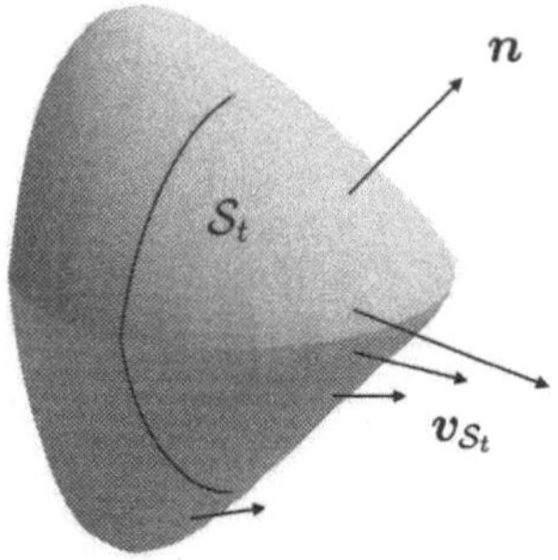

Figure D.4: Closed surface moving and deforming in space.

The last term represents the flux of the quantity f across the moving boundary of the integration domain.

A similar expression holds for the integral of a vector quantity $\boldsymbol{F}(\boldsymbol{x}, t)$ which is defined over a three-dimensional region containing $\mathcal{V}_t$, namely,

$$\frac{d}{dt}\iiint_{\mathcal{V}_t} \boldsymbol{F}(\boldsymbol{x}, t)\, dV = \iiint_{\mathcal{V}_t} \frac{\partial \boldsymbol{F}(\boldsymbol{x}, t)}{\partial t}\, dV + \oiint_{\mathcal{S}_t} \boldsymbol{F}\,(\boldsymbol{v}_{\mathcal{S}_t}\cdot \boldsymbol{n})\, dS.$$

It is to be remarked that these results are obtained without considering any velocity field at points in the interior of the integration domain $\mathcal{V}_t$. In other words, the identities provide the relationship implied by the *kinematics* of the boundary $\mathcal{S}_t$ and by the possible time dependence of the integrand, with no reference to any dynamical aspect associated with the field variable appearing in the integral.

The two vector identities above can be useful to establish the relationship existing between the local and material (substantial) forms of the conservation laws for a fluid. In fact they eliminate the need of Lagrangian coordinates and of the change of variables associated with the flow map which gives the configuration at time t in terms of the initial one. In particular, the analysis of the conservation of momentum will be possible using the following extension of the divergence theorem:

$$\oiint_{\mathcal{S}} \boldsymbol{F}\,(\boldsymbol{G}\cdot \boldsymbol{n})\, dS = \iiint_{\mathcal{V}} \Big[\boldsymbol{F}\,(\boldsymbol{\nabla}\cdot \boldsymbol{G}) + (\boldsymbol{G}\cdot \boldsymbol{\nabla})\boldsymbol{F}\Big]\, dV,$$

which holds for any two differentiable vector fields $\boldsymbol{F}$ and $\boldsymbol{G}$.

[illegible]

Bibliography

GENERAL REFERENCES

[1] Acheson D J *Elementary Fluid Dynamics*, Clarendon Press, Oxford, 1990.

[2] Adams R A *Calculus of Several Variables*, Addison-Wesley, Don Mills, Ontario, 1991.

[3] Batchelor G K *An Introduction to Fluid Dynamics*, Cambridge University Press, Cambridge, 1970.

[4] Bernardi C and Maday Y *Approximations Spectrales des Problèmes aux Limites Elliptiques*, Springer-Verlag, 1992.

[5] Brezis H *Analyse Fonctionnelle. Théorie et Applications*, Masson, Paris, 1983.

[6] Canuto C, Hussaini M Y, Quarteroni A and Zang T A *Spectral Methods in Fluid Mechanics*, Springer-Verlag, New York, 1988.

[7] Carey G F and Oden J T *Finite Elements: Fluid Mechanics*, Prentice-Hall, Englewood Cliffs, New Jersey, 1986.

[8] Chilov G E *Analyse Mathématique. Fonctions d'une Variable*, Vol. 3, Editions Mir, Moscou, 1973.

[9] Chorin A J and Marsden J E *A Mathematical Introduction to Fluid Dynamics*, Second Edition, Springer-Verlag, New York, 1990.

[10] Churchill S W *A Viscous Flows: The Practical Use of Theory*, Butterworth, Boston, 1988.

[11] Currie I G *Fundamentals Mechanics of Fluids*, Mc Graw-Hill, Toronto, 1974.

[12] Dautray R and Lions J-L *Mathematical Analysis and Numerical Methods for Science and Technology*, Vol. 3, Springer-Verlag, Berlin, 1990.

[13] Feynman R P, Leighton R B and Sands M L *The Feynman Lectures*, Vol. II, Commemorative Issue, Addison Wesley, Reading Massachusetts, 1989.

[14] Fletcher C A J *Computational Techniques for Fluid Dynamics*, Vols. I and II, Springer Series in Computational Physics, Springer-Verlag, Berlin, 1988.

[15] Fox L and Parker I *Chebyshev Polynomials in Numerical Analysis*, Oxford University Press, London/New York, 1968.

[16] Girault V and Raviart P-A *Finite Element Methods for Navier–Stokes Equations. Theory and Algorithms*, Springer Series in Computational Mathematics, Vol. 5, Springer-Verlag, Berlin, 1986.

[17] Glowinski R *Numerical Methods for Nonlinear Variational Problems*, Springer Series in Computational Physics, Springer-Verlag, New York, 1984.

[18] Gottlieb D and Orszag S O *Numerical Analysis of Spectral Methods: Theory and Applications*, SIAM, Philadelphia, 1977.

[19] Ingham D B and Kelmanson M A *Boundary Integral Equation Analyses of Singular, Potential and Biharmonic Problems*, Lecture Notes in Engineering **7**, Springer-Verlag, Berlin, 1984.

[20] Jackson J D *Classical Electrodynamics*, Second edition, John Wiley, New York, 1975.

[21] Jaswon M A and Symm G T *Integral Equations Methods in Potential Theory and Elastostatics*, Academic Press, London, 1977.

[22] Karamcheti K *Principles of Ideal–Fluid Aerodynamics*, John Wiley, New York, 1966.

[23] Kreiss H-O and Lorenz J *Initial-Boundary Value Problems and the Navier–Stokes Equations*, Academic Press, San Diego, California, 1989.

[24] Ladyzhenskaya O A *The Mathematical Theory of Viscous Incompressible Flow*, Gordon and Breach, New York, 1969.

[25] Lanczos C *Linear Differential Operators*, Van Nostrand, London, 1961.

[26] Landau L D and Lifshitz E M *Course of Theoretical Physics, Volume 6, Fluid Mechanics*, Pergamon Press, Oxford, 1959.

[27] Lighthill J *An Informal Introduction to Theoretical Fluid Mechanics*, Clarendon Press, Oxford, 1986.

[28] Lugt H J *Vortex Flow in Nature and Technology*, John Wiley, New York, 1983.

[29] Oden J T *Applied Functional Analysis*, Prentice-Hall, Englewood Cliffs, New Jersey, 1979.

[30] Pironneau O *Méthodes des Éléments Finis pour les Fluides*, Masson, Paris, 1988.

[31] Saffman P G *Dynamics of Vorticity*, Cambridge University Press, Cambridge, 1993.

[32] Strang G *Linear Algebra and its Applications*, Mc Graw-Hill, New York, 1977.

[33] Strang G and Fix G J *An Analysis of the Finite Element Method*, Prentice-Hall, Englewood Cliffs, New Jersey, 1973.

[34] Telionis D P *Unsteady Viscous Flows*, Springer Series in Computational Physics, Springer-Verlag, New York, 1981.

[35] Temam R *Theory and Numerical Analysis of the Navier–Stokes Equations*, North-Holland, Amsterdam, 1977.

[36] Thomasset F *Finite Element Methods for Navier–Stokes Equations*, Springer Series in Computational Physics, Springer-Verlag, New York, 1981.

[37] Ting L and Klein R *Viscous Vortical Flows*, Lecture Notes in Physics, No. 374, Springer-Verlag, Berlin, 1991.

[38] Weatherburn C E *Differential Geometry of Three Dimensions*, Cambridge University Press, Cambridge, 1961 (first edition 1921).

ARTICLES

[ ] Abdallah S 1987 "Numerical solution for the pressure Poisson equation with Neumann boundary conditions using a non-staggered grid," *J. Comput. Phys.* **70** 182–192.

[ ] Achard J-L and Canot E 1988 "Formulation integro-différentielle pour les écoulements transitoires d'un fluide isovolume et visqueux," *J. Méca. Théor. Appl.* **7** 645–682.

[ ] Anwar M and Dennis S C R 1988 "Numerical methods for steady viscous flow problems," *Computers and Fluids* **16** 1–20.

[ ] Arakawa A 1966 "Computational design for long-term numerical integration of the equations of fluid motion: two-dimensional incompressible flow. Part I," *J. Comput. Phys.* **1** 119–143.

[ ] Badr H M and Dennis S C R 1985 "Time-dependent viscous flow past an impulsively started rotating and translating circular cylinder," *J. Fluid Mech.* **158** 447–488.

[ ] Basset A B 1887 "On the motion of a sphere in a viscous fluid," *Phil. Trans.* **clxxix** 43; see also *Hydrodynamics*, c. xxii, Cambridge, 1888.

[ ] Bendali A, Dominguez J M and Gallic S 1985 "A variational approach for the vector potential formulation of the Stokes and Navier–Stokes problems in three-dimensional domains," *J. Math. Anal. Appl.* **107** 537–560.

[ ] Bernardi C 1979 "Méthode d'éléments finis mixtes pour les équations de Navier–Stokes," *Thèse de 3ème Cycle*, Université Paris VI.

[ ] Boussinesq J 1885 "Sur la résistance qu'oppose un liquide indéfini en repos, sans pesanteur, au mouvement varié d'une sphere solide qu'il mouille sur toute sa surface, quand les vitesses restent bien continues et assez faibles pour que leur carrés et produits soient négligeables," *Compt. Rend. Acad. Sci. Paris* **100** 935.

[ ] Bristeau M O, Glowinski R and Périaux J 1987 "Numerical methods for the Navier–Stokes equations. Applications to the simulation of compressible and incompressible viscous flows," *Computer Physics Report*, **6** 73–187, North-Holland, Amsterdam, 1987.

[ ] Burggraf O R 1966 "Analytical and numerical studies of the structure of steady separated flows," *J. Fluid Mech.* **24** 113–151.

[ ] Campion-Renson A and Crochet M J 1978 "On the stream function–vorticity finite element solution of the Navier–Stokes equations," *Int. J. Numer. Meth. Eng.* **12** 1809–1818.

[ ] Canuto C and Sacchi Landriani G 1986 "Analysis of the Kleiser–Schumann method," *Numer. Math.* **50** 217–243.

[ ] Cheng Sin-I 1975 "A critical review of numerical solution of Navier–Stokes equations," *Progress in Numerical Fluid Dynamics*, ed. Wirz H J, *Lecture Notes in Physics*, **41** 178–225.

[ ] Chorin A J 1968a "Numerical solution of the Navier–Stokes equations," *Math. Comp.* **22** 745–762.

[ ] Chorin A J 1968b "Numerical solution of incompressible flow problems," *Studies in Numer. Anal.* **2** 64–71.

[ ] Chorin A J 1969 "On the convergence of discrete approximations to the Navier–Stokes equations," *Math. Comp.* **23** 341–353.

[ ] Daube O 1992 "Resolution of the 2D Navier–Stokes equations in velocity–vorticity form by means of an influence matrix technique," *J. Comput. Phys.* **103** 402–414.

[ ] Daube O, Guermond J-L and Sellier A 1991 "Sur la formulation vitesse–tourbillon des équations de Navier–Stokes en écoulement incompressible," *Compt. Rend. Acad. Sci. Paris* **313 II** 377–382.

[ ] Dennis S C R and Chang G-Z 1970 "Numerical solution for steady flow past a circular cylinder at Reynolds numbers up to 100," *J. Fluid Mech.* **42** 471–489.

[ ] Dennis S C R and Dunwoody J 1966 "The steady flow of a viscous fluid past a flat plate," *J. Fluid Mech.* **24** 577–595.

[ ] Dennis S C R, Ingham D B and Cook R N 1979 "Finite difference methods for calculating steady incompressible flows in three dimensions," *J. Comput. Phys.* **33** 325–339.

[ ] Dennis S C R and Kocabiyik S 1990 "The solution of two-dimensional Oseen flow problems using integral conditions," *IMA J. Appl. Math.* **45** 1–31.

[ ] Dennis S C R and Quartapelle L 1983 "Direct solution of the vorticity–stream function ordinary differential equations by a Chebyshev approximation," *J. Comput. Phys.* **52** 448–463.

[ ] Dennis S C R and Quartapelle L 1984 "Finite difference solution to the flow between rotating spheres," *Computers and Fluids* **12** 77–92.

[ ] Dennis S C R and Quartapelle L 1985 "Spectral algorithms for vector elliptic equations in a spherical gap," *J. Comput. Phys.* **61** 218–241.

[ ] Dennis S C R and Quartapelle L 1989 "Some uses of Green's theorem in solving the Navier–Stokes equations," *Int. J. Numer. Meth. Fluids* **9** 871–890.

[ ] Dennis S C R and Walker J D A 1971 "Calculation of the steady flow past a sphere at low and moderate Reynolds numbers," *J. Fluid Mech.* **48** 771–789.

[ ] Donea J 1984 "A Taylor–Galerkin method for convective transport problems," *Int. J. Numer. Meth. Eng.* **20** 101–119.

[ ] Donea J, Giuliani S, Laval H and Quartapelle L 1982 "Finite element solution of the unsteady Navier–Stokes equations by a fractional step method," *Comp. Meths. Appl. Mech. Eng.* **30** 53–73.

[ ] Donea J, Giuliani S, Laval H and Quartapelle L 1984 "Time-accurate solution of advection–diffusion problems by finite elements," *Comp. Meths. Appl. Mech. Eng.* **45** 123–146.

[ ] Donea J and Quartapelle L 1992 "An introduction to finite element methods for transient advection problems," *Comp. Meths. Appl. Mech. Eng.* **95** 169–203.

[ ] Donea J, Quartapelle L and Selmin V 1987 "An analysis of time discretization in the finite element solution of hyperbolic problems," *J. Comput. Phys.* **70** 463–499.

[ ] Ehrenstein U and Peyret R 1989 "A Chebyshev collocation method for the Navier–Stokes equations with application to double-diffusive convection," *Int. J. Numer. Meth. Fluids* **9** 427–452.

[ ] El Dabagi F and Pironneau O 1986 "Stream vectors in the three-dimensional aerodynamics," *Numer. Math.* **48** 561–589.

[ ] Fasel H 1976 "Investigation of the stability of boundary layers by a finite difference model of the Navier–Stokes equations," *J. Fluid Mech.* **78** 355–383.

[ ] Fortin M and Thomasset F 1979 "Mixed finite element methods for incompressible flows problems," *J. Comput. Phys.* **31** 113–145.

[ ] Gadzag J, Takao Y and Fromm J 1982 "Rigorous numerical treatment of the no-slip condition in a vorticity formulation," *Numerical Boundary Condition Procedures*, NASA Report, Moffett Field, California, 367–373.

[ ] Glowinski R and Pironneau O 1978a "Approximation par éléments finis mixtes du problème de Stokes en formulation vitesse–pression. Convergence des solutions approchées," *Compt. Rend. Acad. Sci. Paris* **286 A** 181–183.

[ ] Glowinski R and Pironneau O 1978b "Approximation par éléments finis mixtes du problème de Stokes en formulation vitesse–pression. Résolution des problèmes approchés," *Compt. Rend. Acad. Sci. Paris* **286 A** 225–228.

[ ] Glowinski R and Pironneau O 1979a "Sur la résolution numérique du problème de Dirichlet pour l'opérateur biharmonique par une méthode quasi-directe," *Compt. Rend. Acad. Sci. Paris* **282 A** 223–226.

[ ] Glowinski R and Pironneau O 1979b "Sur la résolution numérique du problème de Dirichlet pour l'opérateur biharmonique par la méthode du gradient conjugué," *Compt. Rend. Acad. Sci. Paris* **282 A** 1315–1318.

[ ] Glowinski R and Pironneau O 1979c "Numerical methods for the first biharmonic equation and for the two-dimensional Stokes problem," *SIAM Review* **12** 167–212.

[ ] Guermond J-L and Quartapelle L 1993 "Equivalence of $\boldsymbol{u}$-p and ζ-ψ formulations of the time-dependent Navier–Stokes equations," submitted for publication.

[ ] Heinrichs W 1992 "A spectral multigrid method for the Stokes problem in streamfunction formulation," *J. Comput. Phys.* **102** 310–318.

[ ] Heywood J G and Rannacher R 1982 "Finite element approximation of the nonstationary Navier–Stokes problem. I. Regularity of solutions and second-order error estimates for spatial discretization," *SIAM J. Numer. Anal.* 19, No. 2, 275–311 (1982).

[ ] Hirasaki G J and Hellums J D 1968 "A general formulation of the boundary conditions on the vector potential in three-dimensional hydrodynamics," *Quart. Appl. Math.* **XXVI** 331–342.

[ ] Hirasaki G J and Hellums J D 1970 "Boundary conditions on the vector and scalar potentials in viscous three-dimensional hydrodynamics," *Quart. Appl. Math.* **XXVIII** 293–296.

[ ] Jespersen D C 1974 "Arakawa's method is a finite-element method," *J. Comput. Phys.* **16** 383–390.

[ ] Kato T 1967 "On classical solutions of the two-dimensional non-stationary Euler equations," *Arch. Rat. Mech. Anal.* **25** 188–200.

[ ] Kleiser L and Schumann U 1980 "Treatment of the incompressibility and boundary conditions in 3-D numerical spectral simulation of plane channel flows," *Notes on Numerical Fluid Mechanics*, ed. Hirschel E H, Vieweg Braunschweig, 1980, 165–173.

[ ] Kocabiyik S 1987 "A study of Oseen flow using integral conditions," *Ph. D. Thesis*, University of Western Ontario, London, Canada.

[ ] Ku H-C, Hirsh R S and Taylor T D 1987 "A pseudospectral method for solution of the three-dimensional incompressible Navier–Stokes equations," *J. Comput. Phys.* **70** 439–462.

[ ] Laval H 1988 "Taylor–Galerkin solution of the time-dependent Navier–Stokes equations," *International Conference on Computational Methods in Flow Analysis*, Okayama, Japan, 5–8 September 1988, eds. Niki H and Kawahara M, Okayama University of Science, 414–421.

[ ] Laval H and Quartapelle L 1990 "A fractional-step Taylor–Galerkin method for unsteady incompressible flows," *Int. J. Numer. Meth. Fluids* **11** 501–513.

[ ] Lighthill M J 1963 "Introduction. Boundary layer theory. Chapter II," *Laminar Boundary Layers*, ed. Rosenhead L, Clarendon Press, Oxford, 1963.

[ ] Le Quéré P and Alziary de Roquefort T 1982, "Sur une méthode semi-implicite pour la résolution des équations de Navier–Stokes d'un écoulement bidimensionnel visqueux incompressible," *Compt. Rend. Acad. Sci. Paris* **294 II** 941–944.

[ ] Le Quéré P and Pécheux J 1989, "A three-dimensional pseudo-spectral algorithm for the computation of convection in a rotating annulus," *Spectral and High Order Methods for Partial Differential Equations*, eds. Canuto C and Quarteroni A, Proceedings of the ICOSAHOM '89 Conference, Como, Italy, 26–29 June 1989, 261–271.

[ ] Marcus P 1984 "Simulation of Taylor–Couette flow. Part 1. Numerical methods and comparison with experiment," *J. Fluid Mech.* **146** 45–64.

[ ] McLaurin J W 1974 "A general coupled equation approach for solving the biharmonic boundary value problem," *SIAM J. Numer. Anal.* **11** 14–33.

[ ] Morino L 1986 "Helmholtz decomposition revisited: Vorticity generation and trailing edge condition. Part 1: Incompressible Flows," *Comput. Mech.* **1** 65–90.

[ ] Napolitano M 1991 "Efficient solution of two-dimensional steady separated flows," *Computers and Fluids* **20** 213–222.

[ ] Napolitano M and Catalano L A 1991 "A multigrid solver for the vorticity–velocity Navier–Stokes equations," *Int. J. Numer. Meth. Fluids* **13** 49–59.

[ ] Napolitano M and Pascazio G 1991 "A numerical method for the vorticity–velocity Navier–Stokes equations in two and three dimensions," *Computers and Fluids* **19** 489–495.

[ ] Nguyen H D, Paik S and Chung J N 1993 "Application of vorticity integral conditioning to Chebyshev pseudospectral formulation for the Navier–Stokes equations," *J. Comput. Phys.* **106** 115–124.

[ ] Olson M D and Tuann S Y 1979 "New finite elements results for the square cavity," *Computers and Fluids* **7** 123–135.

[ ] Orszag S A 1971 "Accurate solution of the Orr–Sommerfeld stability equation," *J. Fluid Mech.* **50** 689–703.

[ ] Picciati G 1907 "Sul moto di una sfera in un fluido viscoso," *Atti Accad. Lincei Rend.* **5** 16-1, 943–951.

[ ] Pironneau O 1986 "Conditions aux limites sur la pression pour les équations de Stokes et de Navier–Stokes," *Compt. Rend. Acad. Sci. Paris* **303 I** 403–406.

[ ] Quartapelle L 1981 "Vorticity conditioning in the calculation of two-dimensional viscous flows," *J. Comput. Phys.* **40** 453–477.

[ ] Quartapelle L 1988 "Factorized formulations of the incompressible Navier–Stokes equations," *International Conference on Computational Methods in Flow Analysis*, Okayama, Japan, 5–8 September 1988, eds. Niki H and Kawahara M, Okayama University of Science, 337–348.

[ ] Quartapelle L and Muzzio A 1988 "Decoupled solution of the vector Poisson equation with boundary condition coupling," *Computational Fluid Dynamics*, eds. de Vahl Davis G and Fletcher C, Elsevier Science Publishers, North-Holland, 1988, 609–619.

[ ] Quartapelle L and Napolitano M 1983 "Numerical solution of natural convection problems by the vorticity projection method," *Numerical Methods in Thermal Problems*, Vol. III, eds. Lewis R W, Johnson J A and Smith W R, Proceeding of the Third International Conference, Seattle, USA, 2–5 August 1983, 382–392.

[ ] Quartapelle L and Napolitano M 1984 "A method for solving the factorized vorticity–stream function equations by finite elements," *Int. J. Numer. Meth. Fluids* **4** 109–125.

[ ] Quartapelle L and Napolitano M 1986 "Integral conditions for the pressure in the computation of incompressible viscous flows," *J. Comput. Phys.* **62** 340–348.

[ ] Quartapelle L and Valz-Gris F 1981 "Projection conditions on the vorticity in viscous incompressible flows," *Int. J. Numer. Meth. Fluids* **1** 129–144.

[ ] Quartapelle L and Valz-Gris F 1990 "Sur l'évolution du champ de pression dans les écoulements visqueux et incompressibles," *Compt. Rend. Acad. Sci. Paris* **310 II** 1381–1386.

[ ] Quartapelle L and Verri M 1992 "On the spectral solution of the Navier–Stokes equations in spherical and cylindrical regions, submitted for publication.

[ ] Quarteroni A, Sacchi Landriani G and Valli A 1991 "Coupling of viscous and inviscid Stokes equations via a domain decomposition method for finite elements," *Numerische Mathematik* **59** 831–859.

[ ] Richardson S M and Cornish A R H 1977 "Solution of three-dimensional incompressible problems," *J. Fluid Mech.* **82** 309–319.

[ ] Rodriguez-Prada H A, Pironti F F and Saez A E 1990 "Fundamental solutions of the streamfunction–vorticity formulation of the Navier–Stokes equations," *Int. J. Numer. Meth. Fluids* **10** 1–12.

[ ] Rogé G 1990 "Sur l'approxmation et l'accélération de la convergence lors de la simulation numérique en éléments finis d'écoulements de fluides visqueux compressibles," *Thèse de Doctorat*, Université Paris VI.

[ ] Ruas V 1991 "On formulations of vorticity systems for viscous incompressible flow with numerical applications," to appear.

[ ] Ruas V and Quartapelle L 1990 "Uncoupled finite element solution of 3D biharmonic problems for vector potentials," *Int. J. Numer. Meth. Fluids* **11** 811–822.

[ ] Saffman P G 1981 "Dynamics of vorticity," *J. Fluid Mech.* **106** 49–58.

[ ] Saiac J-H 1985 "On numerical solutions of time-dependent Euler equations for incompressible flows," *Int. J. Numer. Meth. Fluids* **5** 637–656.

[ ] Schreiber R and Keller H B 1983 "Driven cavity flows by efficient numerical techniques," *J. Comput. Phys.* **49** 310–333.

[ ] Selmin V 1987 "Third-order finite element schemes for the solution of hyperbolic problems," *INRIA Report* **707**.

[ ] Selmin V, Donea J and Quartapelle L 1985 "Finite element methods for nonlinear advection," *Comp. Meths. Appl. Mech. Eng.* **52** 817–845.

[ ] Staniforth A N and Mitchell H L 1978 "A variable-resolution finite-element technique for regional forecasting with the primitive equation," *Mon. Wea. Rev.* **106** 439–447.

[ ] Staniforth A N and Mitchell H L 1987 "Review: Formulating efficient finite-element codes for flows in regular domains," *Int. J. Numer. Meth. Fluids* **7** 1–16.

[ ] Stevens W N R 1982 "Finite element, stream function–vorticity solution of steady laminar natural convection," *Int. J. Numer. Meth. Fluids* **2** 349–366.

[ ] Temam R 1969a "Sur l'approximation de la solution des équations de Navier–Stokes par la méthode des pas fractionnaires (I)," *Arch. Ration. Mech. Anal.* **32** 135–153.

[ ] Temam R 1969b "Sur l'approximation de la solution des équations de Navier–Stokes par la méthode des pas fractionnaires (II)," *Arch. Ration. Mech. Anal.* **33** 377–385.

[ ] Warming R F and Hyett B J 1974 "The modified equation approach to the stability and accuracy analysis of finite-difference method," *J. Comput. Phys.* **14** 159–179.

[ ] Woods L C 1954 "A note on the numerical solution of fourth-order differential equations," *Aero. Q.* **5** 176–182.

Titles previously published in the series

INTERNATIONAL SERIES OF NUMERICAL MATHEMATICS
BIRKHÄUSER VERLAG

ISNM 96 **J. Albrecht, L. Collatz, P. Hagedorn, W. Velte (Eds.):** Numerical Treatment of Eigenvalue Problems, Vol. 5, 1991 (3-7643-2575-5)

ISNM 97 **R.U. Seydel, F.W. Schneider, T.G. Küpper, H. Troger (Eds.):** Bifurcation and Chaos: Analysis, Algorithms, Applications, 1991 (3-7643-2593-3)

ISNM 98 **W. Hackbusch, U. Trottenberg (Eds.):** Multigrid Methods III, 1991 (3-7643-2632-8)

ISNM 99 **P. Neittaanmäki (Ed.):** Numerical Methods for Free Boundary Problems, 1991 (3-7643-2641-7)

ISNM 100 **W. Desch, F. Kappel, K. Kunisch (Eds.):** Estimation and Control of Distributed Parameter Systems, 1991 (3-7643-2676-X)

ISNM 101 **G. Del Piero, F. Maceri (Eds.):** Unilateral Problems in Structural Analysis IV, 1991 (3-7643-2487-2)

ISNM 102 **U. Hornung, P. Kotelenez, G. Papanicolaou (Eds.):** Random Partial Differential Equations, 1991 (3-7643-2688-3)

ISNM 103 **W. Walter (Ed.):** General Inequalities 6, 1992 (3-7643-2737-5)

ISNM 104 **E. Allgower, K. Böhmer, M. Golubitsky (Eds.):** Bifurcation and Symmetry, 1992 (3-7643-2739-1)

ISNM 105 **D. Braess, L.L. Schumaker (Eds.):** Numerical Methods in Approximation Theory, Vol. 9, 1992 (3-7643-2746-4)

ISNM 106 **S.N. Antontsev, K.-H. Hoffmann, A.M. Khludnev (Eds.):** Free Boundary Problems in Continuum Mechanics, 1992 (3-7643-2784-7)

ISNM 107 **V. Barbu, F.J. Bonnans, D. Tiba (Eds.):** Optimization, Optimal Control and Partial Differential Equations, 1992 (3-7643-2788-X)

ISNM 108 **H. Antes, P.D. Panagiotopoulos:** The Boundary Integral Approach to Static and Dynamic Contact Problems. Equality and Inequality Methods, 1992 (3-7643-2592-5)

ISNM 109 **A.G. Kuz'min:** Non-Classical Equations of Mixed Type and their Applications in Gas Dynamics, 1992 (3-7643-2573-9)

ISNM 110 **H.R.E.M. Hörnlein, K. Schittkowski (Eds.):** Software Systems for Structural Optimization, 1992 (3-7643-2836-3)

ISNM 111 **R. Burlisch, A. Miele, J. Stoer, K.H. Well:** Optimal Control, 1993 (3-7643-2887-8)

ISNM 112 **H. Braess, G. Hämmerlin (Eds.):** Numerical Integration IV. Proceedings of the Conference at the Mathematical Research Institute at Oberwolfach, November 8-14, 1992, 1993 (3-7643-2922-X)

ISNM 113 **L. Quartapelle:** Numerical Solution of the Incompressible Navier-Stokes Equations, 1993 (3-7643-2935-1)

MIX
Papier aus verantwortungsvollen Quellen
Paper from responsible sources
FSC® C105338
FSC
www.fsc.org

If you have any concerns about our products,
you can contact us on
ProductSafety@springernature.com

In case Publisher is established outside the EU,
the EU authorized representative is:
Springer Nature Customer Service Center GmbH
Europaplatz 3, 69115 Heidelberg, Germany

Printed by Libri Plureos GmbH
in Hamburg, Germany